UNIFIED EQUILIBRIUM CALCULATIONS

UNIFIED EQUILIBRIUM CALCULATIONS

WILLIAM BENTON GUENTHER
The University of the South,
Sewanee, Tennessee

A Wiley-Interscience Publication
JOHN WILEY & SONS, INC.
New York • Chichester • Brisbane • Toronto • Singapore

Library of Congress Cataloging in Publication Data:

Guenther, William B.
Unified equilibrium calculations / by William B. Guenther.
p. cm.
Includes bibliographical references and index.
1. Chemical equilibrium. I. Title.
QD503.G82 1991
541.3'92--dc20 91-9414
ISBN 0-471-53854-X CIP

Printed in the United States of America

10 9 8 7 6 5 4 3 2 1

CONTENTS

PREFACE

The pervading regularity of step complexing in a wide variety of seemingly chaotic solution equilibrium situations is obfuscated by the rigorous but unchemical ad hoc mathematical description that has come into use in recent decades. I became increasingly unconvinced of the need for the *mass–charge* balance methods in teaching advanced students about 10 years ago. Only, in the 4 years since retirement have I been able to extend my earlier efforts toward a more chemical approach to encompass most of the situations which solution chemistry treats.

This book unifies homonuclear equilibrium calculations in one concept. The α (species fractions) and bound proton (and bound ligand) ratio, $\bar{\mathbf{n}}$, as a function of a single master variable (the unbound **H** or **L**) yield complete balances. A single logic is maintained for all cases by setting equal the chemical binding expressed as an equilibrium condition and as a material balance condition. There is no need to devise and remember multistep rules for deriving complete equations for mixtures since all cases are arrangements of this master balancing equality. A great advantage is that the master $\bar{\mathbf{n}}$ relation is in favorable form for spreadsheet solution of numerical problems of all kinds. The traditional approach has resorted to ad hoc rearrangements and approximations avoided by the clear and direct method proposed here.

Developments that call for changes in the academic treatment of solution equilibria are (1) reduction of time available for teaching this vital topic, (2) the general availability of computers, and (3) the realization that lengthy (and needless) mass–charge balancing derivations can be replaced by a single, more chemically meaningful relation. The book is intended for persons needing to use solution chemistry at a level above the elementary. It provides a clear way for all simple systems and leads to setting up correct equations for less simple multiple equilibria, including solubility, mixed-ligand, and polynuclear systems.

[For brevity, acid–base terms are used here, but analogous metal–ligand terms can often be substituted, as shown in Chapters V and VI.] We demonstrate a unified way to a useful (rapidly converging) complete function for acid–base (weak or strong), buffer, ampholyte, and mixtures of one or many systems. A broader acid–base function, C_{AB}, is introduced to cover almost all mixtures met. It is a bit more versatile and helpful to the learner than the C_H (available protonic acidity) sometimes used. I believe that this unification affords a significant economy of effort and time for students and workers in this field.

Complete equations and high precision numerical results may not often be wanted, but the complete approach using correct chemistry will more likely lead one to valid approximations. Computer use does not replace the chemical reasoning still required to set up the relations but does make the complete approach about as rapid as former approximation guesswork. We try to show application of these ideas to more varied systems than have been treated in teaching in the past. They make chemistry more realistic than the oversimplified ionic reaction approaches of elementary texts.

I have not invented new relations. I have applied known but unused forms to facilitate calculations. Similar equations have been published by Ricci[1] and mentioned by Bruckenstein and Kolthoff[2], and by Butler[3], among others. However, the resulting higher power equations in **H** were declared intractable, except by Ricci, who solved many of them in his book (**H** is used to mean the total hydrated proton concentration sometimes written $[H^+]$ or $[H_3O^+]$ or the true major hydrate $[H_9O_4^+]$). Instead of trying to solve the **H**-polynomial equations in the rather unwieldy arrangements in these references, I show that the bound proton relation makes a readily understandable and computable iterating equation. Some simple ways to numerical results using small computer spreadsheet methods are shown. The proton balance equations (formerly derived by combining charge and material balances) are usually solved in texts and treatises through chemically reasonable approximations that have limitations unclear to students. The latter approach has been widely adopted although it obscures the fundamental proton equilibria which control acidity. The seeming new start for every pH problem is needlessly confusing. Butler's Preface[3] clearly presents the case for mathematically complete treatment of equilibria from the perspective of teaching and computing methods existing about 3 decades ago.

After about 25 years of teaching the long method, I concluded about 10 years ago that the advice to get *n* balancing equations in *n* unknown species concentrations, while mathematically correct, is low in chemical content and is a pointless algebraic exercise after the first time. The student may get the "answer" if the right species are chosen for elimination, by which time little equilibrium information remains. In the method proposed here, one starts with the materials given and enters them into the rigorous binding equations which serve all cases. I am not attacking valid methods used by scientists. I am concerned here with approaches that enlighten, simplify calculations, and reduce time required for learning.

All solutions, whether at equilibrium or not, and all reactions whether acid–base, redox, etc., conform to electroneutrality and conservation of mass. Charge and material balance are not unique to acid–base or any other equilibria. Charge balance is not used here and is never required. As Butler observes[3] (p. 66), it is a device to obtain the proton balance which itself cancels. That is why one obtains the identical Charlot equation for the monoprotic acids HAc, HSO_4^-, and NH_4^+. Their charge balances are quite different but cancel in the traditional derivation. Charge balance is a consequence of proton material balance since there is no reaction that can produce an imbalance of ionic charge on the macrolevel. The balance desired here is that of proton binding to base groups as opposed to water. Those bound to water are the focus of protonic acidity, **pH**.

The initial aspect of the full equations using the $\bar{\mathbf{n}}$ (average proton binding number) relation for polyprotic cases is daunting. However, I have been able to rearrange them in every case to obtain the complete equations derived in the usual ad hoc fashion. But rearrangement is rarely needed or desirable. The $\bar{\mathbf{n}}$ relations make the simplest iterating functions for computing **pH**. The equilibrium $\bar{\mathbf{n}}$ vs. **pH** need be computed only once for a given system. It and the material balance $\bar{\mathbf{n}}'$ always have opposite slopes (vs. **pH**, or more sensitively, vs. **H** or **OH** molarity), making solution facile with a computer spreadsheet or by simple interpolation in a list of computed equilibrium $\bar{\mathbf{n}}$ values ($\bar{\mathbf{n}}'$ is a simply calculated quasi-linear function of **H** and the specific *C*'s put in). Ricci[1] p. 3ff.) chose to use an ionic charge fraction ratio (his β, which is directly proportional to $-\bar{\mathbf{n}}$) to get complete equations identical to the ones in this work. However, he then arranged them in standard polynomial **H** form, which makes for tedious solution. [Examples and alternatives are shown in Appendix B.] Much of his rigorous book is concerned with testing terms for negligibility followed by algebraic toil. Once deciphered (he uses over 200 symbols), his approach is simple and consistent. But the simplifying unity is not apparent without long immersion in his book. I was able to show agreement of $\bar{\mathbf{n}}$ equations only after I had done my derivations and then translated symbols, knowing the result desired! I hope my fewer symbols and single binding focus will convince that equilibrium calculations can be done with much greater simplicity and confidence than has seemed the case. I want to make clear my debt to the pioneering work in Ricci's often praised and little understood book. It is my experience that anything "new" one may find in acid–base relations probably exists in that 1952 monograph. J. N. Butler's 1964 book, *Ionic Equilibrium*,[3] was a major influence converting me, and no doubt other teachers, to better methods of treating equilibria. (See the Annotated Bibliography in the back of this book).

The advantage of the present approach is the unifying of proton equilibria. All single systems (like citrate) can be treated by at most two analytical concentration terms. There is no need to identify the "controlling equilibrium" or to identify major (compatible) and minor acid–base species in each new case since all are handled in the same way in every possible mixture. The $\bar{\mathbf{n}}$ function distributes the protons among species so that we are not obliged to pick dominant ones before setting up the problem as is done in older methods. Although a citrate system, for example, involves a fifth-power **H** equation, the two functions set equal for a given mixture converge rapidly to a p**H** solution via simple spreadsheet calculation. Buffer design, K determination, dilution, titration curves, and Gran plots are all treated by the same $\bar{\mathbf{n}}$ function in this work.

Some applications easily handled rigorously by this method and detailed in the chapters are titration of a mixture of several weak acids, mono- and polyprotic, by NaOH; buffers of citric acid–ethylenediamine; the carbonate problem in NaOH titrating solution for weak acids; the p**H** of diammonium hydrogen phosphate; weak–weak salts like ammonium cyanide and anilinium trichloroacetate; ampholytes; dilution curves of any systems; and calculation of equilibrium constants from p**H** or other electromotive force (emf) data in Chapters II–IV. Applications to solubility, metal–ligand, and mixed equilibria are in Chapters V and VI.

Introductory spreadsheet exercises are given in Chapter I along with reference to a 1990 workbook showing how to use spreadsheets in chemical calculations. Tables of $\bar{\mathbf{n}}$ for some common systems are provided in Appendix D for practice away from the computer.

That the long methods are prevalent can be seen in the group of articles on p**H** calculations in the *Journal of Chemical Education*, the February, March, June, and August 1990 issues. Papers in that Journal over the past 20 or so years try with slight success to clarify and simplify equilibrium mathematics. Looking at the whole forest instead of one tree at a time reveals the unity of the binding concept.

The present treatment differs from my 1975 book[4] most radically in the unified method of arriving at complete equations. First, the idea was there (p. 63–71), but not the universal application with the aid of C_{AB} to incorporate the analytical concentrations of mixtures. Second, it uses solution via computer rather than via approximations aimed at simpler algebra, although with experience one still learns to use approximations and recognize whether they are good or are not valid. Third, elementary chapters (II and III) on equilibrium functions include bar diagrams as well as a variety of other graphical methods to make the relations visually more accessible to students. The bar diagrams help make believable an alternative verbal deduction of rather complex proton balances. The usual α, $\bar{\mathbf{n}}$, and log C diagrams are included because they clarify so well the effects of acid–base strengths and allow prediction of reasonable ranges of solution variables and corroboration of results. Examples are chosen especially to illustrate failure of approximate methods: extremes of high or low concentrations and large and/or closely spaced stepwise constants.

The unified approach should save much learning and teaching time, as seen in the space needed to treat problems in this book. For example, the complete **H** equations for C molar solutions of ampholytes of EDTA, say, Na_2H_2Y, NaH_3Y, or H_5YCl, can be written immediately (Eqs. I-1 to I-6) as $\bar{\mathbf{n}} = 2 - \mathbf{D}/C$, $\bar{\mathbf{n}} = 3 - \mathbf{D}/C$, and $\bar{\mathbf{n}} = 5 - \mathbf{D}/C$. These have the single unknown **H** (in an eighth-power relation if no terms are omitted!). Compare the incomplete equation usually derived by a long process leaving a second unknown, the equilibrium ampholytic species, in the **H** equation [Eq. III-4]. The first of these is almost perfectly handled by the approximate (three-Y-species) equation for C's above 10^{-5} M, while the others benefit by correct inclusion of several more nonnegligible terms for adjacent species. These aspects and the approximate p**H** region are found easily by looking at the $\bar{\mathbf{n}}$ table values for this system at 2, 3, and 5, the values of $\bar{\mathbf{n}}$ and $\bar{\mathbf{n}}'$ if $\mathbf{D}/C$ is negligible.

Rapid computer solutions are described in contrast with the "brute force" usually said to be required. The secant method (straight line interpolation approximations) is advocated rather than the derivative (Newton–Raphson), which is unwieldy for the polyprotic $\bar{\mathbf{n}}$. The favorable slope properties of the $\bar{\mathbf{n}}$ make the secant method admirably convergent and rapid to use. It is described with examples in Chapters II and III and Appendix B. New generations of teachers and students who know the computer well will no doubt develop more varied and efficient numerical methods based on this simple conceptual approach. This is not a book on computer use. Some directions will be suggested and shown, but many variations and improvements are possible. A mathematical appendix (B) on equilibrium-related equations gives some simple problems for the practice of those learning to use a new computing system. Running through the worked examples of this book should provide extensive testing of computer operation. Spreadsheets (or macros for repetition) are sufficient for most computations and graphs in this book. It was written entirely on a microcomputer, Apple Macintosh Plus with the spreadsheet and graphic software of Microsoft EXCEL. Spreadsheet methods are widely applicable to chemical calculations, are generally much simpler to learn and use than BASIC programming, and have similar features in the software available: Macintosh (Excel, Quattro), Apple II (SuperCalc), Commodore (PowerPlan), MS DOS (Lotus1,2,3, Quattro). The similarity allows facile communications among workers on different systems.

A comprehensive survey of all aspects of solution equilibria is not the purpose of this book. It is to illustrate the economy of the method in representative examples. Once grasped, it permits the worker to progress with greater ease and speed than one can by traditional piecemeal approaches. Many references are made for those wanting to study applications in analytical, inorganic, coordination, and biochemistry. See also the Annotated Bibliography.

I hope this book will illuminate a clouded field in chemistry, showing it to be a satisfying example of the major aspect of our science: that seemingly overwhelmingly complex systems are built of simple smaller operating steps of straightforward numerical basis.

REFERENCES

1. Ricci, J. E., *Hydrogen Ion Concentration*, Princeton University Press, Princeton, NJ, 1952.

2. Bruckenstein, S. and Kolthoff, I.M., in *Treatise on Analytical Chemistry*, Wiley Interscience, New York, 1959, Part I, Vol. 1, pp. 421 ff., esp. 454–458.
3. Butler, J. N., *Ionic Equilibrium*, Addison-Wesley, Reading, MA, 1964.
4. Guenther, W. B., *Chemical Equilibrium*, Plenum Press, New York, 1975.

ACKNOWLEDGMENT

I am grateful to my colleagues for help in recent computer applications. Drs. Jeffrey Tassin in Chemistry, John Bordley in Chemistry and Computer Science, and Clay Ross in Mathematics and Computer Science have been invaluable. The helpful macros for calculation and plot of α and $\bar{\mathbf{n}}$ on one graph were written by Dr. Tassin.

I also thank the University of the South for allowing me workspace and access to various facilities since I retired from active teaching.

WILLIAM B. GUENTHER

Sewanee, Tennessee
June 1991

UNIFIED EQUILIBRIUM CALCULATIONS

I

DEFINITIONS AND GENERAL CONCEPTS

Basic chemical information (summarized in Appendix A) is assumed. One needs to know what acids and bases are likely to be strong or weak and must have access to the compilations of equilibrium constants given in chapter references and the Annotated Bibliography at the end of the book. One needs familiarity with the chemistry of Brønsted and Lewis acid–base associations. Formulation of equilibrium constant expressions by the mass action law is assumed. This book is intended for advanced students in chemistry, life, and earth sciences who need facility in rigorous mathematical treatment of solution equilibria.

I-1. SYMBOLS AND BASIC RELATIONS

The symbol C_H is the analytical concentration of total available proton acidity (equivalents per liter) at equilibrium. It is the sum of all weak and strong acids put into a mixture. It includes materials like HCO_3^- that may exist in basic solutions and can supply protons to stronger base. It might be called total titratable acid and has been called BNC (base neutralizing capacity). (A complete function C_{AB} covering mixtures containing a strong base as well is described in what follows.)

Note: Students might wish to skip to Chapters II and III, referring to the following as needed.

Analytical concentrations refer to compounds of the species added: C_0 refers to NaA or Na_2A or Na_3A for mono-, di-, and triprotic acid systems; C_1 refers to HA in any system (NH_4^+ or HA^- if diprotic, HA^{2-} if triprotic, etc.; charges must be adjusted for some cases like the amino acids); C_2 refers to the species having two protons (H_2A in diprotic systems, H_2A- in triprotic systems, etc.); and C_3 refers to H_3A put into a triprotic system mixture. These may give little idea of the *equilibrium* concentrations of species: A solution made from C_0 molar Na_3PO_4 and C_3 molar H_3PO_4 may have major species $H_2PO_4^-$ or HPO_4^{2-} or both, depending upon C values. The C_{AB} method proposed does not require realizing the equilibrium possibilities.

The symbols **H** and **OH** refer to the equilibrium concentration of hydronium and hydroxide ions in a mixture. Ricci's difference function, **H** − **OH**, is represented here by **D**. It appears in every complete equation for acid–base equilibrium so it is convenient to have a symbol for it. (However, Ricci used boldface for activities, Roman for molarities.) This, and omitting ionic charges and the use of brackets for these common items, helps the eye cope with complex relationships such as Eqs. I-3 to I-6. For example, a typical one-step K expression is

$$K_a = \mathbf{H}[A]/[HA]$$

Ratios (R) and fractions (α) of species at equilibrium have especially simple forms in these systems. This ratio dependence makes possible the unified equations of this book because the bound proton level can be expressed as a function of the single master variable **H** alone as follows. Ratios are given as

$$K_a/\mathbf{H} = [A]/[HA] = R_{0-1}$$

the ratio of the unprotonated (0) to the protonated (1) species *at equilibrium*. (This is *not* C_0/C_1.) Species fractions (α) are closely related:

$$\alpha_1 = \frac{[HA]}{[HA]+[A]} = (1 + K_a/\mathbf{H})^{-1} = (1 + R_{0-1})^{-1}$$

$$\alpha_0 = \frac{[A]}{[HA]+[A]} = (1 + \mathbf{H}/K_a)^{-1} = (1 + R_{1-0})^{-1}$$

The plots of these fractions and ratios vs. **H** and p**H** are given in Chapter II, where they are shown to be symmetrical exponential curves when plotted vs. p$\mathbf{H}$. By definition the sum of fractions of all forms is unity, shown for a general H_nA:

$$\alpha_0 + \alpha_1 + \alpha_2 + \alpha_3 + \cdots + \alpha_n = 1 \qquad \text{(I-1)}$$

Fractions of species can be derived from the step equilibrium expressions much like those for the HA case. For example, take a diprotic like H_2CO_3, in general, H_2A. The two constants and K_w control solutions:

$$K_{a1} = \frac{\mathbf{H}[HA]}{[H_2A]} \qquad K_{a2} = \frac{\mathbf{H}[A]}{[HA]} \qquad K_{a1}K_{a2} = \frac{\mathbf{H}^2[A]}{[H_2A]}$$

The third is not independent but is a product of the others. It is useful to express the ratio of A to H_2A. (The use of formation constants inverse to these is now common in reference tables and journals, so it is well to emphasize the subscript, a, for these traditional acidity constants. We shall often drop them for easier reading, but see Section I-3 for the general formulations.) Note that ratios of any two species depend solely on $\mathbf{H}$ and K:

$$R_{1\text{-}2} = \frac{[HA]}{[H_2A]} = \frac{K_1}{\mathbf{H}} \qquad R_{0\text{-}1} = \frac{K_2}{\mathbf{H}} \qquad R_{2\text{-}0} = \frac{\mathbf{H}^2}{K_1K_2}, \qquad \text{etc.}$$

The polyprotic species fractions are simply

$$\alpha_0 = \frac{[A]}{[H_2A] + [HA] + [A]} \quad \text{(denominator is total of A species)}$$

Rearranging and substituting from the K_a expressions yield

$$\alpha_0 = \left(\frac{[H_2A]}{[A]} + \frac{[HA]}{[A]} + 1\right)^{-1} = \left(\frac{\mathbf{H}^2}{K_1K_2} + \frac{\mathbf{H}}{K_2} + 1\right)^{-1}$$

or

$$\alpha_0 = (R_{2\text{-}0} + R_{1\text{-}0} + R_{0\text{-}0})^{-1} \qquad \text{(in ratio terms; } R_{0\text{-}0} = 1\text{)}$$

Then, from similar definitions of the other fractions,

$$\alpha_1 = \alpha_0[HA]/[A] = \alpha_0\mathbf{H}/K_2 \qquad \alpha_2 = \alpha_0[H_2A]/[A] = \alpha_0\mathbf{H}^2/K_1K_2$$

These too can be expressed in R ratio terms. Here we have all the species fractions and ratios in terms of $\mathbf{H}$. This is fundamental: the speciation in mononuclear equilibria depends only on **H** and K values. This does not mean that changing C has no effect. It means that if **H** is adjusted (say, with a strong acid or base) to a given value, the speciation is determined. But changing C may change **H** and thus also α.

A very useful combination of α's is the **n**-bar function, the mean number of protons per A group, or the bound proton number $\bar{\mathbf{n}}$, the equilibrium condition fixed by **H** and K_a values. The $\bar{\mathbf{n}}'$ is that fixed by the C values of a specific mixture and an **H**. It is explained after $\bar{\mathbf{n}}$, which is

$$\bar{\mathbf{n}} = \frac{\text{total bound acidity}}{\text{total A}} = \frac{[\text{HA}] + 2[\text{H}_2\text{A}]}{[\text{H}_2\text{A}] + [\text{HA}] + [\text{A}]} = \alpha_1 + 2\alpha_2 \qquad \text{(I-2a)}$$

An example of the meaning is as follows: If one has $\alpha_0 = 0.1$, $\alpha_1 = 0.6$, and $\alpha_2 = 0.3$, the average number of protons must be 1.2. The same result occurs for the α set 0, 0.8, 0.2 for a different acid. Note that the α's must add to 1. For a chosen acid, there is only one possible set of values of α's and $\bar{\mathbf{n}}$ at a chosen **pH**. The values are determined by the K_a values as the preceding equations show. Typical α and $\bar{\mathbf{n}}$ plots are shown throughout the book.

The general $\bar{\mathbf{n}}$ for any acid H_nA is the weighted average of the number of protonated species fractions. This follows from the preceding α derivations:

$$\bar{\mathbf{n}} = \alpha_1 + 2\alpha_2 + 3\alpha_3 + \cdots + n\alpha_n \qquad \text{(I-2b)}$$

The fraction of species A, α_0, at equilibrium for H_3A is defined as

$$\alpha_0 = \frac{[\text{A}]}{\text{total}} = \frac{[\text{A}]}{[\text{H}_3\text{A}] + [\text{H}_2\text{A}] + [\text{HA}] + [\text{A}]}$$

The species terms can be eliminated as before in favor of K_a and **H** terms using the K_a equilibrium ratio expressions for the H_3A system: e.g., $K_1K_2K_3 = \mathbf{H}^3[\text{A}]/[\text{H}_3\text{A}]$:

$$\alpha_0 = \frac{K_1K_2K_3[\text{H}_3\text{A}]/\mathbf{H}^3}{(1 + K_1/\mathbf{H} + K_1K_2/\mathbf{H}^2 + K_1K_2K_3/\mathbf{H}^3)[\text{H}_3\text{A}]}$$

$$\alpha_0 = \frac{1}{1 + \mathbf{H}/K_3 + \mathbf{H}^2/K_2K_3 + \mathbf{H}^3/K_1K_2K_3} = 1/\mathbf{F}_0 \qquad \text{(I-3)}$$

where $\mathbf{F}_0$ is the polynomial in the denominator. Here $[H_3A]$ cancels. Rearrangement for each α to the form of $\bar{\mathbf{n}}$ given in Eq. I-4 is then done. One may show that

$$\alpha_0 = 1/\mathbf{F}_0 \qquad \alpha_1 = \alpha_0\mathbf{H}/K_3$$

$$\alpha_2 = \alpha_0\mathbf{H}^2/K_2K_3 \qquad \alpha_3 = \alpha_0\mathbf{H}^3/K_1K_2K_3$$

All can be expressed with $\mathbf{F}_0$ or with R ratios as well. The α subscripts are the numbers of protons. That is, α_3 is the fraction as H_3A. This seems to be the present common usage. However, Ricci and the Stumm and Morgan book use the reverse, i.e., numbers of protons lost (this is often the ionic charge). The symbol α has been called the *mole fraction*. However, this term usually has a different meaning in physical chemistry.

The bound proton number expressed in equilibrium terms (for the case H_3A) is (Eq. I-2) (still using the K_a direction)

$$\bar{\mathbf{n}} = \frac{\mathbf{H}/K_3 + 2\mathbf{H}^2/K_2K_3 + 3\mathbf{H}^3/K_1K_2K_3}{1 + \mathbf{H}/K_3 + \mathbf{H}^2/K_2K_3 + \mathbf{H}^3/K_1K_2K_3}$$

or (I-4)

$$\bar{\mathbf{n}} = \frac{3\mathbf{H}^3 + 2K_1\mathbf{H}^2 + K_1K_2\mathbf{H}}{\mathbf{H}^3 + K_1\mathbf{H}^2 + K_1K_2\mathbf{H} + K_1K_2K_3}$$

Extension to any number of protons is analogous following the two cases just shown. Note that speciation (α's) is determined only by $\mathbf{H}$ for any given set of K's. The terms in the first form of I-4 are the ratios $R_{1-0} = \mathbf{H}/K_3$, etc. The acidity constants here are K_{a1}, etc., in dissociation directions and are the conditional molarity constants, sometimes called Q or K'. Conditional constants and species molarities must be used consistently throughout to obtain material (proton) balancing equalities. Thermodynamic K's may be adjusted to ionic strengths, as shown in what follows. The appearance of but three terms in $\bar{\mathbf{n}}$ (in Eq. I-4) makes programming fairly simple, as will be shown. This is a single valued function of $\mathbf{H}$ and has negative slope vs. p$\mathbf{H}$ (positive vs. $\mathbf{H}$). Putting $\mathbf{H}$ as $10^{-\mathrm{p}\mathbf{H}}$ and K as $10^{-\mathrm{p}K}$ shows clearly that the α functions are regular exponential curves with spacing factors pK when plotted vs. p$\mathbf{H}$ as shown in Chapters II and III.

The proton *material balance condition* is $\bar{\mathbf{n}}'$, the C-term ratio of bound protons ($C_H - \mathbf{D}$) to total base A of a single H_nA system. Here C_H is available acidity put in $(3C_3 + 2C_2 + C_1)$, and C_A is the total A from all forms put in $(C_3 + C_2 + C_1 + C_0)$. A discussion of **D**, net **H**, or **OH** formed follows. The bound acid after equilibrium is $C_H - \mathbf{D}$:

$$\bar{\mathbf{n}}' = \frac{C_H - \mathbf{D}}{C_A} = \frac{3C_3 + 2C_2 + C_1 - [H^+] + [OH^-]}{C_3 + C_2 + C_1 + C_0} \qquad \text{(I-5)}$$

where the C_n's are the analytical concentrations of H_nA placed into the mixture. Setting the two $\bar{\mathbf{n}}$-condition expressions (I-4 and I-5) equal gives a complete equation relating equilibrium constants, acidity, and analytical concentrations for a specific mixture. More generally [covering any added amounts of strong acid and strong base (sa, sb)], we can use C_{AB} (explained in what follows) in place of C_H:

$$\bar{\mathbf{n}} = (C_{AB} - \mathbf{D})/C_A$$

or

$$\frac{3\mathbf{H}^3 + 2K_1\mathbf{H}^2 + K_1K_2\mathbf{H}}{\mathbf{H}^3 + K_1\mathbf{H}^2 + K_1K_2\mathbf{H} + K_1K_2K_3} = \frac{3C_3 + 2C_2 + C_1 + C_{sa} - C_{sb} - \mathbf{H} + \mathbf{OH}}{C_3 + C_2 + C_1 + C_0} \qquad \text{(I-6)}$$

Equation I-6 is our master relation for protonic equilibria. The left is the equilibrium proton condition for the chosen acid (or acids), and the right side tells the bound proton portion for any chosen p**H** and analytical concentrations C. Any added strong acid and base ($C_{sa} - C_{sb}$) can be inserted for a more complete equation. Any single system mixture can be expressed by, at most, two C terms, C_{AB}, C_A, or some other acid–base combination, as discussed in what follows, but use of more of the C terms in Eq. I-6 is equivalent: C_{AB} is the sum of the first five terms in the numerator and C_A is the sum in the denominator. Mixtures of systems are easily handled by a summation of system $\bar{\mathbf{n}}$'s. Extensive examples are given in the following chapters. We shall show the superiority of these $\bar{\mathbf{n}}$'s as functions for computing as opposed to condensed polynomial forms. In addition to easier computations, these $\bar{\mathbf{n}}$'s have the advantage of clear chemical significance in the binding, which helps in the writing of complete balances.

I-2. THE NET ACID–BASE FUNCTION C_{AB}

In all acid–base solutions we seek to know the equilibrium distribution of acid protons (if any) on the base positions present. If no proton acidity is put in, any weak base put in reacts with water to remove protons and produce $[OH^-]$. This is controlled by the relations among α's, $\bar{\mathbf{n}}$'s, C_{AB}, C_A, and $\mathbf{H}$ (Eqs. I-2 to I-6). Here C_H is the sum of weak and strong acids (wa, sa) put into the solution. If any strong base (sb) is added, it quantitatively subtracts from the acid to give a resultant C_H. If no acidic materials and no strong base are put in, C_H is zero, and any weak base protonates to equilibrium with water. If excess strong base is added, we can think of the solution as having negative acidity (mathematically correct, the strong acid lacking to make it neutral, in the sense that C_H is zero) and use the term C_{AB}, net acid–base balance, in place of the more limited C_H:

$$C_{AB} = (C_{wa} + C_{sa} - C_{sb}) \qquad \text{(I-7)}$$

This will cover all possible situations. (For many cases, $C_H = C_{AB}$, but not with excess strong base.) The term C_{AB} specifies how to deal with any mixture of weak and strong acids and bases. It does not require decision about what species may be formed in particular equilibria. The master equations take care of the *predominant species* or *controlling equilibria* approaches, which can be mysterious to the beginner. Most texts give no indication of how to deal with mixtures that must react extensively to reach equilibrium (to form compatible species). I believe the $\bar{n}$, C_{AB} formulation deals correctly with all situations in texts and possibly with any that can be devised. I have not yet found exceptions. It is useful for titration calculation with mixtures of strong and weak acids and bases. Numerical examples help clarify C_{AB} for specific cases. These are in Chapters II, III, and IV (using mixtures of up to four systems). See especially the tables of mixtures in monoprotic systems in Chapter II and polyprotic systems in Chapter III.

I-3. PROGRAMMING FUNCTIONS

It is convenient to generate tables (computer spreadsheets) of p$\mathbf{H}$, α_n, and $\bar{\mathbf{n}}$ values for acid–base systems. Using the Eqs. I-2 to I-6, we can program them as follows. Extension to α_6 or higher by analogy is described in what follows. Further computation details are described in the chapters and in Appendix B. A brief introduction to spreadsheet use follows in section I-4. We can use

$$\alpha_0 = \{1 + \mathbf{H}/K_3[1 + \mathbf{H}/K_2(1 + \mathbf{H}/K_1)]\}^{-1}$$

or

$$\alpha_0 = \{1 + 10^{(pK_3 - pH)} * [1 + 10^{(pK_2 - pH)} * (1 + 10^{(pK_1 - pH)})]\}^{-1}$$

$$\alpha_1 = \alpha_0(\mathbf{H}/K_3) \qquad \alpha_2 = \alpha_0(\mathbf{H}^2/K_2K_3) \qquad \alpha_3 = \alpha_0(\mathbf{H}^3/K_1K_2K_3)$$

or the corresponding exponential forms. Then (Eqs. I-1 and I-2)

$$\bar{\mathbf{n}} = \alpha_1 + 2\alpha_2 + 3\alpha_3 + \cdots + n\alpha_n$$

By hand, we may calculate $\mathbf{H}/K_3 = b$, $\mathbf{H}/K_2 = c$, and $\mathbf{H}/K_1 = d$ (for ML or any formation step process, $b = K_1\mathbf{L}$, $c = K_2\mathbf{L}$, etc., as follows) to get

$$\bar{\mathbf{n}} = \frac{b[1 + c(2 + 3d)]}{1 + b[1 + c(1 + d)]} \tag{I-8}$$

These may be calculated from the right side without multiple braces or from the left using memory keys, depending upon convenience for a specific instrument. A programmable calculator with multiple memory keys can greatly speed these calculations. The tables of α and $\bar{\mathbf{n}}$ values in Appendix D can be used to check practice calculations. For example by Eq. I-8, the citrate system at **pH** 3.00 must have $\bar{\mathbf{n}}$,

$$\bar{\mathbf{n}} = \frac{549.5[1 + 22.9(2 + 3 \times 0.851)]}{1 + 549.5[1 + 22.9(1 + 0.851)]} = \frac{57{,}867}{23{,}853} = 2.426$$

$$\alpha_0 = \frac{1}{23{,}853} \qquad \alpha_1 = \frac{549.5}{23{,}853} \qquad \alpha_2 = \frac{12{,}588}{23{,}853} \qquad \alpha_3 = \frac{10{,}715}{23{,}853}$$

For a general equation treating up to six complexes, H_6L or ML_6, we may relate the ratios expressed with acid ("dissociation") constants K_a, which run inversely from 6 to 1, and the ML formation constants K, which run from 1 to 6, to fit in one $\bar{\mathbf{n}}$ equation (for any number of complexes n, use $n, \ldots, 1$ and $1, \ldots, n$ for ranges):

ML	HL or HA
$b = K_1\mathbf{L} = [ML]/[M]$	$b = \mathbf{H}/K_{a6} = [HL]/[L] = R_{1-0}$
$c = K_2\mathbf{L}$	$c = \mathbf{H}/K_{a5}$
$d = K_3\mathbf{L}$	$d = \mathbf{H}/K_{a4}$
$e = K_4\mathbf{L}$	$e = \mathbf{H}/K_{a3}$
$f = K_5\mathbf{L}$	$f = \mathbf{H}/K_{a2}$
$g = K_6\mathbf{L} = [ML_6]/[ML_5]$	$g = \mathbf{H}/K_{a1} = [H_6L]/[H_5L] = R_{6-5}$

Then

$$\bar{\mathbf{n}} = \frac{b[1 + c(2 + d\{3 + e[4 + f(5 + 6g)]\})]}{1 + b[1 + c(1 + d\{1 + e[1 + f(1 + g)]\})]} \qquad \text{(I-9)}$$

Tables having **pH** intervals of 0.2 units suffice for many purposes of graphing and calculation. However, examples will be found in this book where 0.1 and smaller intervals have been calculated. It is convenient to program a column of **H** values for use in interpolations $10^{-\mathbf{pH}}$ since the functions are closer to linear in **H** than in **pH**. Example programs for solutions are given throughout the book. The term $\bar{\mathbf{n}}'$ is entered for specific problems, and its coincidence with $\bar{\mathbf{n}}$ is sought to the desired precision. With a little practice one can recognize when simplifying approximations will serve. For some basic cases where $\bar{\mathbf{n}}$ may be very small, **OH** may be used as the variable. Sometimes log $\bar{\mathbf{n}}$ vs. **pH** or **pOH** is advantageous. Examples of all will be shown.

Note that extension to, e.g., K_4 is simple but that reduction to diprotic systems requires setting $K_{a1} = \infty$ (the formation constant of H_3A is zero), not dropping K_{a3}. The K_{a3} and K_{a2} become the K_{a2} and K_{a1} for the diprotic acid. These problems are avoided if all step equilibrium equations use formation constants, covering acid–base as well as metal ion–ligand relations. For ease of translation to higher polyprotic and ML systems, one may use the formation direction for acidity constants as adopted by the International Union of Pure and Applied Chemistry (IUPAC), as shown in Eqs. I-8 and I-9.

I-4. EXPLICATION OF K_a AND ACTIVITIES: SPREADSHEETS

The ionic strength $\boldsymbol{I}$ of a solution may be thought of as an expression of the charge atmosphere in it. It is the sum of the C of all ions in solution times their charge squared. It arose in the activity coefficient treatment of Debye and Hückel:

$$\boldsymbol{I} = \frac{1}{2} \sum_i C_i z_i^2 \qquad \text{(I-10)}$$

For example, 0.01 M solutions of NaCl, $MgSO_4$, and $AlCl_3$ have ionic strengths 0.01, 0.04, and 0.06 M, respectively. If all three are in one solution, $\boldsymbol{I}$ is the sum, 0.11 M. The activity coefficient of ion i is related to this ionic atmosphere by the Debye–Hückel law:

$$\log f_i = \frac{-0.509 z^2 \sqrt{I}}{1 + \sqrt{I}} \qquad \text{(I-11)}$$

Note that f_i is the coefficient of one specific *ion*, while I must always be the total ionic strength of the *solution*. Elaborations of this approximate equation and their limitations are described in Appendix A.

Spreadsheet Introduction

A simple practice example of spreadsheet use follows.* We enter numbers and formulas for calculating ionic strengths and activity coefficients. This is made general so that one can get the quantities I for the whole solution and f for each ion charge type. First, we show the formula *display* of the spreadsheet. The formulas in columns D through J usually remain hidden and only their numerical result is visible to the operator unless one clicks on a cell to show its formula. Formulas must begin with the equals sign.

The chosen data values are entered in A, B, and C. The first line (4) is for the preceding example where we add up the two 0.01 NaCl ions and the three 0.01 Cl^- in the $AlCl_3$ to get 0.05 M for the total of all positive and negative single-charged ions in the mixture. Column D contains the preceding I formula in terms of relative cell references. For example, A4 refers to the value in column A, row 4. Columns E, F, and G show the preceding formulas for the f's. Column H shows the antilog of E. Columns I and J are similarly antilogs of F and G (not shown for lack of space, but they are in the final values that follow in the spreadsheet).

The last three rows are for 0.01 M solutions of any M^+X^-, $M^{2+}X^{2-}$, and $M^{3+}X^{3-}$ if they are present in a solution of the ionic strength chosen. Remember that here we chose to add all positive and negative ions of the same numerical charge. One can also

*Those new to PC spreadsheets might consult the following introduction to uses in chemistry. In conjunction with the particular software instructions, simple and thorough descriptions with examples are provided: D.E. Atkinson, D.C. Brower, R.W. McClard, and D.S. Barkley (eds.), *Dynamic Models in Chemistry, A Workbook of Computer Simulations Using Electronic Spreadsheets*, N. Simonson & Co., Marina del Rey, CA, 1990.

Formula Display

Ionic Strength and Activity Coefficients

	A *M* tot. 1 ± ions	B *M* of 2 ± ions	C *M* of 3 ± ions	D Ionic strength	E log f_1	F log f_2	G log f_3	H f_1	I	J
4.	0.05	0.02	0.01	= 0.5*(A4 + 4*B4 + 9*C4)	= −0.509*SQRT(D4)/(1 + SQRT(D4))	= 4*E4	= 9*E4	= 10^E4		
5.	0.02	0	0	= 0.5*(A5 + 4*B5 + 9*C5)	= −0.509*SQRT(D5)/(1 + SQRT(D5))	= 4*E5	=9*E5	= 10^E5		
6.	0	0.02	0	= 0.5*(A6 + 4*B6 + 9*C6)	= −0.509*SQRT(D6)/(1 + SQRT(D6))	= 4*E6	= 9*E6	= 10^E6		
7.	0	0	0.02	= 0.5*(A7 + 4*B7 + 9*C7)	= −0.509*SQRT(D7)/(1 + SQRT(D7))	= 4*E7	= 9*E7	= 10^E7		

Spreadsheet as seen in Computer (Columns and Rows are the Same as Before.)

Ionic Strength and Activity Coefficients

M tot. 1 ± ions	*M* of 2 ± ions	*M* of 3 ± ions	Ionic strength	log f_1	log f_2	log f_3	f_1	f_2	f_3
0.0500	0.0200	0.0100	0.1100	−0.1268	−0.5071	−0.1409	0.7468	0.3111	0.0723
0.0200	0	0	0.0100	−0.0463	−0.1851	−0.4165	0.8989	0.6530	0.3833
0	0.02	0	0.0400	−0.0848	−0.3393	−0.7635	0.8226	0.4578	0.1724
0	0	0.02	0.0900	−0.1175	−0.4698	−1.0572	0.7630	0.3390	0.0877

formulate the calculation in other ways that achieve the same results. For convenience we set the *number format* in the spreadsheet to four decimal places to save space. One might add to this spreadsheet the formulas for f's in the Davies and in the Kielland treatments described and graphed in Appendix A.

True thermodynamic equilibrium depends not on concentrations but on activities of species. Activity is not simply related to concentration but may be thought of as an *effective concentration* determined by the free energy of the species in the medium at hand. This conflicts with our need for concentrations: acid–base chemical reactions and balancing conditions involve actual amounts of matter (moles). Therefore, it is convenient to use conditional molar equilibrium constants, K rather than K^0. Activities (designated by parentheses) can be approximated by concentrations (designated by brackets) multiplied by a factor f, called the activity coefficient: $(X) = f_x[X]$. The K_{a1} relation is

$$K^0_{a1} = \frac{f_+[H^+]f_-[A^-]}{f_0[HA]} = \frac{f_+f_-}{f_0}\frac{[H^+][A^-]}{[HA]} = FK_{a1} \qquad \text{(I-12)}$$

Thus, we may obtain estimates of conditional K from the K^0 upon dividing by F. Tables of activity coefficients are described and given in Appendix A. The coefficient for a neutral molecule, f_0, is usually acceptably close to 1 in the dilute solutions we use, while ionic f's are about 0.90, 0.78, and 0.75 for singly charged ions at 0.01, 0.1, and 0.2 M ionic strengths. Thus, noting that HA does not contribute to $\boldsymbol{I}$, which is just 0.10 M to two figures, the corrected K_a for acetic acid for use in the case of 0.100 M buffer of HAc and NaAc (**pH** $\cong$ pK_a) is (take f_0 as 1.0)

$$K_a = K^0_a/(f_+f_-) = 1.764 \times 10^{-5}/(0.78)^2 = 2.88 \times 10^{-5}$$

The ionic medium favors formation of ions. We may calculate that the activity of the hydronium ion a_H ($R \cong 1$) is $2.88 \times 10^{-5}(0.78) = 2.25 \times 10^{-5}$, and the pa$_H$ is 4.649 instead of equal to pK^0_a, 4.756. See the note on **pH** terms at the end of this chapter: List of Symbols. The experimental pa$_H$ on the NBS activity scale of pH reported by Bates (see table and explanation in Appendix C) for this buffer is 4.652. The molarity **pH** is 4.541. Note that the ionic strength of a pure weak acid solution must be calculated from the actual ions formed: 0.1 M HCL (a strong acid) has $\boldsymbol{I}$ of 0.1 M while 0.1 M acetic acid has $\boldsymbol{I}$ of about 0.002 *M* since only about 1.7% is in ionic form. In titrating weak acid with NaOH, large changes in ***I*** occur as the starting, very low value is soon replaced by a strong electrolyte like sodium acetate.

Electrochemical acidity measurements (with pH meter or hydrogen electrode) respond to activity, not to molarity. This is why we are forced to relate the quantities. Further discussion of these problems met in laboratory work are given in Chapter IV and in Appendix C. Further equations and tables are in Appendix A.

The serious problems that arise can be seen in titrations during which the continuous changing in ionic strength (medium) may be large. This causes changes in the value of K_a so that a titration curve cannot be correctly calculated with one K value. Worse, K_a determination cannot be made simply from titrations if ionic strength is not controlled or calculated. Methods of achieving this are given in Chapter IV.

Instead of calculating a K_a from activity coefficients, experimental values are available in standard references (Appendix A), often for several ionic strengths and electrolytes, such as KCl, KNO_3, and $NaClO_4$. It must be stressed that any equilibrium constant is valid only for the medium and temperature attached to it. Any quantities (α's and $\bar{\mathbf{n}}$'s) calculated with K's have similar limitations. Elementary texts often use only K^0 values. This is a poor choice since most solutions have too high ionic strength to allow much closer than an order of magnitude calculations. Note the 64% change in the acetic acid constant when it was adjusted to 0.1 *M* ionic strength (see, p. 12). It is pointless to do calculations even to two significant figures with K^0 for most actual acetate mixtures met in the laboratory.

In general, in this book we use K values for 0.1 *M* ionic strength since solutions can be arranged experimentally to be near this value in many cases, so the results are possible in practice. However, we do adjust to other ionic strengths in cases where that is clearly required. About the only situation allowing K^0 use is dilute solutions of a pure weak acid alone (or an uncharged weak base like NH_3). The major solute is then molecules and the ionic strength is low, about 0.002 *M* in 0.1 *M* acetic acid. Even here the ionic f values are about 0.96. Weaker acids like HCN (pK_a 9.2) have much lower ionic strength. The charged acid NH_4^+ exhibits very low dependence of K_a on ***I*** since the f's nearly cancel in the K expression (but not in pa_H). The K of the monoprotic acid HSO_4^- is very sensitive to ionic strength because of the 2^- sulfate ion. Detailed examples of these effects are given in the next chapter.

I-5. A NOTE ON SYMBOLS, PLOTS, AND EXERCISES

Symbols are a problem in fields where variety still rules worldwide. I have kept the number of special symbols down and tried to use common and suggestive ones as much as possible. But this is subjective. One help to the novice is to keep complex equations clear by reducing characters. Thus, my use of **H** and **OH** for $[H^+]$ and $[OH^-]$ uses three characters instead of nine. The single **H** has been used for a long time in p**H** and is thus reasonable and suggestive. ($[H^+]$ is in some disrepute because the free ion [a proton] does not exist in water and the proton has very special properties far removed from all other ions in chemistry. Often, $[H^+]_{aq}$ used to avoid the confusion, but this adds two more characters to baffle the eye still more in large equations.)

Note in the $\bar{\mathbf{n}} = \bar{\mathbf{n}}'$ that the master equation has about 35 characters on each side, while a more conventional text rendering with $[H^+]$ and K_{a2} etc., would require 62 on the left and 41 on the right. That greatly slows reading comprehension. Throughout I have tried to use the very brief forms first ($\bar{\mathbf{n}} = \bar{\mathbf{n}}'$) to establish the concept used and then print the full relation of entities required for calculations. My list of symbols contains under 20 while Ricci's classic text uses over 200! It is one of the most difficult books in chemistry to decipher. That has effectively limited its use to a few specialists, and its importance is unappreciated. My only composite special symbol is **D**, standing for the difference $\mathbf{H} - \mathbf{OH}$. Ricci also used this because it appears in every complete acid–base equation.

Another feature helpful to the novice is to plot all graphs with quantities increasing upward (ordinate) and to the right (abscissa). Scientists learn to read many inverted and backward graphs, but that is no reason to add a burden to those trying to understand what is being shown (p**H** is an exception since increasing this negative log to the right makes **H** decrease but **OH** increase, so either direction works). However, for log C, log(ligand), and log S graphs, I plot the frequent negative values in the direction that agrees with the increase of the actual physical quantity.

Some ideas for further examination by the reader are suggested within chapters, but they are mainly at the ends of the chapters and in the appendices. They are not purely exercises as in most elementary texts but contain suggestions and answers as guides to fruitful mathematical–graphical investigation of equilibrium behaviors similar to, but often supplementary to, those in the book.

I.6. LIST OF SYMBOLS

The symbols used in this work are substantially those of the major reference tables on stability constants. See references in later chapters and the discussion in Appendix A.

$[x]$	Molar concentration of dissolved species x at equilibrium.
(x)	Activity of species x at equilibrium. Molar scales are used, except for the solvent (water), which is on a mole fraction scale: pure water is 1. The symbol a_H is often used for (H^+_{aq}).
A	Conjugate base of weak acid, H_nA. Charges are not shown.
H_nA	Weak acid species having n protons. Charges are not shown.
C_x	Analytical molar concentration of species or solute x; $x = a$, wa denotes weak acid, $x =$ sa, sb denotes strong acid or base. The term C_3 denotes C of H_3A; C_0 denotes C of unprotonated form. The terms C_L, C_M are the concentrations of a ligand and a metal ion, respectively.
C_{AB}	Net acid–base value, defined as $\Sigma\ C_{wa} + C_{sa} - C_{sb}$.
C_H	Analytical concentration of protonic acidity; the total neutralizable acid.
C_A	Analytical concentration of the total of all species of A regardless of protonation.
D	The acid–base balance (difference) at equilibrium used by Ricci, Bates, and others; **H** − **OH**.
f_z	Ionic activity coefficient; ionic charge is z.
F	Combined activity factor (all f's) for an equilibrium constant expression.
$\mathbf{F}_0$	A denominator of species fractions; the Leden–Froneaus function; $= 1/\alpha_0$.
H, OH	Molar hydronium (total protonated water) or hydroxide at equilibrium; commonly written H, h, $[H^+]$, and $[H_3O^+]$, $[OH^-]$.
I	Ionic strength; $\frac{1}{2}Cz^2$ summed *for all* ions (i) in a solution; z is the ionic charge; $= \frac{1}{2}\ \Sigma_i\ C_i z_i^2$.
K_n^0	Thermodynamic equilibrium constant for single step n in terms of activities; $K_{a,n}^0$ for dissociation direction of weak acids, where n is the number of protons lost

(from the 1st to the *n*th proton); K_f^0 for ML formations. See Sections II-1 and III-1 and the introduction to Chapter V.

K_n Conditional molar equilibrium constant (often called quotient Q or K'). Compare β.

$\bar{\mathbf{n}}$ Average number of protons per base group ($\bar{\mathbf{n}}_H$) or ligands on metal ions ($\bar{\mathbf{n}}_{ML}$) at equilibrium; $\bar{\mathbf{n}}'$ is expressed in terms of analytical C's, the material balance condition, not necessarily at equilibrium; $\bar{\mathbf{n}}$ has been written Z, $\bar{n}$, $\bar{h}$, and *i*-bar by various authors. See Eqs. I-4 to I-6.

α_n Fraction of species *n at equilibrium*; *n* is number of protons or ligands in the species (e.g. α_2 is the fraction of H_2A^- in the H_3A system or of $AgCl_{2^-}$ in the Ag^+ Cl^- system). In some works it is the number of groups *lost* from the highest complex. See the Annotated Bibliography.

β_n Product of conditional K's in the *formation* direction only up to *n* (e.g., $\beta_4 = K_1K_2K_3K_4$ for the ML system forming at least four step complexes).

ϕ Degree of titration of a weak acid; number of equivalents of strong base per mole of acid taken. Thus, it runs 0 to 1 for HA and 0 to 3 for H_3A.

Note: pH terms. The designation p**H** refers to the molarity negative log **H**; pa_H refers to the negative log of the hydronium ion activity and contains some assumption about an ionic activity coefficient; pH_s refers to the value of a certified buffer on the defined NBS standard activity scale; and pH (rarely used here) refers to the general subject of all such quantities, as in "pH meter," where distinction among the others is not needed or not possible. The examples in the chapters and Appendix A should help clarify this topic and the reasons for so many terms.

SUMMARY

1. The term K_a is the proportionality constant between **H** and the *equilibrium* ratio of conjugate species, $R = [HA]/[A]$; $R = \mathbf{H}/K_a$.
2. The value of **H** fixes all the R's in a mixture of systems.
   a. In a solution having boric, formic, and propionic acid

species, each ratio [HA]/[A] has its unique value for the one p**H**. Adding HCl or NaOH changes all *R*'s to new and unique values for the new p**H**.

b. In polyprotic systems, each ratio of species is fixed for each p**H**.

c. Species fractions, α_n's, are related to *R* and so are fixed by **H**.

3. The ratios also determine the average number $\bar{\mathbf{n}}$ of bound protons per A (base group) for every A type in a mixture. This $\bar{\mathbf{n}}$ must also relate to the total available acid–base (C_H or C_{AB}) minus that "free" (**H** − **OH**), or **D** (Eq. I-6). The term C_{AB} is the total protonic acidity (weak plus strong) in equivalents per liter put in minus any *strong* base (Eq. I-7).
4. Molar concentrations of species must be distinguished from their thermodynamic activities at equilibrium. The activity is often the measured property, as in pH readings, but the molarity must be deduced for use in conditional equilibrium relations and in balance equations. The activity and molarity are related approximately by estimated activity coefficients.
5. Spreadsheets (tabular calculation sheets organized in order of the calculation steps) are convenient for many scientific calculations that must be repeated for a set of data points. Their use is not limited to computers. Hand spreadsheets have long been used for laboratory work coupled with any available computation aids: mind and hand, logarithms, the slide rule, electronic calculators, and now computers. Electronic computer spreadsheets are almost perfectly suited to scientific work with shifting values of variables. They are easier to use than linear computer programs and have rapidly become part of scientific work reaching down to elementary levels. The ease of plotting the calculated material in various ways with the computer is a boon.

II

MONOPROTIC ACID–BASES: STRONG AND WEAK ACID–BASE EQUILIBRIA IN WATER AND THE OPERATION OF K_w AND PROTON BALANCING

II-1. DEFINITIONS AND EXPLANATIONS OF TERMS

This chapter presents applications of complete equations using available equilibrium constants. Calculations of constants from experimental data, solution design, and titrations are given in Chapter IV.

Analytical (formal) concentrations, *C*'s, are the molar concentrations placed in solution. These may form significantly different equilibrium concentrations of the same and other species. The term C_{HA} or C_1 (to emphasize the protonation number) is the molarity of the solution made, while [HA] is the molarity present after equilibrium reactions have occurred; C_0 states the molarity of the conjugate base (e.g., sodium acetate). Knowledge of acidity properties and constants of species must be obtained from experimental results. See Appendix A.

The total available proton acidity (equivalents per liter) at equilibrium is designated C_H. It is the sum (in equivalents of **H**) of all weak and strong acids. It includes materials like HCO_3^- which can supply protons to an added strong base and might be called total titratable acid. It has been called BNC (base neutralizing capacity). Any strong base must be subtracted to leave net *available* acidity. The net initially available acid, C_H, if any, will remain at equilibrium but will be redistributed among acceptors.

Example. A mixture having C's of 0.2 M HCl, 0.15 M HA, 0.2 M NaA, and 0.1 M NaOH has a C_H (C_{AB}) of 0.25 M to be distributed among 0.35 M A^- species and water at equilibrium. Here $\bar{\mathbf{n}}'$ is (p. 6):

$$\bar{\mathbf{n}}' = \frac{0.25 - \mathbf{D}}{0.35} = 0.714 - 2.856\mathbf{D}$$

Further examples that include basic cases follow. Refer to the list of symbols in Chapter I or on the back cover while reading. Here C_{AB} is more general ($C_{wa} + C_{sa} - C_{sb}$), the net of all weak and strong acids and strong bases (see Section I-2). It is the same as C_H when a strong base is not in excess of the acids. It can be used for all mixtures, while C_H is less clear in some basic mixtures.

This chapter also presents diagrams to aid the student in understanding equilibrium algebra. It introduces the elements of the proton balancing approach presented in this book but is not essential to the advanced reader surveying that approach. There are also some fundamental properties of acid–base solutions often unmentioned or misstated in texts.

The usual approach to aqueous acid–base problems is to present

$$K_a = [H^+][A^-]/[HA]$$

for the equilibrium,

$$HA + nH_2O \rightleftharpoons H^+(H_2O)_n + A^-$$

Here, the convention of stating the solvent (water) in mole fraction terms makes it 1 in dilute solutions, and this constant value is absorbed into K_a, the conditional molarity acidity constant for a monoprotic acid. The nature of the hydrated proton is mentioned in Chapter I. We symbolize its molarity as **H**. Conversions between conditional and thermodynamic constants (K^0) with activity coefficients are shown in Chapter I and Appendix A.

Elementary treatments try to find approximate values for three of the four quantities in the K_a expression and solve for the fourth. Beyond the elementary level, students should have a firmer grasp of equilibria and see a complete treatment that must include the water ion equilibrium and serves for all cases and mixtures of mono- and polyprotic systems. Diagrams help keep multiple factors clear and reinforce the algebraic derivations that leave some students unconvinced.

For a firm grasp of interacting equilibria, let us start by "talking through" several cases with simple diagrams to demonstrate proton balancing. Then we derive one general algebraic relation that summarizes all the cases.

II-2. INTRODUCTORY DIAGRAMS FOR ACID–BASE RELATIONS

Strong Acid or Base

Any source of protons (acids) or proton binders (bases) will upset the K_w equilibrium condition in water, $[H^+][OH^-] = 1.004 \times 10^{-14}$, at 25° (zero ionic strength). For present purposes, let us write this using the computer multiplication sign (*):

$$\mathbf{H} * \mathbf{OH} = 1.00 * 10^{-14} \quad \text{(II-1)}$$

When a strong acid (HCl) or a strong base (NaOH) is dissolved, the water ionizes it completely. Let us look at what happens if we put just 1.00×10^{-7} mol HCl or NaOH into 1 l of water. This seems to double the **H** (or the **OH**) and causes a Le Chatelier shift, removing H^+ and OH^- in exactly equal numbers (forming water) until the new total **H** and **OH** just satisfies the K_w expression (refer to Fig. II-1 for these steps):

1. Pure water: $K_w = (1.00 \times 10^{-7})(1.00 \times 10^{-7})$.
2. Add 1.00×10^{-7} *M* HCl: $K_w = (2.00 \times 10^{-7})(1.00 \times 10^{-7}) \neq 1.00 \times 10^{-14}$.
3. Make y of water: $K_w = (2.00 \times 10^{-7} - y)(1.00 \times 10^{-7} - y) = 10^{-14}$; solve to get $y = 0.38 \times 10^{-7}$.
4. Thus, at equilibrium, $\mathbf{H} = 1.62 \times 10^{-7}$, $\mathbf{OH} = 0.62 \times 10^{-7}$.

This is shown pictorially to impress this crucial point, $\mathbf{H} - \mathbf{OH} = \mathbf{D} = 1.00 \times 10^{-7}$ (the molarity of strong acid) in this case. This difference is the imbalance introduced by the strong acid. It is fundamental to all water equilibrium situations. Follow the steps, in order, for each case. **Figure II-1** shows water equilibrium upset by 10^{-7} *M* HCl. To attain equilibrium, 0.38×10^{-7} *M* **H** and **OH** ions must combine to form water to satisfy the K_w, relieving the stress of excess amounts in the central mixture.

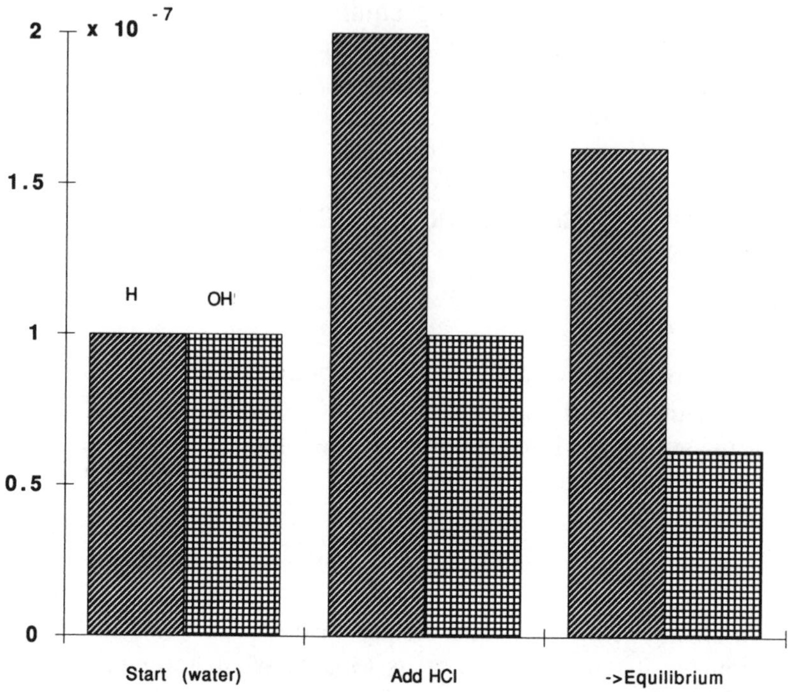

Fig. II-1. Water equilibrium upset by 10^{-7} M HCl. To attain equilibrium, 0.38×10^{-7} M **H** and **OH** ions must combine to form water to satsify the K_w, relieving the stress of excess amounts in the central mixture.

In the schematic drawings that follow (roughly to scale), the bars represent molar quantities of materials. *In thought*, we look at each shift to equilibrium values as if it could occur separately. Of course, the materials interact rapidly and reach their compromise values, satisfying the equilibrium constants.

Figure II-2 shows water upset by addition of 10^{-7} M NaOH. In basic solutions **D** is negative. Note that this is a mirror image of the case in Fig. II-1, as K_w requires.

Alternative Process. We can just as well imagine that we start with pure water that is not yet ionized. First, add 1.00×10^{-7} M HCl. Then let the water form ions just until the right number is achieved to satisfy the K_w. This approach may simplify the pictorial method in more complex, weak cases to come. **Figure II-3** shows an alternative to the method in Fig. II-1 and gives a result identical to the first HCl case.

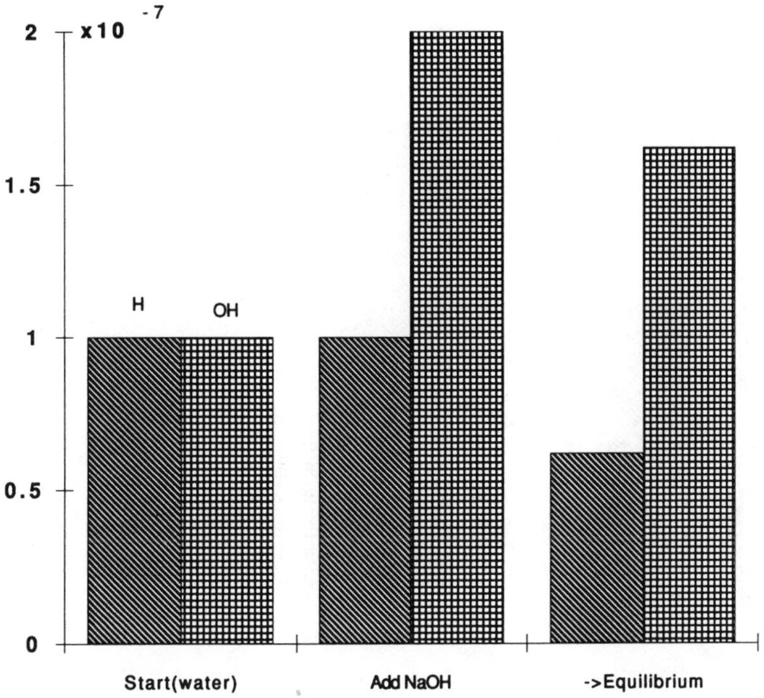

Fig. II-2. Water upset by addition of 10^{-7} *M* NaOH.

Figures II-1 to II-3 should make the algebraic formulation clear:

$$2H_2O \rightleftharpoons H_3O^+ + OH^-$$

which we abbreviate

$$H_2O \rightleftharpoons \mathbf{H} + \mathbf{OH} \qquad y \leftarrow -y - y \quad \text{(Fig. II-1)}$$

For the first two cases some, y, of the starting 10^{-7} M **H** and **OH** had to return to the water to satisfy K_w. Alternatively,

$$-x \rightarrow x + x \quad \text{(Fig. II-3)}$$

for the last case, where we let the x of water make the ions from a zero ion start. Both give the result $x = 10^{-7} - y$)

$$K_w = \mathbf{H} \cdot \mathbf{OH} = \begin{cases} (M + x)(x) & \text{for the HCl case} \\ (x)(M + x) & \text{for the NaOH case} \end{cases}$$

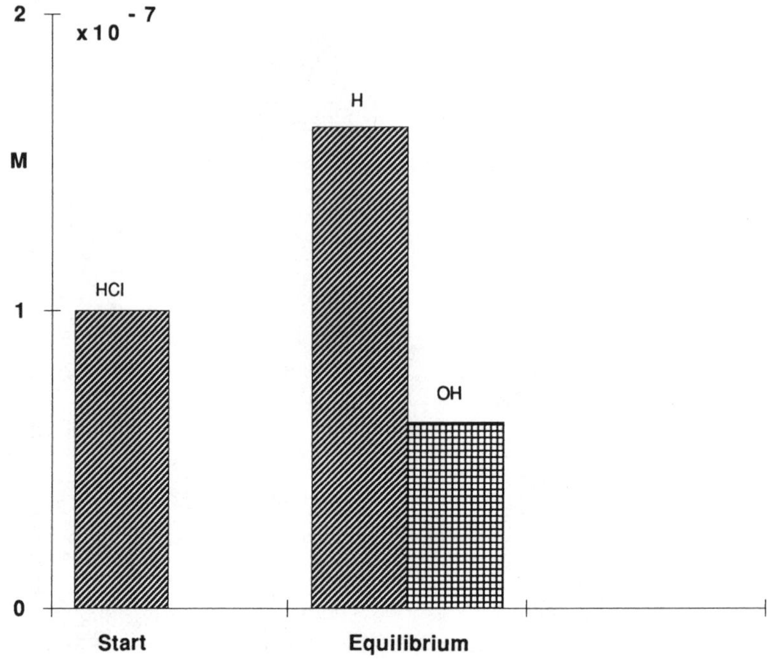

Fig. II-3. Alternative method for the case in Fig. II-1 with result identical to the first HCl case.

The important generalization, which will hold for every solution we study, is $\mathbf{H} - \mathbf{OH} = \mathbf{D}$, the difference, the imbalance of acid and base produced by proton transfers no matter how many or complex they are. In the first two cases,

HCl: $\mathbf{H} - \mathbf{OH}$ ($\mathbf{OH} = H^+$ from water, x) $= \mathbf{D} = [H^+_{aq}]$ from the acid; $1.62 \times 10^{-7} - 0.62 \times 10^{-7} = 1.00 \times 10^{-7}$.

NaOH: $\mathbf{OH} - \mathbf{H}$ ($\mathbf{H} = OH$ from water, x) $= -\mathbf{D} = [OH]^-$ from NaOH.

The reader should try the same diagrams, Figs. II-1 and II-3, using 2×10^{-7} *M* HCl, which becomes the **D** value.

In acidic solution **D** is positive ($\mathbf{H} > \mathbf{OH}$), in basic solution **D** is negative ($\mathbf{H} < \mathbf{OH}$), and in neutral solution **D** is zero. Examine Figs. II-1 to II-3 and the algebra to clarify the meanings and that no individual *concentration* is ever negative.

Bar Diagrams for Weak Acid–Base Cases

To the water equilibrium in K_w, we now add the equilibrium condition for the weak acid, HA, system:

$$K_a = \mathbf{H}[A]/[HA] \qquad \text{(II-2)}$$

We use the symbols given previously and unit activity for the solvent water, which limits eq. II-2 to dilute solutions where water is constant to two to three significant figures or to a certain medium, say, 1 M $NaClO_4$, that may be needed for the experimental system. [We omit the ion charges since they have no mathematical effect but differ for HA cases like NH_4^+ and HSO_4^-, which are correctly described by the same relations we use here. They do have an activity effect.]

(i) HA alone: The protons from this and from water will interact to reach a compromise **H** that will satisfy both equilibrium constant expressions. Some, x, of the acidity will come from the HA, and some, y, will come from the water. Just as the water must make y of OH^-, the HA must make x of A^-. So, we get equilibrium expressions

$$[HA] = C_1 - x \qquad [A] = x \qquad \mathbf{H} = x + y \qquad \mathbf{OH} = y$$

These relations are shown in **Fig. II-4**. They constitute a proton balance, a statement that species formed from initially added species

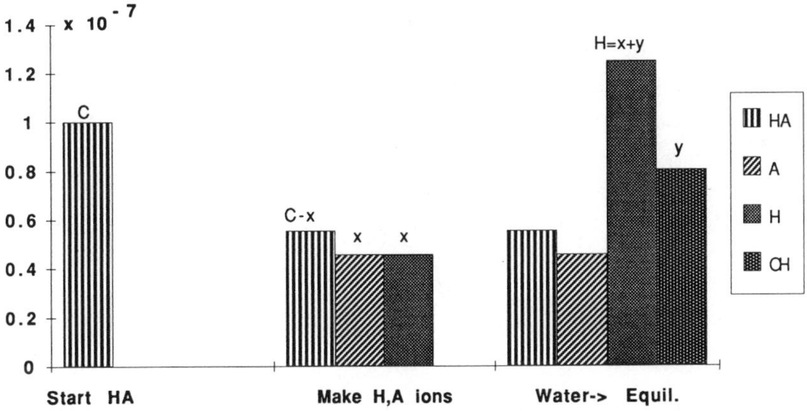

Fig. II-4. Weak acid. First HA forms ions, x; then water adds y more ions to satisfy K_w.

by proton loss must equal those formed by proton gain. Here, H_2O and HA can lose and only water can gain:

$$\text{(formed by proton loss)}\ \mathbf{OH} + [A^-] = \mathbf{H}\ \text{(formed by water + a proton)}$$

$$y + x = x + y$$

(This also happens to be a charge balance. It is not a charge balance for charged acids, HA's such as NH_4^+ or HSO_4^-, for which it is still the correct proton balance needed for equilibrium calculations. For this reason, we prefer not to use charge balancing methods, which turn out never to be required.)

The relations can be rearranged to give

$$\mathbf{H} - \mathbf{OH} = x + y - y = x = [A] = \mathbf{D}$$

Putting these into K_a (Eq. II-2) gives

$$K_a = \mathbf{H}(\mathbf{D})/(C_1 - \mathbf{D}) \tag{II-3}$$

Using $\mathbf{OH} = K_w/\mathbf{H}$ (Eq. II-1) in the $\mathbf{D}$ terms allows one to obtain a cubic equation in one unknown $\mathbf{H}$, if K_a and C_1 are given [part (iii)].

(This case is for $C_1 = 10^{-7}\ M$ with pK_a 10^{-8} giving $x = 0.455$, $y = 0.802$, and $\mathbf{H} = 1.25$, all times $10^{-7}\ M$. But the specific case is not the point here.)

(ii) Next we examine the conjugate base, A, alone. (Charge has no effect on the algebra, so we omit it.) Let it have starting concentration C_0. It must take some protons from water to make equal amounts, y, of HA and OH^-. Again the water must form a bit more H^+ and OH^-, x. So we have the equilibrium relations $[A] = C_0 - y$, $[HA] = y$, $\mathbf{H} = x$, and $\mathbf{OH} = x + y$. Thus, $\mathbf{D} = -y$. The picture can be left to the reader. [See (iii).] The resulting equations are

$$K_a = x(C_0 - y)/y \qquad K_w = x(x + y)$$

or

$$\mathbf{H} = K_a(-\mathbf{D})/(C_0 + \mathbf{D})$$

$$\mathbf{H} = K_a(\mathbf{OH} - \mathbf{H})/(C_0 + \mathbf{H} - \mathbf{OH}) \tag{II-4}$$

A traditional K_b form is usually derived from this equation. Since the solutions are basic, **OH** > **H**, so that **D** ≅ −**OH** on the right. However, the lone **H** on the left cannot be dropped. Making it K_w/**OH** gives an *approximate* weak base equation in the single unknown:

$$\mathbf{OH}^2 = \frac{K_w}{K_a}(C_0 - \mathbf{OH}) \quad \text{where the traditional } K_b \text{ is } K_w/K_a$$

Methods to follow show that this form is rarely required; the general Charlot equation covers all cases.

(iii) Next we examine the general case, mixtures of any concentrations of the acid (HA) and the base (A), C_1 and C_0. If either is zero, we use Eq. I-3 or I-4 for one of them alone. So, only the general equation is really needed. With both present, the mixture is a buffer. Here we add the shifted amount of HA to C_0 and subtract it from C_1 or vice-versa. Either can happen, and it turns out that they give the identical relation, the signs taking care of the direction of the shift. See **Fig. II-5** to visualize these relations. In the figure a dilute HA–NaA buffer is described. The HA has pK_a 7, $C_1 = 3 \times 10^{-7}$ M and $C_0 = 2 \times 10^{-7}$ M.

Let us choose the first (acidic) case. At equilibrium, [HA] = $C_1 - x$, [A] = $C_0 + x$, and an equal amount, x, of H^+ must form. Then y more H^+ and OH^- from water form, **H** = $x + y$, **D** = x. The relations here, for Eqs. II-1 and II-2, are

$$K_a = (x + y)(C_0 + x)/(C_1 - x) \quad \text{and} \quad K_w = (x + y)(y)$$

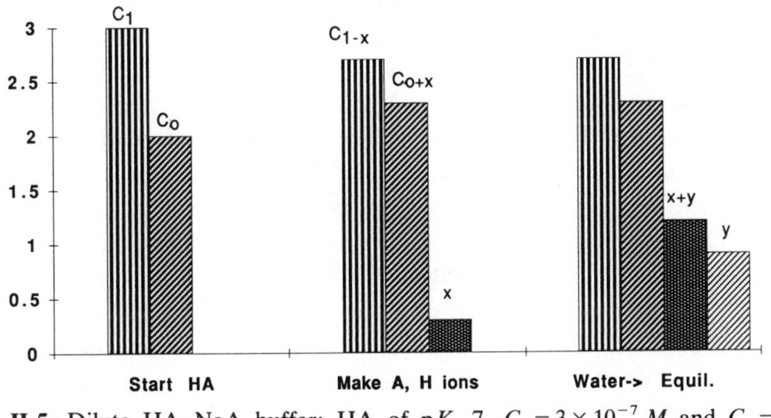

Fig. II-5. Dilute HA–NaA buffer: HA of pK_a 7, $C_1 = 3 \times 10^{-7}$ M and $C_0 = 2 \times 10^{-7}$ M; then ($x = 0.32$, $y = 0.86$) $\times 10^{-7}$ M.

Note this is not the common approximation with $y = 0$ in K_a. Eliminate x and y to give

$$\mathbf{H} = K_a(C_1 - \mathbf{D})/(C_0 + \mathbf{D}) \tag{II-5}$$

Figure II-5 shows species as they occur for HA of pK_a 7, $C_1 = 3 \times 10^{-7}\,M$ and $C_0 = 2 \times 10^{-7}\,M$. Then $(x = 0.32,\ y = 0.86) \times 10^{-7}\,M$. For a less acidic system like HCN, boric acid, or NH_4^+, which all have pK_a near 9, buffers will usually be basic. Thus, some A takes protons from water, as for the base alone, and we get $C_1 + y$ for the [HA] and $C_0 - y$ for the [A^-] left. Water makes x more H^+ and OH^- to give $\mathbf{H} = x$ and $\mathbf{OH} = x + y$. It is left for the reader to see that the changes in direction and signs (the proton balance) give exactly the same **H** relation (Eq. II-5) as the acidic case. The reader should draw a picture for the basic case and derive the equation to see the identical result. (See the next section for algebraic details.) This is sometimes called the Charlot equation, a complete, mathematically exact relation for any mixture we can make with a single weak acid–base system: any combination of C_1 and C_0 values, including zero, that give an acid, or base, alone, as shown in the preceding. (Numerical precision is, of course, limited by that of the molarity constant (K). It is not always the best relation for complete calculations as will be shown. Treatments of mixtures of systems and additions of strong acids and bases follow.

II-3. GENERAL RELATIONS: α AND CHARLOT EQUATIONS

In all monoprotic acid–base solutions, the equilibrium distribution of acid protons on the base positions present is specified by the α_1 relation and the analytical concentrations C_H, C_A and the equilibrium **H**. The concentration C_H is the sum of the weak and strong acids put into the solution. If strong base is added, it quantitatively subtracts from the acid to give the resultant C_H. If no acidic materials and no strong base are added, C_H is zero, and the weak base reacts to equilibrium with water. If excess strong base is added, we can think of the solution as having negative acidity (the acid lacking to make the C_H zero) and use the term C_{AB}, the net acid–base balance, which is $C_{wa} + C_{sa} - C_{sb}$, in place of the more limited C_H. This will cover all possible situations. (For most cases, $C_H = C_{AB}$, but not those with excess strong base.) Some example solutions explain the cases: weak acid alone, weak base alone, buffer, etc. in Table II-1.

TABLE II-1. Terms for Monoprotic Mixtures[a]

Given	C_H	C_{AB}	Sign of **H − OH = D**
1. C_{wa}	C_{wa}	C_{wa}	+
2. C_{wb}	0	0	−
3. C_{wa}, C_{wb}	C_{wa}	C_{wa}	(buffer) +, −, 0 (all possible, depending on K_a)
4. $C_{wa} > C_{sb}$	C_{wa}	C_{wa}	
5. $C_{wa} < C_{sb}$	Not applicable	$(C_{wa} - C_{sb}) < 0$	−
6. All four C's	Reduces to one of the above depending on C and K_a		

[a]Abbreviations: wa, weak acid; wb, weak base; sb, strong base.

Examples. a. Mix conjugates to get C_{wa} and C_{wb} both equal to 0.10 M, a buffer of HA and NaA. By reacting, this mixture is identical to one made with 0.20 mol HA and 0.10 mol NaOH per liter. Both have $C_H(C_{AB})$ of 0.10 M and C_A of 0.20 M. The equilibrium **H** depends on K_a and is approximately equal to K_a. See items 3, 4, 6, in Table II-1.

b. See item 5 in Table II-1. 0.10 M HA, 0.10 M NaA, and 0.20 M NaOH. (This is identical to 0.20 M HA and 0.30 M NaOH.) Here C_{AB} is −0.10 M, i.e., an excess strong base; C_{wb} is C_A, i.e., 0.20 M. A specific example is shown for HCN with an excess strong base, (see p. 46).

Algebraic Derivations

Now we derive the complete monoprotic relations algebraically to compare them with the diagrammatic chemical arguments presented so far in this chapter.

For monoprotic systems, $\bar{\mathbf{n}} = \alpha_1$, the bound protons per base group. We use C_1 (C_{wa}) for the analytical concentration of the acid (HA) and C_0 (C_{wb}) for A^-, as from sodium acetate, and $\bar{\mathbf{n}} = \alpha_1 = 1 - \alpha_0$. For the case of no strong acid or base and a single weak HA system,

$$C_H = C_{AB} = C_1 \quad \text{and} \quad C_A = C_0 + C_1$$

The bound proton condition (the master equation; review Eqs. I-3 to I-5 and II-5) becomes

$$\begin{array}{c}\text{Material balance for} \\ \text{the mixture given}\end{array} = \begin{array}{c}\text{equilibrium condition for} \\ \text{any mixture of this system}\end{array}$$

$$\alpha_1' \qquad = \alpha_1$$
$$(C_{AB} - \mathbf{D})/C_A = [HA]/([HA] + [A])$$

Using C_1 and C_0 mixtures on the left and $K_a/\mathbf{H} = [A]/[HA]$ on the right gives

$$(C_1 - \mathbf{D})/(C_1 + C_0) = (1 + K_a/\mathbf{H})^{-1} \qquad \text{(II-6)}$$

A form of the master equation, the preceding equation can be rearranged to yield the well-known (Charlot) equation:

$$\mathbf{H} = K_a \frac{C_1 - \mathbf{D}}{C_0 + \mathbf{D}} \qquad \text{(II-7)}$$

This is cubic in **H** (since $\mathbf{OH} = K_w/\mathbf{H}$) and reduces to common approximate forms (by taking **D** to be small relative to the C terms) for acid ($C_0 = 0$), base ($C_1 = 0$), or buffer cases. Complete solutions will be illustrated in examples following. (Bar diagrams that help visualize the derivation are Figs. II-4 and II-5.) Similarly

$$\mathbf{OH} = \frac{K_w(C_0 - \mathbf{OH} + K_w/\mathbf{OH})}{K_a(C_1 + \mathbf{OH} - K_w/\mathbf{OH})} \qquad \text{(II-8)}$$

The traditional $K_b = K_w/K_a$. This is equivalent to the **H** form of the complete monoprotic **H** equation. It is sometimes convenient especially in approximate forms obtained by omitting $K_w/\mathbf{OH}$ in basic solutions. (To add confusion, K_b is sometimes called a "hydrolysis" constant of the ion A^-.)

Note that only acidity balancing and equilibrium conditions (K_a and K_w expressions) are needed to derive Eqs. II-7 and II-8 that may be derived in many ways. Another method that does not use charge balancing uses A and acid balances:

A balance: $C_1 + C_0 = [HA] + [A]$ $\qquad$ C_H balance: $C_1 = [HA] + \mathbf{D}$

Divide the two equations and eliminate the expressions in brackets with the K_a expression to get only the **H** and C terms (Eq. II-7).

A broader equation uses C_{AB} instead of C_H in deriving Eqs. II-6 and II-7, noting that $C_{wa} = C_1$, $C_{wb} = C_0 + (C_{sb} - C_{sa})$. (Any net strong base is converted to a weak base or excess strong acid is converted to HA. An excess of either is subtracted out correctly by the **D** term.) Examples to follow will clarify uses of C_{AB}. The following form allows exact treatment of all the mixtures in the examples that follow:

$$\mathbf{H} = K_a \frac{C_{AB} - \mathbf{D}}{(C_0 - C_{sa} + C_{sb} + \mathbf{D})} \tag{II-9}$$

that is, all mixtures of strong acids and bases with one weak HA system. No approximations have yet been made in these derivations. The α, α' form is often easier to use than the Charlot rearrangement (Eq. II-7). The following graphical solutions demonstrate what numerical solutions are about, and do not suggest that graphing is the preferred method.

The α_0 and α_1 functions plotted vs. **pH** (Figs. II-6, II-10, and II-11) are useful for a wide **pH** range. Note that $\alpha_0 = 1 - \alpha_1$. These symmetrical curves are the same shape for any weak acid–base system, HA. They are placed along the **pH** axis to cross at **pH** = pK_a (**Fig. II-6**). In the figure for pK_a 6, the **pH** coordinates at the bottom apply. For any other pK_a system, the top coordinate for **pH** should be used.

The nature of these equilibrium curves may be clearer if we substitute k for K_a and x for **pH** in the preceding α expressions. This

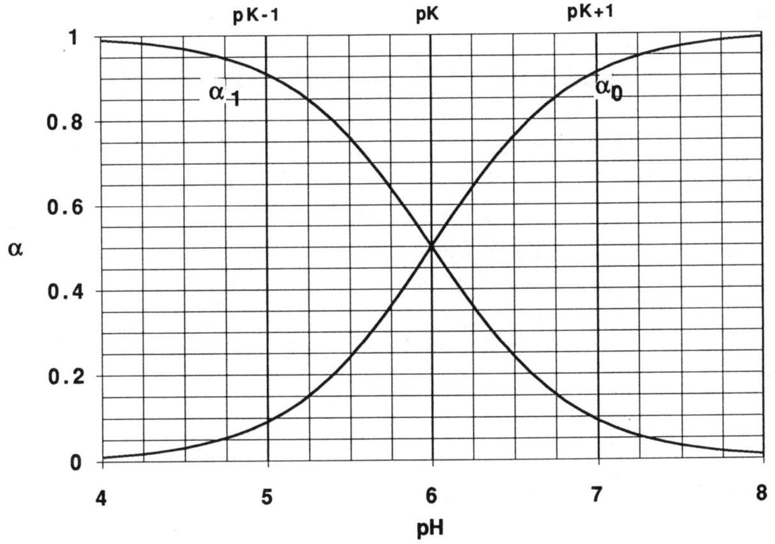

Fig. II-6. Plots of α vs. pH. For HA of pK_a6, the **pH** scale at the bottom applies. For any other pK_a system, use the top scale for **pH**.

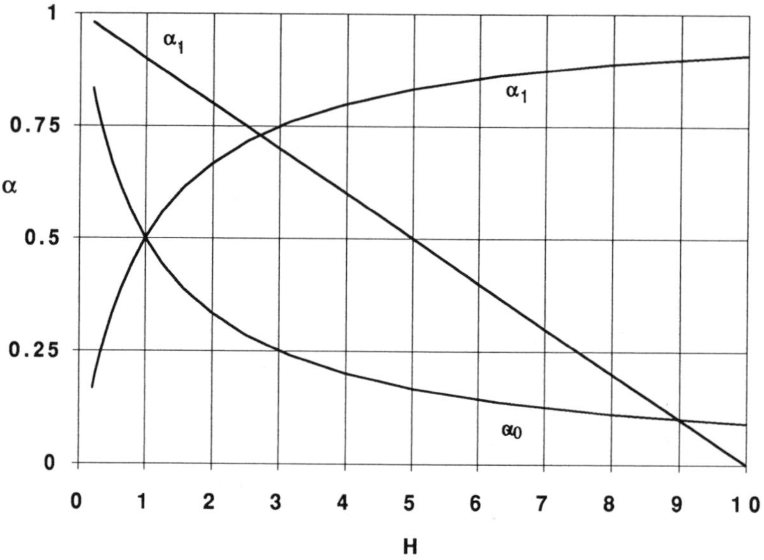

Fig. II-7. Plots of α vs. **H** (**H** $\times$ 1000, i.e., 1 means 0.001 M).

shows that they are simple exponential curves (in the **pH** plot):

$$\alpha_1 = \frac{1}{1 + k \times 10^x} \qquad \alpha_0 = \frac{1}{1 + (10^{-x})/k} \tag{II-10}$$

For the pK_a 4.00 plot,

$$\alpha_1 = (1 + 10^{x-4})^{-1} \quad \text{and} \quad \alpha_0 = (1 + 10^{4-x})^{-1}$$

Inserting x (**pH**) values at 0.1 unit intervals gave the plot in Fig. II-6. Try the simple values $x = 3$, 4, 5 to see how these functions behave.

Linear plots of **H** vs. α for pK_a 3.00 and the α_1' line for a 0.010 M HA solution are shown in **Fig. II-7** ($\alpha_1' = 1 - \mathbf{D}/0.01$), where 1 means 0.001 M. Thus, pK_a is 3. The α_1' line is for a 0.01 M HA solution and shows the same result as the ratio plots in Figs. II-8 and II-9. Note that **H** increases to the right here so that the α's are reversed from the **pH** plots.

II-4. RATIO OF CONJUGATES

The ratio form $\mathbf{H}/K_a = [\text{HA}]/[\text{A}] = R_{1\text{-}0}$ (or the inverse $R_{0\text{-}1}$) helps stress the dependence of the two A species upon **H**. Fixing **H**

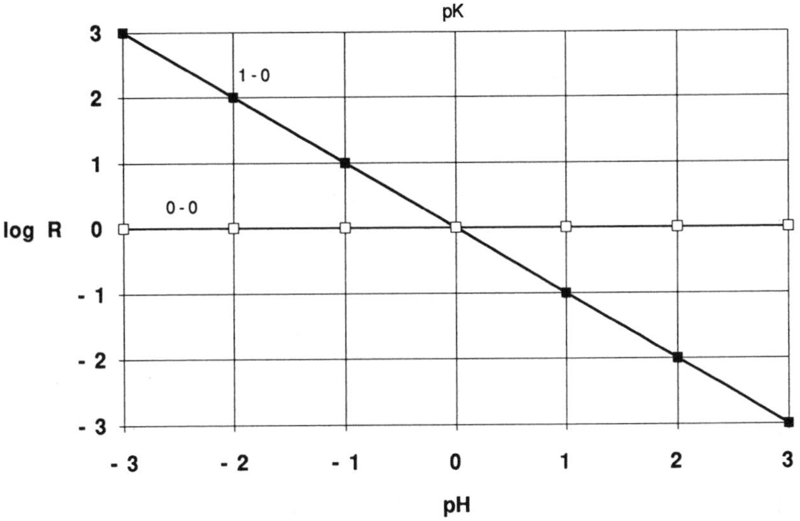

Fig. II-8. Log ratio [HA]/[A] (R_{1-0}) for any acid HA system mixture centered on its pK_a.

determines the ratio and the fractions (α's) as well. For a wider range of **pH**, we must use the log plot shown in **Fig. II-8** for any acid HA, where the **pH** scale is related to pK_a. Plots of log R' will be curves. The R_{0-0} merely emphasizes the relation of A to HA already included in the coordinates.

The K_a *of an acid is numerically the* **H** *at which the equilibrium base–acid (or acid–base) ratio is 1.* Note that these ratios are *not* the ratios of the analytical concentrations C_1 and C_0 (unless the solution is neutral). *The* K_a *of an acid is numerically the ratio of conjugate base to acid (at equilibrium) when* **H** *is 1.*

Figure II-9 shows a linear coordinate plot of $R_{1-0} = \mathbf{H}/K_a$ for an acid of pK_a 3. $R' \cong (0.01 - \mathbf{H})/\mathbf{H}$.

II-5. TREATMENT OF VARIOUS MONOPROTIC SYSTEMS

In the example problems in this section, assume M and K_s *are given to three significant figures.*

Solution of a Buffer Case

The α_1' line in Fig. II-10 is for a 1:1 buffer, 0.001 M HA and A (total acidity is 0.001 M, and the total concentration of species A, $C_A =$

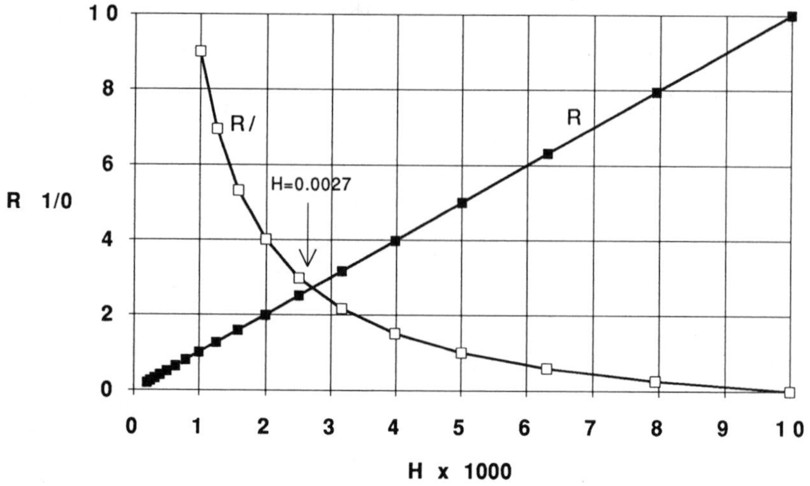

Fig. II-9. Linear coordinate plot of $R_{1-0} = \mathbf{H}/K_a$ for an acid of pK_a 3 [$R' \cong (0.01 - \mathbf{H})/\mathbf{H}$ for 0.01 *M* HA alone). Simultaneous solution is shown (R' □). On abscissa, 1 means 0.001 *M* **H**. The **pH** is 2.57 instead of the approximate 2.50, from $\sqrt{K_a C_a}$. The R_{1-0} ratio dependence is also linear vs. **H** as shown. The ratio R' is expressed in material balance terms: [HA] = $C_1 - \mathbf{D}$ and [A] = **D** for the 0.010 *M* HA case. (Compare Fig. II-4.)

0.002 *M* for Eq. II-6). The expression for this case is then

$$\alpha_1' = \frac{\text{bound H}}{\text{total A's}} = \frac{C_1 - \mathbf{D}}{C_1 + C_0} = (0.001 - \mathbf{H})/0.002 = 0.5 - 500\mathbf{H}$$

Thus, we plot 0.5–500**H** vs. **pH** to find its intersection with α_1 (not with α_0). **Figure II-10** shows the intersection of α_1 and α_1' from Eq. II-6 with a **pH** of about 4.1, not the approximate 4.0, the pK_a.

We use **D** = **H** in the acid range where **OH** is negligible. The intersection shows the true **pH** of about 4.1. To see how this was plotted, substitute **pH** = 3, 3.3, 3.7, 4, . . . and get the α_1' line. (This is the acidic half of the α_1' line. Full plots of α' lines for various mixtures are shown in Figs. III-8 to III-10.) Examination of a rough α diagram allows the choice of **pH** region for a solution of the cubic equation by successive approximation on a computer or calculator. Quadratic approximation is usually possible for monoprotic cases (either **H** or **OH** is negligible), but for polyprotic cases a complete method is needed.

Spreadsheet Computer Method. See the activity coefficient spreadsheet exercise in Chapter I and Appendix B. Make a table with

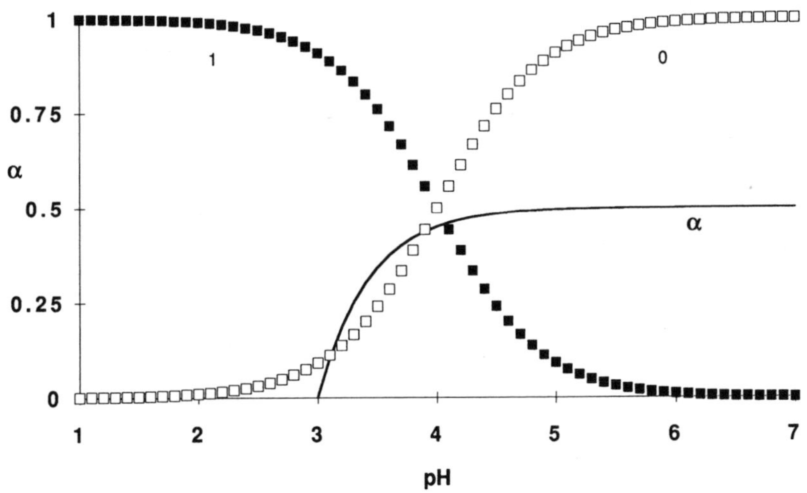

Fig. II-10. Intersection of α_1 and α_1' from Eq. II-6 for 1:1 buffer, 0.001 *M* each, pH 4.1.

columns for **H**, α_1, α_1' and Δ their difference (see the example spreadsheet that follows). Use the preceding formulas for α. Put a numerical guess for $\mathbf{H}_1$ in the first **H** cell. In the second **H** cell, $\mathbf{H}_2$, use*

$$= \mathrm{IF}(\Delta < 0,\ 1.2 * H_1,\ H_1/1.2)\ ,$$

meaning that if Δ is negative, increase $\mathbf{H}_1$ by 1.2; if not, decrease it. Here it is decreased. Then $\mathbf{H}_x$ (interpolated) is derived from a linear extrapolation (or interpolation) to a Δ of zero:

$$\mathbf{H}_x = \mathbf{H}_2 + (\mathbf{H}_1 - \mathbf{H}_2)/(1 - \Delta_1/\Delta_2)$$

This takes the first two points to give the slope of a straight line (for the difference, Δ, line) and finds an approximate **H** value where Δ will hit *zero*. This value, $\mathbf{H}_x$, is then used as $\mathbf{H}_3$ in the calculation of the third line with $\mathbf{H}_2$. Successive values converge rapidly to a constant value. (The arbitrary factor 1.2 may need to be changed to 1.5, 2, 3, etc., if negative **H** or oscillation occurs when the first guess is far off.) Note how the **H** column changes and the α's converge. Experimenting with several larger and smaller guesses of $\mathbf{H}_1$ will demonstrate the method and the invariant final result. Extend the

*(Note that formulas in EXCEL must start with the equal sign.)

columns if low Δ and convergence are not obtained with about six rows. (FILL DOWN, being careful to start the fill-down of **H** from the *third*, or lower, **H**.) Any number of other problems can be solved by appropriate changes of the K_a and C values in formulas. (Refer to Appendix B for details of spreadsheets.) An iteration method or one based upon ratios can also be used, as shown in Chapter III.

In the spreadsheet that follows $K_a = 0.0001$, $\mathbf{D} \cong \mathbf{H}$, $\alpha_1' = 0.5 - 500\mathbf{D}$, and both the HA acid and the NaA are 0.001 M:

H	α_1	α_1'	Δ	$\mathbf{H}_x$ **H** (interp.)	pH
2.000E − 04	0.6667	0.4000	2.67E − 01		3.699
0.0001667	0.6250	0.4167	2.08E − 01	4.7619E − 05	3.778
4.7619E − 05	0.3226	0.4762	−1.54E − 01	9.8143E − 05	4.322
9.8143E − 05	0.4953	0.4509	4.44E − 02	8.6817E − 05	4.008
8.6817E − 05	0.4647	0.4566	8.12E − 03	8.4279E − 05	4.061
8.4279E − 05	0.4573	0.4579	−5.16E − 04	8.4431E − 05	4.074
8.4431E − 05	0.4578	0.4578	5.69E − 06	8.4429E − 05	4.074

Note that after Δ becomes negative, the $\mathbf{H}_x$ is an interpolation that is quite close to the root. The next step also changes sign and makes an even closer interpolation. Note that the final Δ has become zero to the precision shown in the α's. The true **H** is only 84% of the approximate value, $\mathbf{H} = K_a$.

Calculator Methods. Trial and error or approximations by a quadratic in **H** are suitable with monoprotics:

(i) From the Charlot equation (II-7) for the preceding buffer,

$$\mathbf{H} = K_a \frac{C_1 - \mathbf{D}}{C_0 + \mathbf{D}} = 10^{-4} \frac{0.001 - \mathbf{D}}{0.001 + \mathbf{D}} \cong 10^{-4} \frac{0.001 - \mathbf{H}}{0.001 + \mathbf{H}}$$

First approximation: If $\mathbf{D} \ll 0.001$, $\mathbf{H} = 10^{-4}$. Check this on the right side to get

$$\mathbf{H} = 10^{-4}(0.0009)/(0.0011) = 8.2 \times 10^{-5}$$

Continuing shows slow oscillation converging to the root.

(ii) *Quadratic*: Solve the preceding **H** approximation to get

$$\mathbf{H}^2 + 0.0011\mathbf{H} - 10^{-7} = 0 \quad \text{which gives } \mathbf{H} = 8.45 \times 10^{-5} \quad \mathbf{pH} = 4.073$$

A Full Cubic Example

(i) Find the **pH** of 10^{-7} *M* acetic acid. No term is negligible since **H** is near 10^{-7}. We use K_a^0 and K_w^0 values at this dilution:

$$\alpha_1' = 1 - 10^7\mathbf{D}, \quad \text{and} \quad \alpha_1 = (1 + 1.75 \times 10^{-5}/\mathbf{H})^{-1}$$

for use in Eq. II-6 or in the preceding spreadsheet calculation. Instead, one may try the Charlot equation (II-7):

$$\mathbf{H} = 1.75 \times 10^{-5}(10^{-7} - \mathbf{H} + K_w/\mathbf{H})/(\mathbf{H} - K_w/\mathbf{H})$$

It is usual practice to make approximations and iterate the latter to find **H**. However, for some values of K_a and C_1, this form diverges, oscillating about the true root.[1] The first form, α_1, is always soluble and makes good iterating functions since the α_1 values need be obtained but once, all monoprotic systems having the same values just shifted to a central $\mathbf{pH} = \mathrm{p}K_a$. This method is shown since it is the unified method of this book. It is only a slight help for these simple monoprotic cases but becomes far more useful in the mixtures of systems and polyprotic cases to follow. (See also Appendix B.)

A computer readout of α values is given in Table II-2. These values can be used for any HA if a computer spreadsheet method is not being used by setting $\mathrm{p}K_a$ at 6.000, taking the **pH** distance from the desired $\mathrm{p}K_a$, measured from 6.000. See the Fig. II-11 and following examples.

To find the α's for acetic acid, for example, place 4.756 at the **pH** 6 line. Then, α_1 at **pH** 6.756 is 0.009901, and α_0 is 0.9901. These are found, for example, 2 units above $\mathrm{p}K_a$ (6). To solve our problem, see that **H** must be between 2×10^{-7} and 1×10^{-7} (all or no protons lost from the 10^{-7} *M* acetic acid). Calculate $\alpha_1' = 1 - 10^7\mathbf{D}$ in this region and make a table with the α_1 values to find the interval of the root. Then program or calculate finer intervals to the desired precision. (Appendix B suggests spreadsheets and various other treatments of cubic and higher equations that chemists have sometimes used.)

pH	$\mathbf{H} \times 10^7$	$\mathbf{OH} \times 10^7$	$\mathbf{D} \times 10^7$	α_1'	α_1 (table)
6.756	1.754	0.5702	1.184	−0.1838	0.009901
6.956	1.1066	0.9037	0.203	0.7971	0.00627

Table II-2. α Values for Monoprotic Acids Centered on pK_a = 6

pH	α_1	α_0	pH	α_1	α_0
0.0	0.99999901	1E − 06	6.2	0.38686318	0.61313682
0.2	0.99999842	1.5849E − 06	6.4	0.28474725	0.71525275
0.4	0.99999749	2.5119E − 06	6.6	0.20076001	0.79923999
0.6	0.99999602	3.9811E − 06	6.8	0.13680689	0.86319311
0.8	0.99999369	6.3095E − 06	7.0	0.09090909	0.90909091
1.0	0.99999	9.9999E − 06	7.2	0.05935094	0.94064906
1.2	0.99998415	1.5849E − 05	7.4	0.0382865	0.9617135
1.4	0.99997488	2.5118E − 05	7.6	0.02450337	0.97549663
1.6	0.99996019	3.9809E − 05	7.8	0.01560166	0.98439834
1.8	0.99993691	6.3092E − 05	8.0	0.00990099	0.99009901
2.0	0.99990001	9.999E − 05	8.2	0.00627001	0.99372999
2.2	0.99984154	0.00015846	8.4	0.00396529	0.99603471
2.4	0.99974887	0.00025113	8.6	0.00250559	0.99749441
2.6	0.99960205	0.00039795	8.8	0.00158239	0.99841761
2.8	0.99936944	0.00063056	9.0	0.000999	0.999001
3.0	0.999001	0.000999	9.2	0.00063056	0.99936944
3.2	0.99841761	0.00158239	9.4	0.00039795	0.99960205
3.4	0.99749441	0.00250559	9.6	0.00025113	0.99974887
3.6	0.99603471	0.00396529	9.8	0.00015846	0.99984154
3.8	0.99372999	0.00627001	10.0	9.999E − 05	0.99990001
4.0	0.99009901	0.00990099	10.2	6.3092E − 05	0.99993691
4.2	0.98439834	0.01560166	10.4	3.9809E − 05	0.99996019
4.4	0.97549663	0.02450337	10.6	2.5118E − 05	0.99997488
4.6	0.9617135	0.0382865	10.8	1.5849E − 05	0.99998415
4.8	0.94064906	0.05935094	11.0	9.9999E − 06	0.99999
5.0	0.90909091	0.09090909	11.2	6.3095E − 06	0.99999369
5.2	0.86319311	0.13680689	11.4	3.9811E − 06	0.99999602
5.4	0.79923999	0.20076001	11.6	2.5119E − 06	0.99999749
5.6	0.71525275	0.28474725	11.8	1.5849E − 06	0.99999842
5.8	0.61313682	0.38686318	12.0	1E − 06	0.999999
6.0	0.5	0.5			

Then by hand or with the preceding spreadsheet method:

6.856	1.3932	0.7178	0.6754	0.3246	0.00788
6.793	1.61145	0.6205	0.99096	0.00904	0.009104

We see that α_1' has a high slope while α_1 changes little. Hand calculations are simple with monoprotics. Note the approximation (with **D** negligible) $\mathbf{H} = \sqrt{K_a C_a} = 13 \times 10^{-7}$ (p**H** 5.88) compared with 1.61×10^{-7} and 6.793) from the complete equation. Another approximate form taking **OH** negligible (in **D** of the Charlot form) yields a

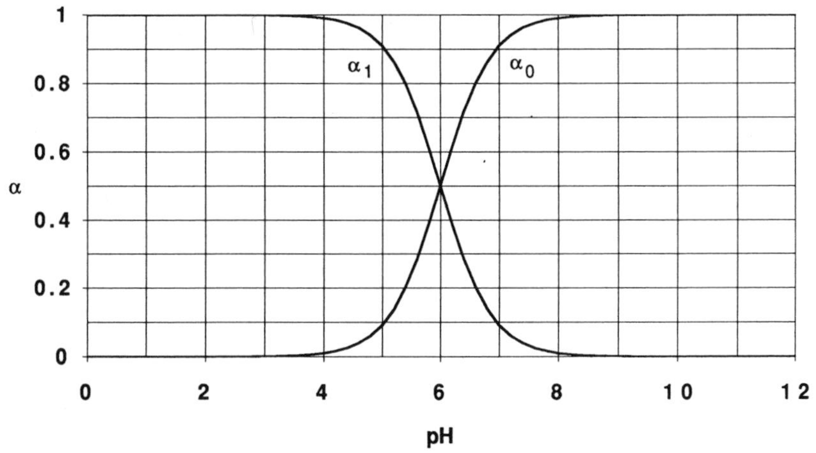

Fig. II-11. Plot of α values in Table II-2. The slopes of α are identical at corresponding distances from $\mathrm{p}K_a$ for every K_a case. The apparent differences are caused by the spread of **pH** scales. Compare Figs. II-6 and II-10.

pH of 3.5! Examination of the terms in the equations (Table II-2) shows the reason. Here **H** is larger than C_1 while **D** is smaller. The ratio function can be used as well as $\boldsymbol{\alpha}$, as in Fig. II-8, but may have more convergence problems in very dilute solutions or if one of the C terms is *zero*.

The expanded central region of the general HA curve and the logarithmic curve follow. The latter is useful in regions far from the $\mathrm{p}K_a$ value where the α values cannot be read precisely on the other plots. **Figure II-12*a*** is a log α plot of a general HA, for example, $\alpha_1 = 10^{-5}$ at a **pH** five units above the $\mathrm{p}K_a$ value of the system. This cannot be seen in Fig. II-11.

(ii) A buffer made 10^{-7} M in each HAc and NaAc. As before use Table II-2 and the new $\alpha_1' = 0.5 - 5 \times 10^6\mathbf{D}$ or $\alpha_2' = 0.5(1 - 10^7\mathbf{D})$ so that the previous values times 0.5 are applicable. A **pH** of 6.795 is found showing that the very weak base hardly has any effect at this dilution. The approximate **pH** $= \mathrm{p}K_a$ is far off because of the high α_0. Similarly, 10^{-7} M NaAc alone has $\alpha_1' = 0 - 10^7\mathbf{D}$ and the **pH** is 7.001.

In effect, we solved Eq. II-7 by using the form of Eq. II-6 as iterating functions. Several other rearrangements would serve here (R, the species ratio), but for the more complex mixture that follows and polyprotic cases the α or **n**-bar form is most efficient and offers a single valid way to obtain correct results.

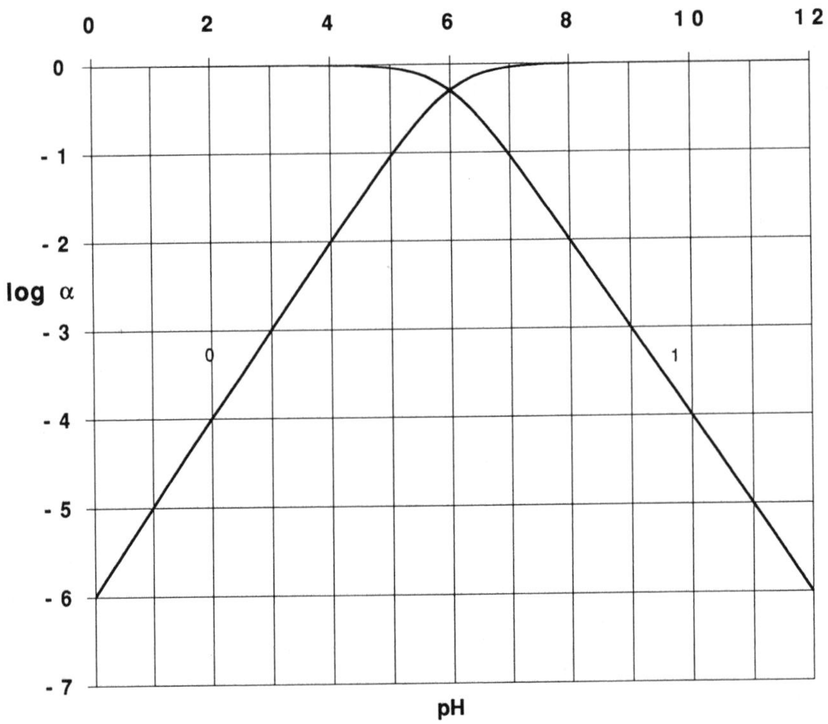

Fig. II-12*a*. Log α plot of a general HA.

Further Numerical Calculations

Use of the Charlot equation for weak acids, conjugate bases, and their mixtures (buffers) is well known and is treated in elementary texts. Let us look at more revealing uses of the master relations (see Eqs. II-4 to II-9).

II-6. MIXTURES OF SYSTEMS

(i) What is the pH of a mixture made 0.0200 M in each: formic, acetic, and propionic acids (designated with subscripts f, a, p)?

The total acidity is the total bound proton concentration plus **D** (net "free" or unbound). Here $\alpha_1 C$ gives the true equilibrium concentration of each HA:

$$C_{AB} = C_H = \alpha_f C_f + \alpha_a C_a + \alpha_p C_p + \mathbf{D} \quad \text{(total acidity)}$$

We insert each α_1 as $(1 + K_a/\mathbf{H})^{-1}$ for each of the acids. We choose 0.1 M ionic strength, 25°C, pK_a values of 3.56, 4.56, and 4.67 for formic, acetic, and propionic acids, respectively, C_H is 0.0600 M, so we have an equation in the single unknown **H**. One solution is to try several **pH** values and compute or plot the right side of the equation to find its crossing of 0.06 M. The preceding spreadsheet method works well here. The function Δ is just the whole right side minus 0.06. Since we know the answer is a **pH** between 2 and 3, this is not lengthy. This may be done for 0.1 or 0.05 pH unit intervals on the computer and the crossing interpolated. A spreadsheet for this case with the formulas is displayed in Appendix B. (A calculator with several memory keys also serves.) The reader may check that the answer, **pH** 2.6135* gives the following α_1 values: formic 0.8984, acetic 0.9888, and propionic 0.9913 for a bound proton level of 0.05757 M. The **H** value is 0.00244 M, which gives a total acidity, C_H, of 0.06001, which is probably close enough. If the pK_a's have uncertainty 0.01, the result may be recalculated at these limits and expressed as **pH** 2.614 ± 0.004.

Compare the longer, and approximate, methods used in advanced texts, except for the one by Ricci,[2] who does it in detail. A monoprotic muxture done by a log C diagram method is given by Hogfeld.[3] See, also, Butler[4] and the Kolthoff–Elving *Treatise*[5] for a variety of ad hoc approaches. In general, any mixed system **pH** problem can be handled rigorously and easily by the summation of **H** binding shown here.

The same setup solves related problems:

(ii) What is the **pH** after adding NaOH? Subtract the C_{sb} added from the 0.0600 M solution to get a new C_H. The reader can practice the method to check the following results. NaOH was added to make the lowered C_H values (or HCl for the first higher value):

C_H	0.065	0.06	0.05	0.04	0.03	0.02	0.01	0.005
pH	2.219	2.614	3.394	3.896	4.302	4.678	5.126	5.487

*The α values may be read from Table II-2. For example, α's for **pH** 2.614 are found (pK_a 2.614) units below 6.000 (e.g., 3.56–2.614 = 0.946. Subtracting 0.946 from 6 gives the α line in the Table for formate at pH 2.614. These come at "**pH**" 5.054, 4.054, and 3.944 for formic, acetic, and propionic systems in Table II-2. A large graph of that table can serve to yield α values for all monoprotic acid systems (see Figs. II-11 and II-12*a*.

These are points along a titration curve of the mixed acids at constant volume and I. The closely related equation for a conventional titration curve is given in Chapter IV.

(iii) What concentration of NaOH, C_{sb}, is needed to make the **pH** to a desired value? Insert the α values calculated from the given **pH** and get the new C_{AB}, which is $0.0600 - C_{sb}$. The reader may check that **pH** 4.000 is obtained if 0.02251 mol NaOH is added per liter of the mixture in (i).

(iv) A mixture that is awkward in the traditional treatments is easily handled by the α method. Find the **pH** and other species in a mixture of 0.0400 M acetic acid and 0.0200 M sodium formate.

First obtain valid constants: I is approximately 0.020 M. The Davies activity coefficient $f = 0.871$. Subtract $0.120(\log f^2)$ from the pK_a^0 of the systems to get acetic pK_a, $4.756 - 0.120 = 4.636$, and formic pK_a, $3.75 - 0.12 = 3.63$.

Next the total acid is

$$C_H = 0.0400 = C_f \alpha_{1f} + C_a \alpha_{1a} + \mathbf{D}$$

$$0.0400 = 0.02(1 + 10^{\mathbf{pH}-3.63})^{-1} + 0.04(1 + 10^{\mathbf{pH}-4.636})^{-1} + \mathbf{H}$$

The right side gives, for **pH** 3.9, $C_H - 0.04091$, and for **pH** 4.0, $C_H = 0.03857$. Linear, not log, interpolation to $C_H = 0.04$ (via **H**, 1.259×10^{-4} and 10^{-4}) gives $\mathbf{H} = 1.158 \times 10^{-4}$ M, or **pH** 3.936. Then, returning this to the α expressions yields M values at equilibrium:

Formic acid	0.0066
Formate	0.0134
Acetic acid	0.0333
Acetate	0.0067

Note that one need not decide in advance whether major proton transfer occurs. The α methods give a correct proton distribution. The identical result is found for a mixture 0.02 M in formic acid, 0.02 M in acetic acid, and 0.02 M in sodium acetate. Note that the proton acidity balance (0.04 M) holds when the 0.0001 M **H** is added in with the HAc and HForm.

A helpful illustration is a logarithmic concentration diagram for this problem (**Fig. II-12*b***). The species concentrations are $\alpha_n C$, or $\log C_n = \log \alpha_n + \log C$. Thus, log C species diagrams are just the log α diagrams shifted down the log C axis. In this case we place two

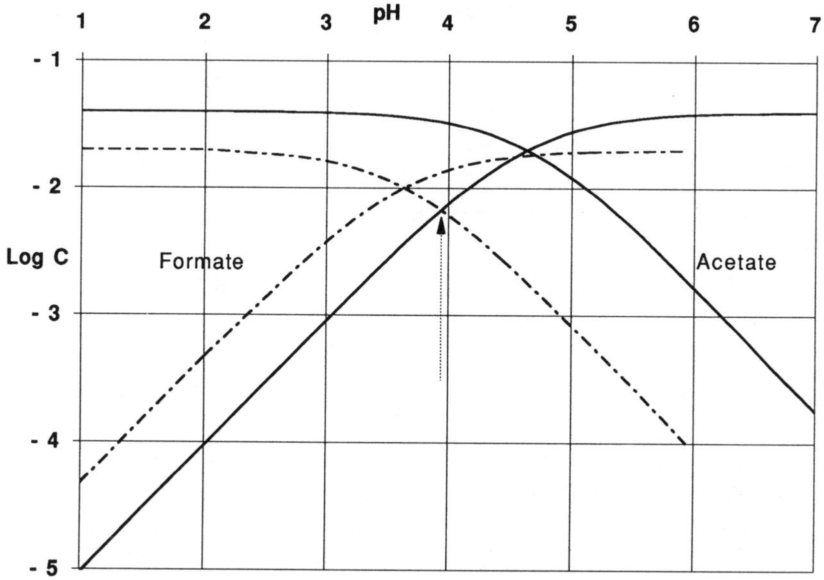

Fig. II-12*b*. Logarithmic concentration diagram for the mixture 0.04 *M* HAc + 0.02 *M* Na formate. The arrow shows the approximate solution at log(0.0066 *M*) or −2.18 at **p**H near 4.

log α monoprotic curves (Fig. II-12*a*) with crossings at **pH** 3.63 and 4.64 and with maxima at log 0.02 and log 0.04 (−1.7 and −1.4). The solution to the problem occurs approximately where [HForm] = [Acetate$^-$], implied by the reaction,

HAc	+ Formate$^-$ ⇔	HForm +	Acetate$^-$	
0.0333	0.0134	0.0066	0.0067	(equilibrium *M*)

Figure II-12*b* is a logarithmic concentration diagram for the mixture 0.04 *M* HAc + 0.02 *M* Na formate.

Salt of a Weak Acid and a Weak Base. Much effort has been put into complex methods for explaining how to find the **p**H of solutions of compounds like ammonium formate or ammonium cyanide where we have a weak proton donor and a weak acceptor, not conjugates. Now one can see that it is just another example of a two-system mixture with equal *C*'s of the systems so that $C(\alpha_{1x} + \alpha_{1y}) + \mathbf{D}$ constitutes the total acidity balance and gives the complete **H** equation as used in the preceding example.

Much confusion has arisen in trying to explain the "degree of reaction," meaning protonation or proton transfer. In the ammonium formate, 99 + % of the ions remain as they started, while for the cyanide, about half changes to NH_3 and HCN (both pK_a's near 9.2) owing to the values of the K_a's and the balancing **pH**. Looking at rough sketches of the log $C(\log \alpha)$ diagrams should clarify this. Examples are shown in **Fig. II-12*c*** and in Butler's book. The balancing **pH** is compatible with high α's of the starting species in the first but not the second case. The α diagrams make this immediately apparent. There is no **pH** at which both NH_4^+ and CN^- have α's near 1, while NH_4^+ and formate$^-$ have their α's near 1 over the **pH** range 6–8: no large proton transfer is needed to reach an equilibrium satisfying the pK_a's and the K_W. The opposite is true with ammonium cyanide.

The alpha diagrams in Fig. II-12*c* are of three monoprotic acids for weak–weak salt illustrations. Visualize central equilibrium for salts: ammonium acetate, anilinium acetate. The same results are seen starting with the conjugates of each case, e.g., acetic acid plus ammonia. The cyanide system was omitted for clarity. It is close to the ammonia, making a pair like the acid systems.

Returning to the **H** balance,

$$C_H = C(\alpha_{1x} + \alpha_{1y}) + \mathbf{D}$$

For the 1 : 1 salt cases, the single C value describes both systems and the total available protons. This produces complete equations for weak–weak monoprotic systems of this type:

$$\alpha_{1x} + \alpha_{1y} = 1 - \mathbf{D}/C$$

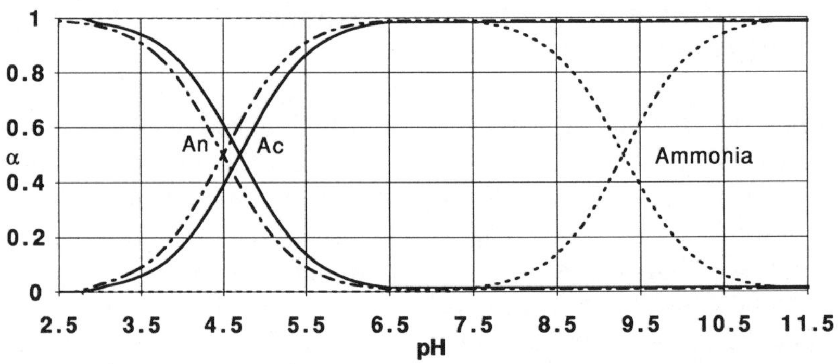

Fig. II-12*c*. Alpha diagrams of three monoprotic systems for weak–weak salt illustrations: An, anilinium system; Ac, acetate; basic is ammonia system.

or, using $1 - \alpha_1 = \alpha_0$,

$$\alpha_{0y} - \alpha_{1x} = \mathbf{D}/C \quad \text{or} \quad \alpha_{0x} - \alpha_{1y} = \mathbf{D}/C \qquad \text{(II-11)}$$

These balances are seen on the α diagrams (Fig. II-2*c*) where two-system intersections occur above and below the intermediate point.

Note the significance that the imbalance of protonation of the two systems gives the **D** imbalance of the solution. Putting in the α expressions for the two systems, one obtains a multiterm cubic equation in **H**. As usual, it is easier to use the α forms in a spreadsheet to find the **H** that satisfies specific cases, as was done in the multisystem problems given in the preceding. Generally, without the 1 factor and with the larger α terms (not those approaching zero), the forms II-11 give less rounding error and better convergence in computation.

In this case, some insight is gained from the approximation found by looking at a high C (above 0.1 M) and low **D** (**pH** 4–10), which make $\mathbf{D}/C$ much smaller than 1. This shows that the conjugate α's must correspond:

$$\alpha_{1x} \cong 1 - \alpha_{1y} = \alpha_{0y} \quad \text{and} \quad \alpha_{1y} \cong 1 - \alpha_{1x} = \alpha_{0x} \quad \text{and} \quad \mathbf{H} \cong \sqrt{K_{a,x}K_{a,y}}$$

This implies that all protons lost by the acidic species are gained by the acceptor. For example, in ammonium acetate, approximately 1% of the protons of NH_4^+ are lost and 1% of acetate is protonated to give α values of 0.99 NH_4^+, 0.01 NH_3, 0.01 HAc, and 0.99 Ac^-. This clearly happens if the final **pH** lies halfway between the pK_a values of the two systems (Fig. II-12*c*). But remember that this is only valid in the approximation region used (high C and **pH** 4–10; see the examples that follow that compare 0.01 M and 0.0001 M). Thus, ammonium acetate has a predicted **pH** 7.0 (6.9), which is neutral, because of the fortuitous symmetry of these pK_a values: ammonia is as strong a base as acetic is an acid (in very dilute water solutions). Whereas NH_4CN has $\mathbf{pH} \cong \frac{1}{2}(9.3 + 9.1) = 9.2$, anilinium acetate has $\mathbf{pH} = \frac{1}{2}(4.65 + 4.56) \cong 4.6$. The last two require α values near 0.5 since the systems are near their pK_a crossings. The reader can sketch the α curves to demonstrate that the systems move toward each other to obtain the compromise **pH** (in the high C limit cases). Looked at in this manner, it is not so mysterious that proton transfer occurs until $K_a/\mathbf{H}$ is satisfied for both systems.

The reader might use the spreadsheet method to calculate the **pH** and α values of varied C's of ammonium cyanide and of anilinium trichloroacetate (pK_a 0.7). Start with the approximate pH as before (9.2 and 2.675) and obtain new **H** trials, with the complete **H** equation (II-11; α from II-6) moving toward neutral until unchanging α's and **D**/C are obtained. Using the pK_a values mentioned here, with pK_w 13.80 yields the following:

C Values (M)	pH	α_{1x}	α_{0y}	**D**/C
	Final Values			
NH_4CN				
0.01	9.197	0.559	0.556	−0.003
0.0001	9.044	0.643	0.468	−0.175
Anilinium trichloroacetate				
0.01	3.344	0.953	0.998	0.045
0.0001	4.427	0.626	0.9998	0.374

Note that Eq. II-11 is satisfied, but the α equality approximations become poorer with dilution. The approximate values are the high C limiting ones. See how they are approached by 0.1 and 1 M for C. (The other α's are obtained if desired as with any monoprotics by subtraction from 1.)

A paper treating this topic with a complete equation but with different methods is that of Cardinalli et al., *Journal of Chemical Education*, 1990, Vol. 67, p. 221. Note their Eq. 1 is our II-11. Other computer methods may be found in the continuing "Computer Series" articles of the *Journal of Chemical Education*, for example, no. 111 in February 1990, p. 150, and no. 115 in June 1990, p. 500. I believe that $\bar{\mathbf{n}}$ derivations of equations and spreadsheet computations are more efficient.

II-7. WEAK–STRONG MIXTURES

(a) What is the pH after titrating HCN and adding 1% excess base? Take the final C_0 of CN^- to be 0.200 M, which means $C_{sb} = 0.002\ M$. (Note that this is equivalent to making a mixture with analytical concentrations 0.2 M HCN and 0.202 M KOH for Eq. II-6.) Use pK_a 8.96 and pK_w 13.75 for ionic strength 0.2 (Davies equation). Note that C_{AB} is negative in this mixture. In Eq. II-9 we get,

$$\mathbf{H} = 10^{-8.96}\,\frac{-0.002 - \mathbf{H} + \mathbf{OH}}{0.200 + 0.002 + \mathbf{H} - \mathbf{OH}}$$

(*Logic*: The numerator, [HCN], is also the $[OH^-]$ in excess of the NaOH and that from water, **H**, and thus is that produced by protonation of CN^-. The denominator is the total CN^- minus that lost to the numerator as HCN. The master equation with C_{AB} always gives these correctly.)

With the **H** terms negligible on the right and $\mathbf{H} = K_w/\mathbf{OH}$ on the left, solving the quadratic gives **OH**, 0.00306 *M* (p**H** 11.235). The CN^- has produced an extra 0.00106 *M* OH^- (the HCN formed). Note $C_1 = 0$, but $[HCN]_{eq} = 0.00106$.

Alternatively, use Table II-2 and Fig. II-11. For this case, $\alpha_1' = (-0.002 + \mathbf{OH})/0.200$. Here **OH** must be above 0.002 *M* (p**H** 11.050). Working up from this leads to $\alpha_1' = \alpha_1 = 0.0053$ at 2.28 p**H** units above pK_a.

(b) What is the pH of 0.100 *M* H_2SO_4? This constitutes 0.1 *M* strong acid (C_{sa}, the first-step proton) followed by a weak equilibrium step. Use K_a 0.029, pK_a 1.538, $C_{AB} = 0.200$ *M*, $C_0 = 0$, and $C_{sa} =$ 0.100 in Eq. II-7 ($\mathbf{D} \cong \mathbf{H}$ in this acidity). Note that taking **H** or **OH** as negligible in a calculation always requires those terms added to or subtracted from other larger terms so that omitting them has an effect smaller than the precision desired. Other **H** and **OH** terms cannot be omitted with small effect.)

Chemistry: Start from 0.1 *M* H^+ and HSO_4^-. Then go to equilibrium:

$$\begin{array}{ccccc} HSO_4^- & \rightleftharpoons & H^+ & + & SO_4^{2-} \\ 0.1 - x & & x & & x \end{array}$$

so that total $\mathbf{H} = 0.1 + x$. Then, in Eq. II-9,

$$\mathbf{H} = K_a\,\frac{C_{AB} - \mathbf{H}}{C_0 - C_{sa} + \mathbf{H}} = 0.029\,\frac{0.2 - \mathbf{H}}{\mathbf{H} - 0.1}$$

(*Term logic*. Numerator: $C_{AB} - \mathbf{D}$ [or **H** here] = bound protons = $[HSO_4^-]$. Denominator: $x = [SO_4^{2-}] = \mathbf{D} - 0.1 \cong \mathbf{H} - 0.1$; acidity in excess of the first step (0.1 *M*). No salt like K_2SO_4 was added, so $C_0 = 0$. Logical analysis of the terms obtained from the master equation helps see why it always works and why rederivation for each case is not required.)

Solve the quadratic. Alternatively, use the general HA Table II-2 or Fig. II-11 and $\alpha_1' = (0.2 - \mathbf{D})/0.1$ to find their simultaneous **pH**. Note that the highest possible **pH** is 1.0; there is no H^+ from HSO_4^-. Start from this (0.538 units below pK_a) and work down to find $\alpha_1 = \alpha_1' = 0.80$ at $\mathbf{H} = 0.120\ M$ (**pH** 0.92). Thus, 20% of the HSO_4^- is ionized. This would be about 40% if the strong acid (0.1 *M* of the first step) was absent, as in a 0.1 *M* $KHSO_4$ solution.

Initially, iteration may be required to adjust to the proper ionic strength. The constant used is for $\boldsymbol{I} = 0.14\ M$ and Davies equation activity coefficients, $K_a^0 = 0.010$. The ionic strength of the equilibrium mixture is about 0.14 *M*, which often must be found by successive trials. The species inventory at equilibrium is as follows: $\mathbf{H} = 0.120\ M$, $HSO_4^- = 0.080\ M$, and $SO_4^{2-} = 0.020\ M$. But if K_a^0 is used, $\mathbf{H} = 0.108$, $HSO_4^- = 0.092$, and $SO_4^{2-} = 0.008\ M$. Here K_a^0 gives 9% error in **H** but over 50% error in the weak acid ionized.

Realistic I Iteration Method. The effect of ionic strength on the acid HSO_4^- is large owing to the 2− ion charge. In general, one would not know the final ***I***, and iteration would be required using K_a^0 and *f*. The effect of ***I*** on log K_a is shown in **Fig. II-13**. The Davies equation predicts better than the DH equation without the added 0.3***I*** term

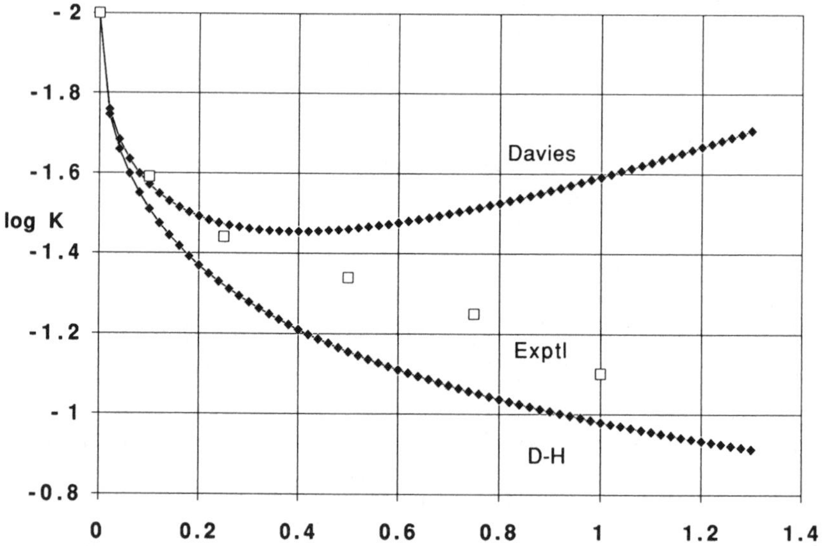

Fig. II-13. Variation of log K_a of HSO_4^- with ionic strength, log $K_a^0 = -2.00$. Some experimentally determined pK_a values are shown (□).

(Guntelberg form) up to about $0.4\boldsymbol{I}$ (see Appendix A). The experimental values are for low concentrations of the acid in the presence of electrolyte $NaClO_4$ to set the desired ionic strengths. Put *all* the ions present of charge z and concentration c in the usual $\boldsymbol{I}$ expression:

$$\boldsymbol{I} = \frac{1}{2} \sum_{i=1}^{n} c_i z_i^2$$

$$K_a = K_a^0 / f_{2-}$$

as the f_+ and f_- (of H^+ and HSO_4^-) approximately cancel in this K expression.

We need the usual α, α' expressions but inserting for K_a, K_a^0/f_{2-}. Using $K_a^0 = 0.0100$ and the Davies equation (I-10) with an added term (Appendix A), this becomes

$$K_a = K_a^0/f_{2-} = \frac{0.0100}{E[-4 \times 0.509[\sqrt{\boldsymbol{I}}/(1+\sqrt{\boldsymbol{I}})] + 0.61\boldsymbol{I}]}$$

The E term is the antilog of the Davies log f_{2-} expression. This is used with $\boldsymbol{I} = 0.1 + 2x$. (Prove with the preceding $\boldsymbol{I}$ expression.) Here $\alpha_1' = 2 - 10\mathbf{H}$ and $\alpha_1 = (1 + K_a/\mathbf{H})^{-1}$. A program starting with trial values of x quickly iterates through $\boldsymbol{I}$ and the α's by trial to precise values of **H**. [Choose ITERATE and MANUAL CALCULATION in the Calculation menu.] The value 0.11975 was obtained and the corresponding $\boldsymbol{I}$ was 0.1396 *M*. This is a more realistic approach to the problem since one usually does not know the final ionic strength for such highly ionized acids. Once the program is set up, other mixtures of the acid and its salts can be solved as well as other acid systems by changing the K and C parameters. See exercises at the end of this chapter.

II-8. DILUTION OF WEAK ACIDS

This topic corrects a common misconception (that weak acids increase steadily in ionization, α_0, when diluted) and illustrates an algebraic simplification useful to know. It also is an extreme example of the wrong results approximate equations give when used beyond their limitations. Let us examine the **H** and the α_0 (fraction ionized) of acids of various strengths, shown in **Fig. II-14**, for pK_a 5, 7, and 9. Simple Le Chatelier reasoning in Eq. II-2 suggests that dilution of

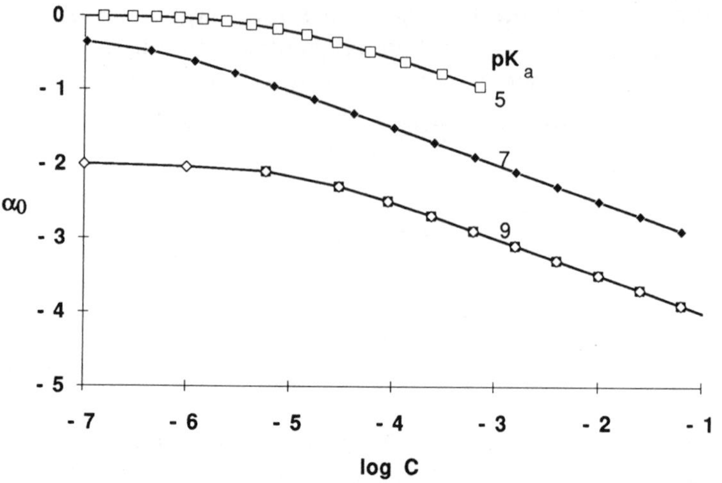

Fig. II-14. Ion fraction α_0 upon dilution of three pure weak acid solutions showing that nearly complete ionization does not occur for weaker cases.

HA solutions must increase the fraction ionized. The operation of K_w (in the complete equation, II-7) show that this is not so simple. For the acid alone, we have

$$K_a = \mathbf{H}(\mathbf{D})/(C_1 - \mathbf{D}) \qquad \mathbf{D} = \mathbf{H} - K_w/\mathbf{H} \tag{II-12}$$

and

$$\alpha_0 = (1 + \mathbf{H}/K_a)^{-1}$$

For a given acid K_a over a wide range of C_1 values, we must calculate **H** and then α_0. This looks like a long job since Eq. II-12 is cubic in **H**. The trick is to invert the variables and find C_1 for a range of chosen **H** values:

$$C_1 = \mathbf{D}(1 + \mathbf{H}/K_a) \quad \text{and then} \quad \alpha_0 = \mathbf{D}/C_1 \tag{II-13}$$

For example, to have p**H** 5, an acid of pK_a 9 must be 0.1 M, giving $\alpha_0 = 10^{-4}$. The graph shows that such a weak acid (such as HCN or NH_4^+) never rises above 1% ionization ($\alpha_0 = 0.01$, or log $\alpha_0 = -2$) because all solutions approach p**H** 7 as C_1 goes below 10^{-7}. Thus, $\alpha_0 \rightarrow (1 + 10^{pK_a - 7})^{-1}$ as a limit. For a stronger acid like acetic (pK_a 4.7), α_0 approaches the constant value 1/1.005, or 99.5% (but never reaches 100%). For the pK_a 7 acid, the limit is just 50% ionized ($\log 0.5 = -0.3$).

Dilution of a Pure Acid HA to Produce 50% Ionic Form

Only acids of pK_a < 7 can be ionized to more than 50% by dilution of the pure acid in water. These stronger acids are 50% ionized at $C_1 = 2\mathbf{D}$ (Eq. II-13) (approximately at $C_1 = 2\mathbf{H} = 2K_a$). This is where $\alpha_0 = 0.5$. For example, an acid of pK_a^0 2.0 (e.g., HSO_4^-) would be 50% ionic at 0.02 M if the zero ionic strength K were valid. Acetic acid is 50% ionized at 5.5×10^{-5} M (0.1 M ionic strength, pK_a 4.56). Note that this rules out 50% ions in weaker acids for which $\mathbf{H} = K_a$ would require a basic solution. (No dilution of acid, no matter how weak, can produce a basic solution!)

There may seem to be discrepancy between Eqs. II-12 and II-13 for α_0. However, it is resolved when we realize that for finite (not zero) C_1, the **pH** never gets to 7 (**D** cannot be zero). For pH 7 α_0 can be evaluated, while C_1 cannot, for solutions of HA alone. If $C_1 = 0$, α_0 becomes indefinite (in II-13). Buffers are different. (The points in the graph of Fig. II-14 were calculated at **pH** increments of 0.2 units from about pH 4.1–6.9.)

The idea that diluted weak acids approach 100% ionization rests on an approximate form of Eq. II-12 that holds only at high concentration. Assuming $x = [A] = \mathbf{H}$ gives:

$$K_a = x^2/C_1$$

A further approximation, $\alpha_0 = x/C_1$, produces (after dividing both sides of the first equation by C_1 and eliminating x)

$$\alpha_0 \cong \sqrt{K_a/C_1}$$

This predicts that α_0 grows larger as C_1 gets smaller. The contrast with the limiting behavior of the complete equation is presented in **Fig. II-15**. The figure shows fractions ionized in the pK_a 9 acid by an approximation valid only in high concentrations and by the complete equation.

The approximations may even predict basic solutions and α_0's above 1. For 10^{-6} M HA with pK_a 5 we get $\alpha_0 = \sqrt{10}$. For 10^{-7} M HA of pK_a 9, we get a **pH** of 8 with the approximate equation. (No matter how weak an acid may be, its pure solutions cannot be basic.) We repeat that the trouble here is omission of $[H^+]$ from water (the K_w term in correct equations), which becomes larger than that from HA at high dilution and represses its ionization.

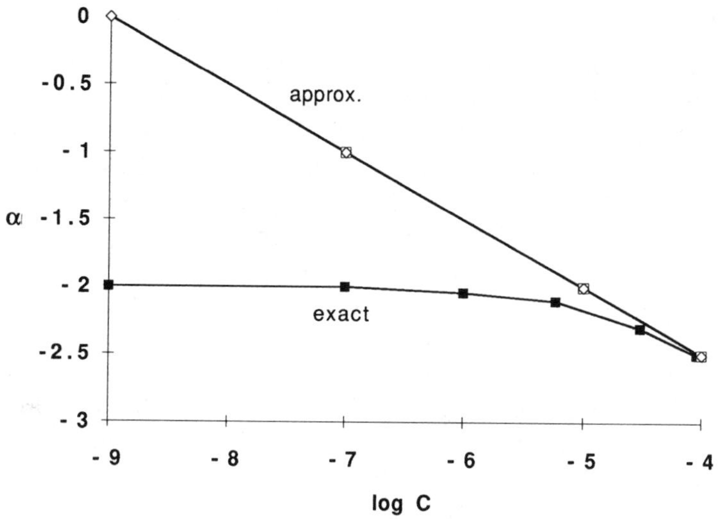

Fig. II-15. Fractions ionized in pK_a 9 acid of Fig. I-14 by an approximation valid only in high concentrations and by the complete equation.

II-9. CONJUGATE BASE SOLUTIONS AND COMPLETE CUBIC EQUATIONS

Equation II-7, with **D** in terms of **H**, becomes

$$\mathbf{H} = K_a \frac{C_1 - \mathbf{H} + K_w/\mathbf{H}}{C_0 + \mathbf{H} - K_w/\mathbf{H}}$$

[Iterating by trial-and-error guesses for **H** on the right side diverges in some cases. For instance, if a trial **H** is $<\sqrt{K_a C_1}$ for the acid alone ($C_0 = 0$), trials do not converge. See ref. 1.] Letting $\mathbf{H} = x$ and collecting terms yields the conventional cubic form

$$x^3 + (C_0 + K_a)x^2 - (K_a C_1 + K_w)x - K_a K_w = 0 \qquad \text{(II-14)}$$

The question arises as to which form of the equation (II-6 or II-7) is desirable for computer (or graphical) solution when no terms are clearly negligible. The cubic form can have at most one positive real root (Descartes' rule of signs). A computer calculation at **H** intervals over a small range should lead easily to the root. However, the cubic equation coefficients must be recalculated with new C values in every new problem. (See Appendix B.)

A more useful form is the bound proton function, α_1 (Eq. II-6), the master relation for a monoprotic system. This turns out to be a natural iterating function for equilibrium problems, and it converges. The α_1 term is a unique curve for each system and need only be evaluated once while the left material balance function can easily be evaluated with the specific C's of the case and trial **H** values to find a solution. The slope of α_1 is always down while the other (α_1') slopes up vs. pH or vs. **OH** (or the opposites vs. **H** and **pOH**). This makes their crossing well defined and easily found. The plot vs. **OH** (or **H**) (or a log-log plot) gives more clearly defined solutions at the extremes where the semilog slopes are smaller. Figures II-7 and II-10 illustrate the intersection of a typical α_1, α_1' case.

Figure II-16 shows the α_1 lines of four weak acids (pK_a's of 2, 4, 6, and 8). The α_1 is fixed by the K_a value, whereas the α_1' line depends only on C values of a specific problem. Here C_1 and C_0 are 0.0100 M, a buffer having **pH** $\cong$ pK_a only where **H** is negligible compared to C_1 so that the α's are both 0.5. And $\alpha_1' = 0.5 - 50$**D**. [The value 0.5 is the limit at **pH** near 7. In a base, the α' line is an inverted mirror image of the acid part (Figs. III-9 and III-10). Thus, the A takes on protons in the pK_a 8 case to make basic solutions.] Dilution, making the C's smaller (α curves unaffected) shifts α_1' to the right. The second case has both C's equaling 0.001 M ($\alpha_1' = 0.5 - 500$**D**). When the intersection occurs on the horizontal portion of α_1' (as for the acid of pK_a 6), the approximate buffer equation is valid. For the others a quadratic or cubic relation is required. Study of this graph can explain what

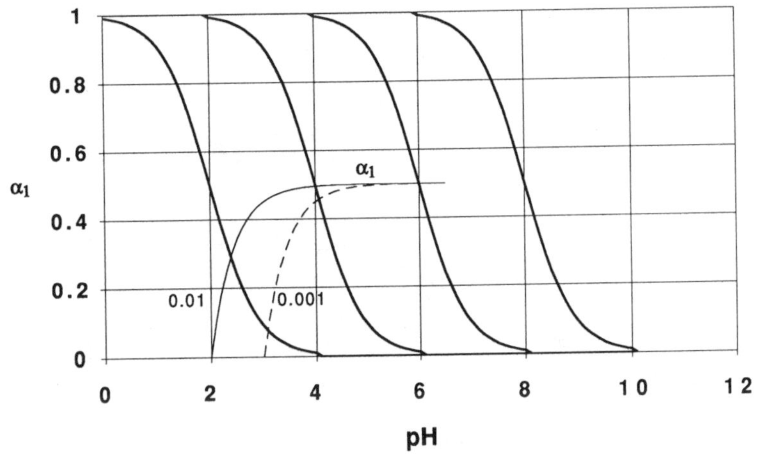

Fig. II-16. The α_1 lines of four weak acids, pK_a's 2, 4, 6, and 8, with intersections of α_1' for two 1:1 buffers having total C_A of 0.01 and 0.001. Note again the shift of α_1' one unit right for a 10-fold decrease in C.

occurs in the computer solutions illustrated in the preceding. (Log–log plots are shown in Figs. II-17b, III-17 and in Figs. III-16.)

Example. Figure II-17*a* repeats some previous work to compare two ways to solve the problem: Find the **H** in a solution of 10^{-5} *M* in NaA alone of a system of pK_a 6. The cubic form above gives

$$x^3 + 1.1 \times 10^{-5}x^2 - 10^{-14}x - 10^{-20} = 0$$

Letting $x = y(10^{-7})$ and multiplying through by 10^{21} gives

$$y^3 + 110y^2 - y - 10 = 0 = \boldsymbol{d}$$

Figure II-17*a* shows the plot ***d*** of this cubic crossing zero and the plot of Δ of the α's, the difference of the two sides of Eq. II-6($\times 100$) crossing zero as they become equal at a **pH** 7.514 or $\mathbf{H} = 3.06 \times 10^{-8}$.

A variety of computer iteration programs and/or plots can be used to solve such problems. In this case, for example, either function could be computed at **H** intervals until the function changes sign. Then smaller intervals can be used until desired precision is obtained. Zero ionic strength is used here. This is not an intersection method! The solution occurs at zero for both methods plotted.

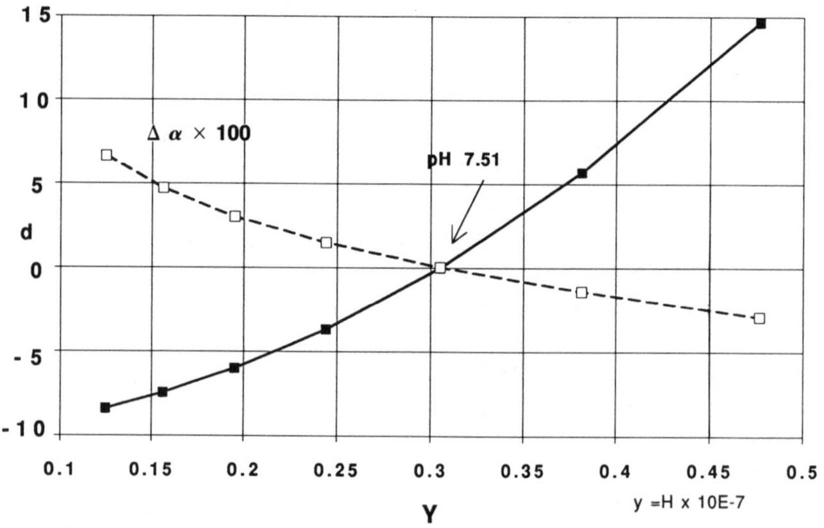

Fig. II-17*a*. Plot ***d*** of cubic crossing zero and plot of Δ of α's.

Calculation of some α values should clarify the approach for the reader. See the α's given in Table II-2 for this acid. Note $C_1 = 0$ and $C_0 = 10^{-5}$. Examples near the solution are as follows:

pH	H	OH	α_1	α_1'	Difference, Δ
8	10^{-8}	10^{-6}	0.0099	0.099	−0.089
7.5	3.16×10^{-8}	3.16×10^{-7}	0.0306	0.0285	−0.0021
7.514	3.06×10^{-8}	3.27×10^{-7}	0.0297	0.0296	+0.0001
7	10^{-7}	10^{-7}	0.0909	0	+0.0909

Note the opposite trends of the α's. The reader should ponder the meaning of the α values. For example, at **pH** 7, $\alpha_1 = 0.0909$, or $\frac{1}{11}$ is the fraction protonation in any HA system when the **pH** is $pK_a + 1$. An α_1' of zero signifies that a pure A^- solution can have neutral **pH** only if no proton transfer takes place (regardless of pK_a). That these are irreconcilable means that this is not a possible equilibrium **pH** for this case. Only the true **pH** makes the α's equal, satisfying the equilibrium and the material balance conditions simultaneously.

Log-Log Method for Basic Cases

The curvature of the α lines in basic cases when their values are small can be reduced by plots (and spreadsheet calculations) using log α_1 and log α_1' vs. **pH**. An example for 0.100 *M* sodium acetate and 0.1 *M*

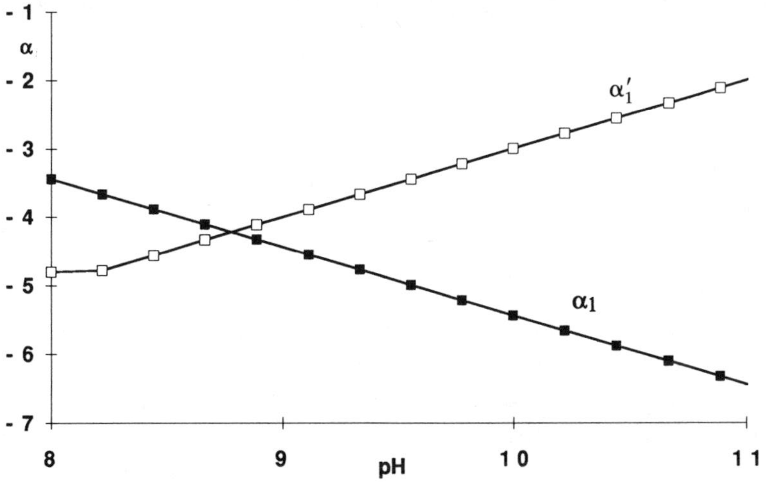

Fig. II-17*b*. 0.1 *M* sodium acetate log α plots.

I value constants is shown in **Fig. II-17*b***. The spreadsheet interpolation formula using logs gives the intersection at **pH** 8.77. With functions so nearly linear, one such calculation is correct since convergence is assured for straight lines. Compare Figure II-17*a*.

II-10. APPROXIMATE pH CALCULATIONS: GETTING ESTIMATES OF pH REGION TO USE IN RIGOROUS CALCULATIONS

See the α in Fig. II-11 diagram for the correlations with **pH** approximations.

The Charlot relation (Eq. II-7) is used. Letting **D** be negligible compared to C terms and taking **D** as **H** in acid and as **OH** in basic cases leads to **H** equations of the first or second power:

(a) Weak acid solutions alone: C_1 (molar), $\mathbf{H} = K_a(C_H - \mathbf{D})/\mathbf{D} \cong K_aC_1/\mathbf{H}$. Thus,

$$\mathbf{H} \cong \sqrt{K_aC_1}$$

is the approximation. It may be good at high C values and low K_a values. It gives worthless results at the opposite extremes. Compare 0.1 M and 10^{-6} M acetic acid (pK_a 4.7). Compare 0.1 M trichloroacetic acid (pK_a 0.7). The complete cubic is required only near neutral **pH**.

(b) Weak base alone: C_0 (molar), $\mathbf{H} \cong K_a\,\mathbf{OH}/C_0$. Using $\mathbf{OH} = K_w/\mathbf{H}$ gives

$$\mathbf{H} \cong \sqrt{K_aK_w/C_0}$$

(c) Buffers (mixtures of C_1 and C_0):

$$\mathbf{H} \cong K_aC_0/C_1$$

Conclusion. For monoprotic cases, the Charlot equation can often be solved with obvious approximations to get satisfactory results. With mixtures of systems and with polyprotic acids, however, the complete forms are more satisfactory and lead to correct results without guesswork. The complete approach for these monoprotic cases was given to show the essence of the methods in this work. Once a spreadsheet is prepared, solving by the complete α method (by entering the new C and K parameters) is faster than the quadratic. The time saving and power of the method are more clearly seen in the following chapters.

SUMMARY

1. Equivalent solutions and proton balance: Protonically equivalent mixtures (those having the same **pH** and acid–base active species but possibly differing in inert ions, like Na^+ and Cl^-) can be made by mixing appropriate quantities from any of the following pairs: (a) HAc + NaAc, (b) HAc + NaOH, (c) HCl + NaAc.
2. At a chosen **pH**, the K_a expression, the α_n relations, and the conjugate ratios R contain the same information. They are arrangements of the equilibrium condition.
3. Proton balancing using the bound proton condition sets equal the sum of bound acidity, $\Sigma\, \alpha_1 C_A$, for all HA systems and the material balance (total $C_{AB} - \mathbf{D}$). This produces a complete equation in a single variable **H** for mixed systems (Section II-6). This handles correctly any mixtures of weak and strong systems including the interesting special cases of salts of a weak acid and a weak base like NH_4CN and anilinium trichloracetate. These have C (molar) of each system and also C_H all equal to C.
4. In solutions of a weak acid, HA, alone, the fraction ionized, α_0, does *not* increase steadily upon dilution. It approaches a limit that is the value of α_0 at neutral **pH**. The C needed to give a desired fraction ionized can be calculated only for certain allowed cases (Section II-8).

EXERCISES

1. For a buffer made 0.01 M in HA and 0.001 M in NaA, use Eq. II-6 to determine what terms are needed for precise **H** calculation of pK_a cases 2, 6, 7, and 12.

2. Set up a complete equation relating **H**, K_a's, and C's for any mixture of four weak monoprotic acids and their sodium salts. Assume the K_a's are closely spaced near 10^{-3}.

3. At what C is dilute acetic acid 10.00% ionized? Assume pK_a 4.56.

4. (a) Find C_{AB} and C_A for mixture made 0.0200 M in HCl, 0.700 M in Na acetate and 0.0060 M in NaOH. (b) How can the same resultant **H** solution be obtained using NaOH and acetic acid? Take 0.5000 M of each and make 1 l.

5. Use the trial iteration scheme in the sulfuric acid example (Section II-7*b*) to find accurate **H** in the following mixtures using K_a^0 0.0100 for HSO_4^-. Set up the computer for trial values of **H**, $\boldsymbol{I}$, the Davies f_{2-}, α_1, α_1', and their difference. When the difference changes sign, subdivide the trial intervals until the desired precision (three figures) is obtained.

a. 0.0100 M H_2SO_4. Compare with the 0.100 M result in the chapter. (*Ans*: $\mathbf{H} = 0.0150_5$ M.)

b. 0.0200 M $NaHSO_4$. (*Ans*.: $\mathbf{H} = 0.0126$; $\boldsymbol{I} = 0.02 + 2\mathbf{H}$.)

c. 0.0100 M each of $NaHSO_4$ and Na_2SO_4. (*Ans*.: $\mathbf{H} = 0.00583$; $\boldsymbol{I} = 0.04 + 2\mathbf{H}$.)

6. Find the C of $NaHSO_4$ that will be 50 and then 90% ionized. First use a 0.1 M ionic strength value for K_a and then try to program the iterations for the case where the $NaHSO_4$ is the only material added (find the $\boldsymbol{I}$ function in C terms).

REFERENCES

1. Norris, A. C., *Computational Chemistry*, Wiley, New York, pp. 79–80. Various methods of solving equations and avoiding divergence are given in this book.
2. Ricci, J. E., *Hydrogen Ion Concentration*, Princeton University Press, Princeton, NJ, 1952.
3. Hogfeldt, E., in *Treatise on Analytical Chemistry*, 2nd ed., Vol. 2, Kolthoff, I. M. and Elving, P. J., Eds., Wiley-Interscience, New York, 1979, Part I, p. 10.
4. Butler, J. N., *Ionic Equilibrium*, Addison-Wesley, Reading, MA, 1964, p. 229.
5. Bruckenstein, S. and Kolthoff, I. M., in *Treatise on Analytical Chemistry*, Vol. 1, Wiley-Interscience: New York, 1959, Part I, Ch. 12, p. 446.

III

POLYPROTIC ACID–BASE EQUILIBRIUM

III-1. H RELATIONS

For symbols and fundamental relationshps, refer to Chapter I. The chemical background outlined in Appendix A is assumed. The master relation expressible in several arrangements, from Eqs. I-3 to I-6 gives proton binding expressed in equilibrium and in analytical concentration (material balance) terms (bound H/total base ligand). For example, in a system from H_3A, the terms are as follows:

Equilibrium	Material Balance
$f(\mathbf{H}) = \bar{\mathbf{n}}$	$= \bar{\mathbf{n}}'$
$\alpha_1 + 2\alpha_2 + 3\alpha_3 = \bar{\mathbf{n}}$	

$$\frac{\mathbf{H}/K_3 + 2\mathbf{H}^2/K_2K_3 + 3\mathbf{H}^3/K_1K_2K_3}{1 + \mathbf{H}/K_3 + \mathbf{H}^2/K_2K_3 + \mathbf{H}^3/K_1K_2K_3} = (C_{\mathrm{AB}} - \mathbf{D})/C_A \qquad \text{(III-1)}$$

Using the various interelations of $\bar{\mathbf{n}}$ and the α's (Eqs. I-1 to I-3), relations among **H** and the analytical concentrations of any mixtures can be found. We proceed to demonstrate that mononuclear* acid–

*Mononuclear refers to acids having a single central base coordinating n protons (e.g., H_2S, H_3PO_4, H_6EDTA^{2+}) as opposed to polynuclear acids like H_2F_2, $H_4P_2O_7$, and $H_2B_4O_7$. The latter give more complex algebra from fragmentation of the initial base group, which will be described later.

base equilibrium relations can be obtained and problems solved numerically starting with the single relation $\bar{\mathbf{n}} = \bar{\mathbf{n}}'$. This is not to deny the widespread text statement that about $n + 2$ equations in as many unknowns may be required to set up the **H** relation for H_nA systems from basics. Here, the $\bar{\mathbf{n}}$ function has combined all the K_a equations to eliminate equilibrium concentrations of species by using ratio forms to get a general **H** equation (Eq. III-1) (**D** is a function of **H** and K_w). Depending on the problem, any of the terms may be unknown. There is no need to rederive the left side, equilibrium $\bar{\mathbf{n}}$, which is the same curve for all mixtures containing the system.

A computer printout of α and $\bar{\mathbf{n}}$ values for citric acid at 0.1 unit intervals of p**H** from 0 to 12 (see Appendix D) will serve for most problems in the citrate system at the same conditions of temperature and ionic strength; see the graphs, in Figs. III-14*a*,*b*, Figs. III-15 and III-16. Alternatively, a spreadsheet program for $\bar{\mathbf{n}}$ and $\bar{\mathbf{n}}'$ can be used with trial p**H** values until satisfactory agreement (Eq. III-1) is reached (examples follow here and in Appendix B). Thus the chemistry involved lies in deducing the C term description of the specific problem, usually C_{AB} and C_A. Inserting **H** into the individual α expressions may be used to find species concentrations. The equilibrium concentration of the n species must be $C_A(\alpha_n)$ from the definition of α. Designing a solution to have any required p**H** is especially simple since the master relation is linear in the unknown C terms. Various types of calculations follow. Various computer strategies can be applied. A few simple ones are shown. Spreadsheet software will sufficice for most work (see Appendix B), while advanced mathematical applications will also solve the polynomial equations since the complete **H** equations (but not some approximations) all have only one real positive root.

[Note that the K_a values in Eq. III-1 are conditional molarity constants for one temperature and medium (ionic strength). We often use 0.1 M ionic strength and the value 13.80 for pK_w. The K_a^0 thermodynamic constants are used only for high dilution cases. Conversion of K_a^0 to conditional K_a's using activity coefficients is described in Chapter I and Appendix A.]

Typical plots of α and $\bar{\mathbf{n}}$ vs. p**H** for an idealized H_3A of pK_a values 2, 3, and 4 (close to citric acid) are shown in Figs. III-1*a*,*b*. Contrast phosphoric and citric acids in Figs. III-6 and III-14. Note the addition of weighted α's to give $\bar{\mathbf{n}}$: At p**H** 2, α_3 and α_2 are about 0.5, giving $\bar{\mathbf{n}} \cong 3 \times 0.5 + 2 \times 0.5 = 2.5$. The other α's are small here.

Note the α crossings exactly at the p**H** = pK_a values. If one knows the total of all forms of A (the C_A) in a solution and calculates α's,

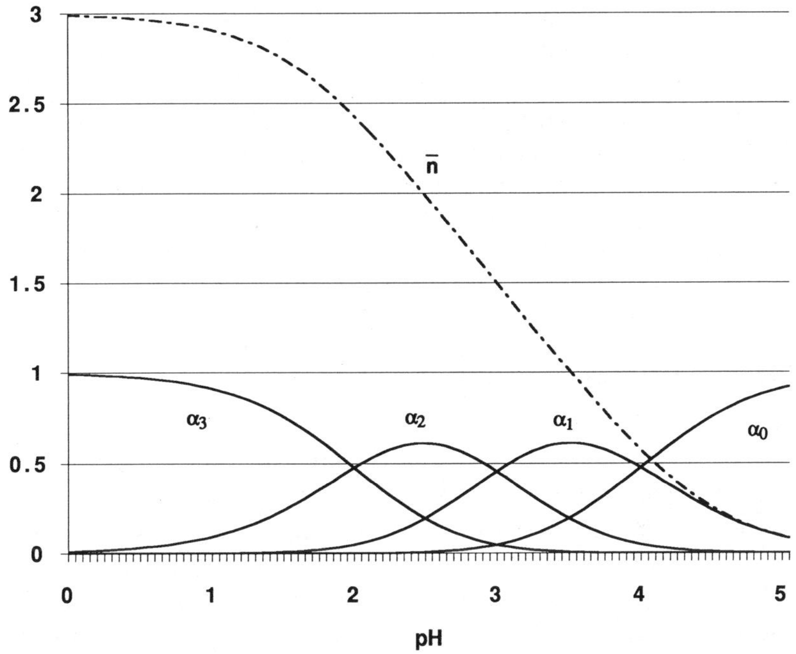

Fig. III-1*a*. Plots of α, $\bar{\mathbf{n}}$ vs. **pH** for a triprotic acid (pK_a 2, 3, 4).

one can find the precise concentration of any species, $[H_nA] = \alpha_n C_A$, for any chosen **pH** value. **Figure III-1*b*** clarifies the four α curves in this system. Only interior species have maxima. The curves shown are simple exponential functions of **pH**. (See Chapter II.)

The **H** equations for polyprotic systems can be obtained most simply by the general proton balance condition (Eq. III-1) $\bar{\mathbf{n}} = \bar{\mathbf{n}}'$,

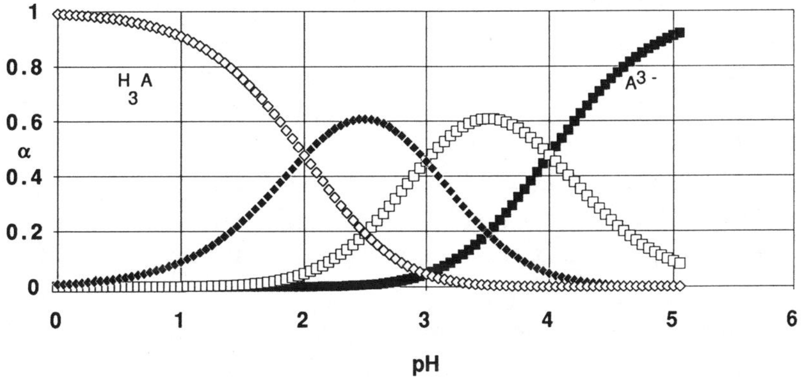

Fig. III-1*b*. Plots of α species vs. **pH**.

where the first is the equilibrium expression (above) and the second is the acid proton material balance condition, $(C_{AB} - \mathbf{D})/C_A$. [Here C_{AB} is a broader term extending to strong–weak acid–base mixtures; details are in Chapters I and II and the examples that follow.] Learning to solve **pH** problems is a matter of converting any given mixture to the two C parameters just given (it is always possible in single-acid systems) and then solving the high power of the **H** equation that results. (The method is summarized in Fig. III-2.) (Mixtures of systems will be treated with summations of the $\bar{\mathbf{n}}$'s.)

Significance of $\bar{\mathbf{n}}' = (C_{AB} - \mathbf{D})/C_A$

Note that $\bar{\mathbf{n}}'$ contains no term pertaining to a specific acid–base. It tells what value $\bar{\mathbf{n}}$ must have for the given C's (of any system: acetate, citrate, tartrate, etc.) if a chosen **pH** is to result. Plots of $\bar{\mathbf{n}}'$ curves vs. **pH** depend only on the C of the mixture. Partial curves are seen in Fig. III-2, and both acid and basic portions are seen in Figs. III-9 and III-10. Those experiencing difficulty with this function might look at those plots now and calculate some points for the examples in Fig. III-2 and Table III-1 (see Section III-8). The $\bar{\mathbf{n}}$, in contrast, contains only the conditional K's of one specific system and gives the unique $\bar{\mathbf{n}}$ value at each **H**. For a mixture, only one point on the $\bar{\mathbf{n}}'$ curve is also an equilibrium point on the $\bar{\mathbf{n}}$ curve of a specific acid–base system. **Figure III-2** shows six of the unlimited number of $\bar{\mathbf{n}}'$ mixtures for the single acid–base $\bar{\mathbf{n}}$ system chosen. The figure shows the equilibrium $\bar{\mathbf{n}}$ curve for an H_3A acid having pK_a's = 2, 3, and 4 (Fig. III-1*a*, similar to citric acid) and $\bar{\mathbf{n}}'$ plots for six different mixtures. The $\bar{\mathbf{n}}'$ functions are: ($\mathbf{D} \cong \mathbf{H}$, for these acidic cases)

Mixture	$\bar{\mathbf{n}}' = (C_{AB} - \mathbf{D})/C_A$	$\bar{\mathbf{n}}'$ function
0.01 M H_3A alone	$(0.03 - \mathbf{D})/0.01$	$3 - 100\mathbf{D}$
Ampholytes: 0.01 M H_2A^-	$(0.02 - \mathbf{D})/0.01$	$2 - 100\mathbf{D}$
0.01 M HA^{2-}	$(0.01 - \mathbf{D})/0.01$	$1 - 100\mathbf{D}$
1:1 Buffers:		
0.005 M H_3A, and H_2A^-	$(0.025 - \mathbf{D})/0.01$	$2.5 - 100\mathbf{D}$
0.005 M H_2A^-, and HA^{2-}	$(0.015 - \mathbf{D})/0.01$	$1.5 - 100\mathbf{D}$
0.005 M HA^{2-}, and A^{3-}	$(0.005 - \mathbf{D})/0.01$	$0.5 - 100\mathbf{D}$

The first term in the last column is $\bar{\mathbf{n}}'$ if **D** is negligible. This illustrates the use of the simple proton balancing equality above for solving **H** problems. The solution is at the crossing, where the two $\bar{\mathbf{n}}$ expres-

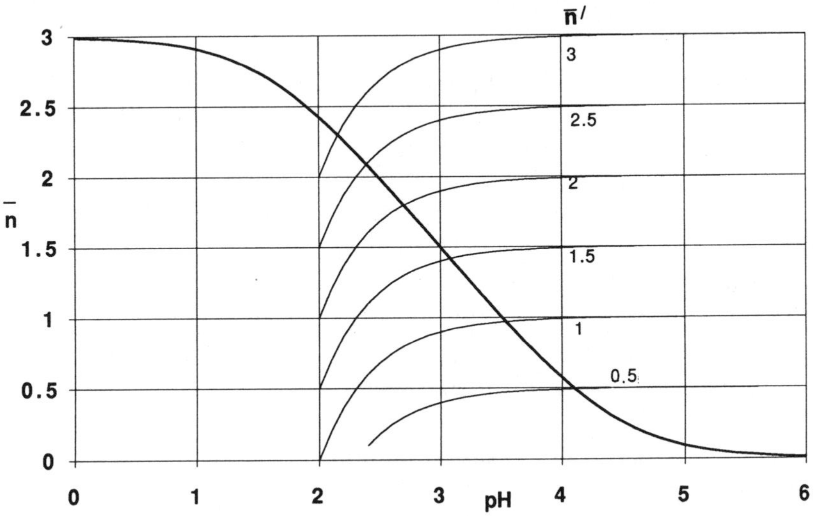

Fig. III-2. Equilibrium $\bar{\mathbf{n}}$ curve for an H_3A acid having pK_a's 2, 3, and 4 (heavy line) and $\bar{\mathbf{n}}'$ plots for six different mixtures (thin lines).

sions are simultaneously satisfied. Computer plots (or value tables) of $\bar{\mathbf{n}}$'s are easily made once one decides (from α diagrams) the region of **H** to use. The process is described later in this chapter. (Base A^{3-} alone is described in Section III-12 and Table III-1. $\mathbf{D} \cong -\mathbf{OH}$.)

An arbitrary decision about counting a very weak proton will not cause error if it is treated consistently. For example, the third proton in H_3PO_4 can be ignored if the solutions are treated mathematically as diprotic acids in a **pH** range in which the third deprotonation is negligible, say in acid solutions. We do this with citric acid; the fourth proton (an **OH** group) has pK_a of about 16. We would get identical results treating it as a tetraprotic acid with maximum $\bar{\mathbf{n}} = 4$. In other words, the *reference acidity level* is usually chosen as the completely deprotonated base, but this is not the only choice possible. It is a simple, consistent approach for avoiding mistakes.

Next we look at the meanings of these relations and at diagrammatic derivations to clarify them. We try to show the chemical logic of the master relations. However, the bar diagrams that follow (Figs. III-3 to III-5) are not an integral part of the master equation method for solving problems. They help some students grasp the proton balancing relations and their rigorous mathematical completeness. They derive from α diagrams like Fig. III-1, although the chemical argument here stands alone.

III-2. POLYPROTIC DIAGRAMS AND BALANCES

Let us look first at the ampholyte of a diprotic acid and then go to H_3A for general cases. Here we show, as in Chapter II, that proton balancing by chemical sense produces the master $\bar{\mathbf{n}}$ relation in every case. Thus, with experience, one may use the master relation without deriving an equation for every weak acid–base situation. These illustrations are thus given for clarification purposes, not as the necessary approach to problems. Instead of our usual **H**, $[H^+]$ and $[H_3O^+]$ are used in several spots to emphasize proton gain–loss balances. Rather than C_{AB}, C_H is used since no mixtures with excess strong base are included in this section.

1. Take C_1 as 20 units (say, millimolar) NaHA. Start with the HA^- given and water (reference level species). Take C_1 as C, C_A since there is only one starting substance. [One may see the relation of the final bar heights (dark) to the α diagram of oxalic acid (Fig. III-11), at about p**H** 2.5, but the verbal argument here does not depend on that diagram.]

Produce x of H_2A and y of A^{2-} and subtract these from C. Make a of H^+ and OH^- from water as before (Section II-2). Proton gain products are H_2A and **H** (H_3O^+), and loss products are A^{2-} and OH^-, made from the starters HA^- and H_2O (**Fig. III-3**, which shows ampholyte HA equilibrium). Here, $[H^+] = y - x + a$, or $\mathbf{D} = y - x$, is the proton balance, which we write as

$$[H_2A] + [H^+] = [A^{2-}] + [OH^-]$$
$$x + (y - x + a) = y + a$$

We insert the α fraction expressions to get the equilibrium concentrations: (at equilibrium, any species n has concentration $C\alpha_n$ by definition)

$$\mathbf{D} = [A^{2-}] - [H_2A] = \alpha_0 C - \alpha_2 C = C(\alpha_0 - \alpha_2)$$

Using the definitions $\alpha_0 + \alpha_1 + \alpha_2 = 1$ and $\bar{\mathbf{n}} = \alpha_1 + 2\alpha_2$ combined as

$$1 - \bar{\mathbf{n}} = \alpha_0 - \alpha_2$$

yields

$$\mathbf{D} = C(1 - \bar{\mathbf{n}}) \quad \text{or} \quad \bar{\mathbf{n}} = (1 - \mathbf{D}/C) = \bar{\mathbf{n}}' \qquad C = C_A$$

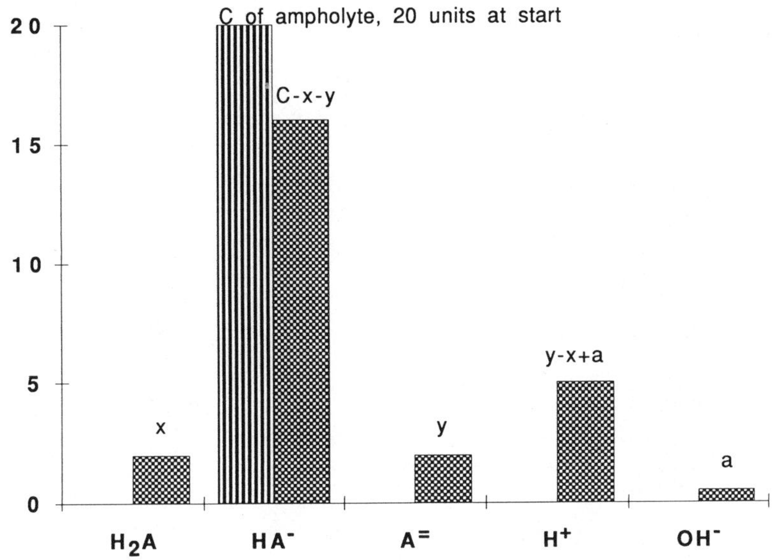

Fig. III-3. Ampholyte HA equilibrium: stripes, start; dark, final equilibrium solution.

This is the general proton balance equation (I-5 and III-1) for this case. Here, $\bar{\mathbf{n}}$ is a function of K and $[H^+]$, while $\bar{\mathbf{n}}'$ is a function of C and $[H^+]$.

2. Polyprotic acid H_3A alone. (a) Start with 20 units C_3 acid in water. (b) Make equilibrium amounts: some H_3A must form H_2A^-, HA^{2-}, and A^{3-}, say, y, z, and w, of these species, Make a of water ions for K_w. Assume C_3 is C or C_A since there is but one initial species. We get **Fig. III-4**:

$$[H^+] = y + 2z + 3w + a$$

Then, using the $\alpha_n C_n$ for each species,

$$\mathbf{H} = C(\alpha_2 + 2\alpha_1 + 3\alpha_0) + [OH^-]$$

The α sum $(3\alpha_3 + 3\alpha_2 + 3\alpha_1 + 3\alpha_0 = 3)$ and $\bar{\mathbf{n}} = \alpha_1 + 2\alpha_2 + 3\alpha_3$ are used as before to get

$$\mathbf{H} = C(3 - \bar{\mathbf{n}}) + \mathbf{OH} \quad \text{or} \quad \bar{\mathbf{n}} = 3 - \mathbf{D}/C$$

This is the general bound proton equation intersection (for pure H_3A solutions, C_3 alone) shown graphically in Fig. III-2. Here H_3O^+ is the

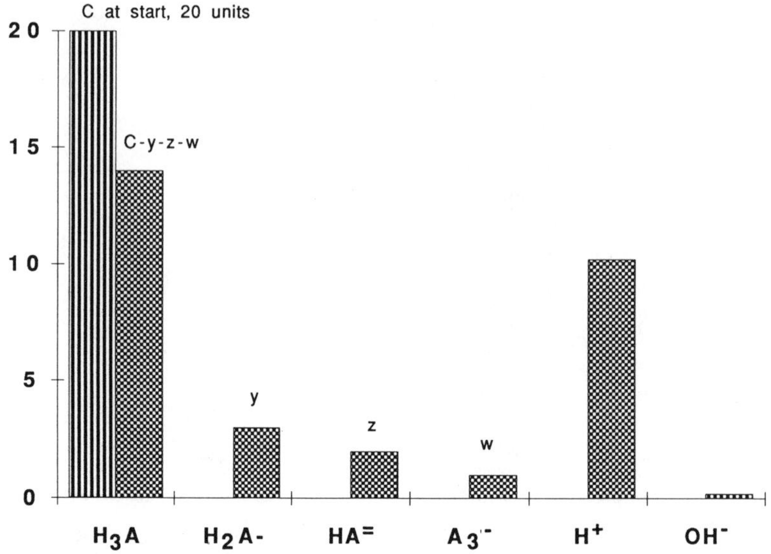

Fig. III-4. Stripes, C (molar) H_3A acid alone to start; dark, other A species.

only proton gainer. Loss of one proton from water makes OH^- and, from H_3A, y, two-proton loss, loss from H_3A gives z, and three-proton gives w of A^{3-}.

3. A buffer, C_3 H_3A and C_2 H_2A^- are the starting species. (a) Take $C_3 = 0.13\ M$ and (b) $C_2 = 0.100\ M$, expressed as millimolar in **Fig. III-5**.

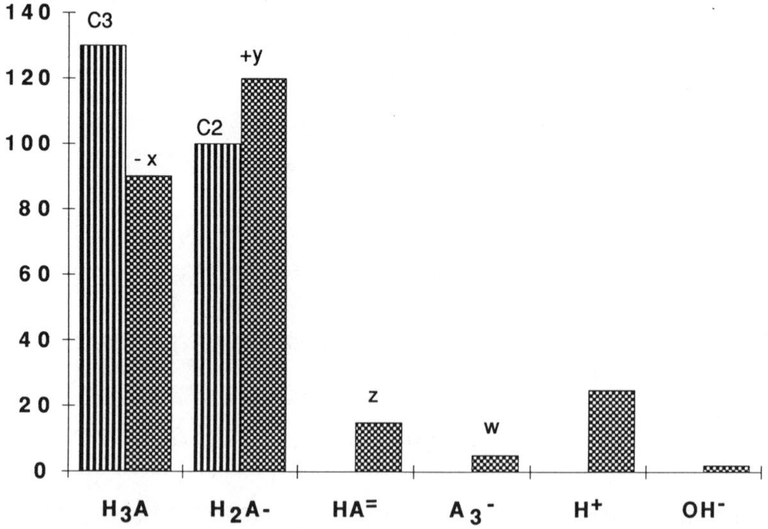

Fig. III-5. Buffer equilibrium shifts for the buffer mixture 130 mM H_3A and 100 mM H_2A^- (striped to start).

For a first case, assume that x of H_3A forms some of each of the other species, y, z, and w, to reach equilibrium (dark) and water forms a of H^+ and OH^-. The six items in the diagram at equilibrium are then

$$C_3 - x \quad C_2 + y \quad z \quad w \quad \mathbf{H} \quad a$$

From two ways of bookkeeping protons,

$$[H^+] = \mathbf{H} = 3x - 2y - z + a = 3w + 2z + y + a$$

The first is the total number of protons removed from H_3A and water $(3x + a)$ minus those put into other species; the second is the sum of protons lost in forming new species. They can be interconverted by the relation $x = y + z + w$ (a material balance on A, the loss and gain in species A), which shows them to be identical. They make a proton balance as in all the cases.* Putting in α's as before gives

$$[H^+] = 3(C_3 - \alpha_3 C_A) - 2(\alpha_2 C_A - C_2) - \alpha_1 C_A + [OH^-] \quad \text{where} \quad C_A = C_3 + C_2$$

*Compare the common use of material ("mass") and charge balances to obtain the **H** equation (C_2, NaH_2A is used for $[Na^+]$ in the charge balance; one must sum all conceivable ions in a mixture):

Material: $C_A = C_3 + C_2 = [H_3A] + [H_2A^-] + [HA^{2-}] + [A^{3-}]$

Charge, total positive = total negative: $C_2 + \mathbf{H} = \mathbf{OH} + [H_2A^-] + 2[HA^{2-}] + 3[A^{3-}]$

Subtracting the second relation from 3 times the first, we get one possible relation:

$$3C_3 + 2C_2 = 3[H_3A] + 2[H_2A^-] + [HA^{2-}] + \mathbf{H} - \mathbf{OH}$$

or

$$\mathbf{D} = 3(C_3 - [H_3A]) + 2(C_2 - [H_2A^-]) - [HA^{2-}]$$

which is the complete equation above when the α equivalents are entered ($\alpha_3 C_A = [H_3A]$, etc.). The same method can be used with the previous examples. Numerous other correct but not very helpful equations can be obtained by eliminations of other H_nA terms instead of the $[A^{3-}]$ chosen here. That and the problem of solving this equation in four unknown equilibrium concentrations makes the $\bar{\mathbf{n}}$ method with functions of **H** a more straightforward technique. Application to ampholytes is shown in Section III-9.

As in the previous cases, this transforms to $\bar{\mathbf{n}} = (C_H - \mathbf{D})/C_A$, which is equal to $\bar{\mathbf{n}}$. Here, $C_H = 3C_3 + 2C_2$. Signs of the changes in buffer species can be positive or negative as in the monoprotic case (Chapter II) for various values of constants. The algebra will take care of this: just realize that an opposite sign (if you get finally that $[H_2A] < C_2$) means an opposite change from the one assumed in setting up the equations. To test this, try a second case, letting both H_3A and H_2A decrease. One gets the same algebraic result.

Thus, all cases lead to the general $\bar{\mathbf{n}}$, $\bar{\mathbf{n}}'$ balance condition, and may use this for any mixture. There is no need to go through the balance and derivation of an equation for each new **pH** problem. Just find C_A and C_H and set $\bar{\mathbf{n}}$ equal to $\bar{\mathbf{n}}'$. Using $[OH^-] = K_w/[H^+]$ provides an equation relating K's, C's, and $[H^+]$, which we call **H**.

III-3. SUMMARY OF METHODS

Three ways of doing pH problems are described, the first being advocated here.

1. Use the $\bar{\mathbf{n}}$ method detailed in this book. One general acidity balancing equation gives a complete relation among K's, C's, and **H**. The chemical part is deducing the C's from the specific starting materials given. (These may be the unknowns with **H** given.) A computer can generate the $\bar{\mathbf{n}}$ functions in seconds, which allows a solution by inspection, or with a spreadsheet convergence plan for interpolation and iteration as necessary to locate the equality of the two $\bar{\mathbf{n}}$'s, shown in what follows.

2. Instead of the simple concept described in 1, material and charge balances for each case are used to get a proton balance in terms of **H**, K's, and [X]'s, the equilibrium concentrations of "major" species. Approximations of the last are used to solve the problem. Or, $[X] = \alpha_x C$ can be used for the "major" species to get complete equations identical to 1. These are in a form that is less clear when used to solve a problem, which is usually done by brute force methods for high order equations. Our $\bar{\mathbf{n}}$, $\bar{\mathbf{n}}'$ setup is well poised for successive approximation or iteration solutions and makes chemical and graphical sense. Texts and treatises have multiple-page derivations for each case, which are not needed in our method. See the previous footnote.

3. Elementary texts use approximate equations and chemical intuition (largely unavailable to the beginner!) from the start. Guess which polyprotic step(s) is (are) the major equilibrium(s) and omit the others. Results may be nearly correct or very far off. Without the full equation, much experience, or use of α diagrams, there is no clear and simple way to judge in many situations.

III-4. FEATURES AND USES OF α-pH DIAGRAMS

1. Monoprotics all have the same shape, α diagram, shifted along the **pH** axis according to pK_a (see Fig. II-6). At **pH** = pK_a, $\alpha_0 = \alpha_1 =$ 0.5. For given C_1 and C_0, test whether approximate or full equations are needed. Is $\mathbf{D} \ll C$? Look at the α diagram for the system to see if $\mathbf{H} - \mathbf{OH}$ is small at the α_1 implied by the C values given (details in Chapter II).

2. Polyprotic α shapes depend on the *spacings* of the pK_a values. For example, H_3PO_4 (pK_a of about 2, 7, and 12) gives three α crossings at **pH** values of pK_a that almost resemble three separate monoprotic HA curve for pK_a of about 2, 7, and 12 (**Fig. III-6**). Note that *four* separate curves appear on such a diagram. *Two* curves cross near zero at **pH** 4.5 (α_3, α_1) and 9.5 (α_2, α_0) (compare Figs. III-1 and III-2. The maxima above these are single lines for α_2 and α_1. The $\bar{\mathbf{n}}$ line is in Fig. III-10, and the α equations are shown in Chapter I.

However, closer spacing like the 2, 3, 4, acid shown in Figs. III-1 and III-2 makes overlapping curves. They cross below $\alpha = 0.5$. The

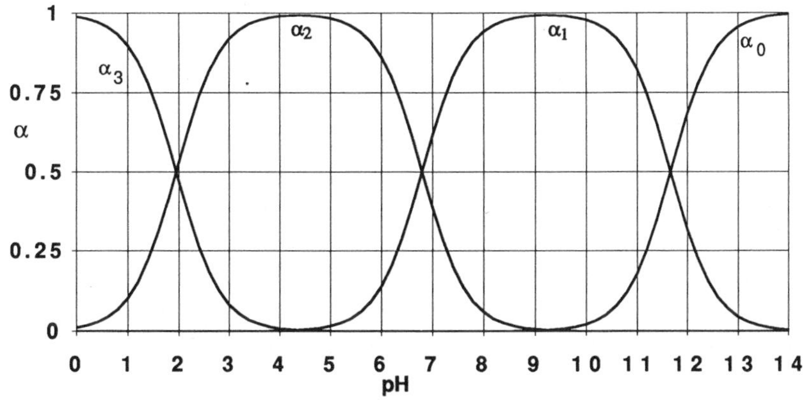

Fig. III-6. The α_3, α_2, α_1, α_0 lines for phosphoric acid at 25°C and 0.1 M ionic strength (pK_a 1.95, 6.80, and 11.67).

sum of all the α's at a **pH** must be 1. They are symmetrical about a center. An example, for the 2, 3, 4 acid follows:

pH	α_0	α_1	α_2	α_3	$\bar{\mathbf{n}}$
2.00	0.000476	0.0476	0.476	0.476	2.428
3.00	0.0455	0.455	0.455	0.0455	1.500
4.00	0.476	0.476	0.0476	0.000476	0.5726

Note the symmetry around the central **pH** 3. The **pH** of crossings are exact, but overlap reduces α levels, the most at **pH** 3. (Compare phosphoric acid in the preceding with crossings almost up to $\alpha = 0.5$.) As an exercise, enter these and a few other **pH** values into the α and **n**-bar expressions to see what happens. It is easy for these round K values. For nonintegral pK values like the citrate system that follows (Fig. III-14), be thankful for computers and their graphics.

Quick Sketch Method for α Curves. With practice it becomes easy to sketch rough α's to use as a guide in problem solving. Place an $\times$ at each pK_n value at $\alpha = 0.5$ (or below if pK's are closer than 2 units apart). Connect α lines through the $\times$ that have maxima halfway between the pK_a = **pH** values. Ratios of adjacent species change 10-fold for each **pH** unit (plus or minus as required by each K_a/**H** term). Note this and the pK_a = **pH** effects in the preceding table for the 2, 3, 4 acid system (e.g., α_3/α_2 at **pH** 2, 3, and 4 is 1:1, 1:10, and 1:100). Examination of the α graphs in this chapter and α tables in Appendix D will make clear the process for these similar exponential curves. Then, $\bar{\mathbf{n}} = \alpha_1 + 2\alpha_2 + 3\alpha_3 + \cdots$. It is helpful to sketch in the $\bar{\mathbf{n}}$ on an α diagram by summing the α's and comparing the differences between citric and phosphoric systems. At the first crossing (α_3 and α_2) of H_3PO_4 (see Figs. III-6 and III-10), the α's are very close to 0.5 so that $\bar{\mathbf{n}} = 2.5$. At the following level section of α_2, $\bar{\mathbf{n}} = 2.0$. At the last crossing $\bar{\mathbf{n}} = 0.5$, etc. For citric acid, one must estimate all visible α's to sum to $\bar{\mathbf{n}}$.

Polyprotic acids having the same spacings do have the same shapes. For example, the 2, 3, 4 acid diagram is just shifted for diagrams for acids of pK_a's 4, 5, 6 or 9, 10, 11 (similar to some triaza bases). The $\bar{\mathbf{n}}'$ curve, however, changes with **pH** and C's so that corresponding mixtures may not have relatively the same algebraic solutions (intersections).

III-5. REVIEW OF APPROXIMATE pH CALCULATIONS: ESTIMATING THE pH

See the α diagrams of various species for the correlations with these approximations.

Monoprotic

In Chapter II, the Charlot relation (Eq. II-7) is used. Letting **D** be negligible compared to C terms and taking **D** as **H** in acid and as **OH** in basic cases leads to **H** equations of the first or second power.

(a) Weak acid solutions alone, C_1 (molar): $\mathbf{H} = K_a(C_H - \mathbf{D})/\mathbf{D} \cong K_a C_1/\mathbf{H}$. Thus,

$$\mathbf{H} = \sqrt{K_a C_1}$$

is the approximation. It may be good at high C values and low K_a values. It gives worthless results at the opposite extremes. Compare 0.1 M and 10^{-6} M acetic acid (pK_a 4.7). Compare 0.1 M trichloroacetic acid (pK_a 0.7). The complete cubic is required only near neutral p**H**.

(b) Weak base alone, C_0 (molar): $\mathbf{H} \cong K_a\mathbf{OH}/C_0$. Using $\mathbf{OH} = K_w/\mathbf{H}$ gives

$$\mathbf{H} \cong \sqrt{K_a K_w / C_0}$$

(c) Buffers (mixtures of C_1 and C_0): $\mathbf{H} \cong K_a C_0/C_1$. This (or its logs) is the famous (infamous) Henderson–Hasselbalch (He–Ha) equation beloved of premedical chemistry teaching. Try it for buffers of trichloroacetic, citric (Fig. III-16, see next section), and other fairly strong acids. It gives p**H** that is off by several tenths of a unit (in usual high C). At low C, the He–Ha equation is worthless, as shown in the dilution curves in Chapter II and in Figs. III-11 to III-13, which follow.

Polyprotic

Divide C_H by C_A to obtain $\bar{\mathbf{n}}'_{max}$, that obtained if all protons are bound. [If a strong base is present in excess, the C_{AB} approach, described in Chapter II and in what follows (Eq. III-2), is required.]

A weak acid, say H_nA, or a weak base, A^{n-}, alone are treated just like monoprotics with K_1 or K_n, as if one proton step predominates. For these two cases $\bar{\mathbf{n}}'_{max}$ is n or zero.

Buffers are treated with the single K relating the two conjugate species put into solution, as in the preceding (He–Ha equation). The $\bar{\mathbf{n}}'_{max}$ is fractional between the species proton numbers: $\bar{\mathbf{n}}'_{max} = 2.278$ signifies a buffer of H_3A and H_2A^- in analytical ratio 0.278/0.722; $\bar{\mathbf{n}}'_{max} = 0.661$ signifies a buffer of analytical ratio 0.661/0.339 for HA/A.

Ampholytes are treated with the two K's controlling the proton addition or loss on each side of the ampholyte. The complete equations, omitting $\mathbf{D}$ and other small-species terms, yield $\mathbf{H} \cong \sqrt{K_1K_2}$ for the first ampholyte (H_2A^- in the H_3A system or HA^- in the H_2A system) and $\mathbf{H} \cong \sqrt{K_2K_3}$ for HA^{2-} in the H_3A system. Later extensive examples in Section III-9 show degrees of agreement found with these equations.

Older texts and handbooks list these and hybridized equations containing some of the omitted terms. They can be used at one's own risk! The one-significant-figure (or worse) results serve us as starting points for successive approximation or iteration solutions of the complete $\bar{\mathbf{n}}$ equations.

The poor agreement of the He–Ha equation at extreme buffer ratios for the overlapping acid, citric, is shown in Fig. III-16 (see Section III-12).

III-6. CHOICE OF STARTING CONCENTRATION TERMS: EQUIVALENT SOLUTIONS

In applying Eqs. I-3, I-4, and III-1, it should be clear that more than two C terms are redundant but not wrong. Any combination of species C's can be expressed in terms of the two parameters C_{AB} and C_A (Eq. I-6) for any H_nA, here illustrated for H_3A:

$$\bar{\mathbf{n}}' = (C_{AB} - \mathbf{D})/C_A = \frac{3C_3 + 2C_2 + C_1 + C_{sa} - C_{sb} - \mathbf{H} + \mathbf{OH}}{C_3 + C_2 + C_1 + C_0} \qquad \text{(III-2)}$$

For example, if all four species were put into solution (C_3 of H_3A, C_2 of NaH_2A, C_1 of Na_2HA, and C_0 of Na_3A), C_A is the sum of all four and C_{AB} is the numerator sum of the five C terms including any strong acid and base added.

Three protonically equivalent solutions illustrate this:

(a) 0.1 *M* NaH_2A with 0.2 *M* Na_2HA; C_2, C_1 given; C_0, $C_3 = 0$.
(b) 0.05 *M* H_3A with 0.05 *M* NaH_2A, 0.15 *M* Na_2HA, and 0.05 *M* Na_3A; four *C*'s given and C_H can be deduced.
(c) 0.1 *M* Na_3A with 0.2 *M* H_3A and 0.2 *M* NaOH; C_0, C_3 given; C_1, $C_2 = 0$.

Each has $C_{AB} = 0.4$ and $C_A = 0.3$. Using Eq. III-2 and $\bar{\mathbf{n}} = \bar{\mathbf{n}}'$ we get

$$0.4 - \bar{\mathbf{n}}(0.3) = \mathbf{D}$$

for all three. This reads: The total acidity, C_{AB}, minus the bound equals the "free," **D**. Then, using Eq. III-1, yields an equation in one variable, **H**. Note that Eqs. I-4 to I-6 (and III-2) will also give the same result with the three different sets of *C* values. The maximum value of $\bar{\mathbf{n}}'$ is $0.4/0.3 = 1.333$. This occurs as $\mathbf{D} \rightarrow 0$.

Two *C* terms can describe any mixture possible in a single weak system. One can be the net acid–base value, C_{AB}, and the other the total of base groups, C_A. Ricci[1] used a different pair, which we shall find useful later in treating titrations. He thinks of any mixture as derived from the pure acid (H_3A) *C*, and C_{sb}, strong base needed to make the mixture. So $C_H = 3C - C_{sb}$ and $C_A = C$. (C_{sa} can be added when present.) Looking at the preceding examples from both viewpoints should clarify this bivariant system model. Ricci's method gives $0.9 - 0.5 = 0.4 = C_H$ by adding up the H missing on species below H_3A (removed by NaOH). Thus, our mixtures (a, b, and c) could also be analytically described as:

(d) 0.3 *M* in H_3A and 0.5 *M* in NaOH; all have $C_{Na} = 0.5$ *M* and are indistinguishable.

This method does not cover a mixture with a strong acid also added as clearly as does our C_{AB} method. See the examples in Table III-1 in Section III-7.

General Polyprotic Mixtures. The two-term description might be called the protonic phase rule. There are two degrees of freedom, protonic acidity, C_{AB} (including its negative, excess strong base), and total weak base acceptors, C_A. One of the *C* terms may be zero for certain cases described in this chapter. Examining the chemical source of **H** or **OH** and the α, $\bar{\mathbf{n}}$ diagrams show the following generalizations about the three kinds of solutions possible in single polyprotic systems. (We use H_3A as an example.) (All solutions approach neutrality with increasing dilution, $C \rightarrow 0$.)

(i) Pure acid or base solutions ($\bar{\mathbf{n}}'_{max} = 3$ or 0): H_3A solutions (can be described by a single $C_A = C$ and $C_H = 3C$) have **H** increase with C. Pure A^{3-} solutions give the mirror image variation vs. **OH**.

(ii) Ampholytes, H_2A^- or HA^{2-} alone ($\bar{\mathbf{n}}'_{max}$ of 2 or 1), are also described by one C term and must have **H** closer to neutral than the $\mathbf{H}_{max}$ at their species maximum (α_2 or α_1) and approach $\mathbf{H}_{max}$ with increasing C.

(iii) Buffers (given or calculable C's of two stable conjugate or *compatible* species, so that $\bar{\mathbf{n}}'_{max}$ has nonintegral values like the 1.333 example in the preceding or 2.5, 1.5, or 0.5 for 1 : 1 buffers. These have **H** closer to neutral than the high concentration **H** limit of their buffer ratio. At neutral p**H**, the buffer has this limiting value.

Example. H_3A and HA^{2-} are *incompatible* as major species: they transfer protons to make some H_2A^-, depending on the relative proportions mixed. However, the result has the correct C_H and C_A (as in the preceding equivalent solutions) so that one need not make any change in applying the $\bar{\mathbf{n}}$ methods. This avoids a major source of confusion about the treatment of incompatible mixtures. The result and α values will clarify which are the true major species. The two C values will lead to correct results even for incompatible mixtures. Numerous examples follow.

The **H** approaches limits in the cases i and ii because $\bar{\mathbf{n}}$ cannot exceed $\bar{\mathbf{n}}'_{max}$ in acidic solution and must go above that maximum to produce a basic solution. [Thus, $\bar{\mathbf{n}}_{neu}$ is a better description: it is the value in neutral solution, the maximum in acidic cases and the minimum in basic cases (Fig. III-9 in Section III-8).] In basic solutions of buffers and ampholytes, $\bar{\mathbf{n}}$ has increased (from $\bar{\mathbf{n}}'$) as starting species become further protonated. See the diagrams that follow under dilution curves, Figs. III-12 and III-13.

These can be confirmed algebraically, or more simply by sliding the $\bar{\mathbf{n}}'$ lines in Fig. III-2 to the left and observing the change in p**H** of the intersection with $\bar{\mathbf{n}}$. The similar situation in basic solution, inverted and vs. p**OH**, is shown later in Fig. III-9. The H_3A and A^{3-} ($\bar{\mathbf{n}}'_{neu}$ of 3 and 0) have no limit, while ampholytes and buffers ($\bar{\mathbf{n}}'$

between 3 and 0) approach the $\bar{\mathbf{n}}'_{\text{neu}}$ value intersection with $\bar{\mathbf{n}}$ at high C. In general, the log C_A value is seen to be the **pH** value (in acidic cases) at which the $\bar{\mathbf{n}}'$ line is one unit below its value in neutral solution. All the solutions in Fig. III-2 are 0.01 M. Sliding them all to **pH** 1 gives the **pH** for 0.1 M solutions. The inverted analog holds in basic solutions (Section III-8).

Pondering the properties illustrated in Figs. III-1 and III-2 can reveal all the behaviors of solutions in the H_3A system, including difficult or improbable cases like close, equal, or inverse order pK_a values. Numerical computer solutions can then be arranged to any desired precision. Contrast this with approximate approaches that require checking complete equations to determine validity.

III-7. NUMERICAL SOLUTIONS TO pH PROBLEMS

Rigorous Methods: Programs

From Eqs. I-5 and III-2 and earlier expressions for terms, the plots of $\bar{\mathbf{n}}$ and $\bar{\mathbf{n}}'$ vs. **H** will intersect at the **H** value that satisfies both $\bar{\mathbf{n}}$'s simultaneously. [The alternative of inserting all **H** terms (K_w/**H** in **D**) and solving the high order equations as Ricci chose is also correct, but it is laborious and less instructive as to the chemistry occurring.] The plot vs. **pH** sometimes serves, but in some regions the intersection is hard to find because slopes are near zero. Plotted vs. **H** or **OH**, over the short range used in these probems, $\bar{\mathbf{n}}'$ is straight and $\bar{\mathbf{n}}$ is straight to good approximation (the slope changes slowly with **H**). Log–log plots may also be used, as shown later. The region of **pH** to use is found by the usual approximations or by trial and error. But since the slopes of the two $\bar{\mathbf{n}}$'s are of opposite sign, it is a simple matter to program the intersection to the precision needed. When $\bar{\mathbf{n}} < \bar{\mathbf{n}}'$, the next trial **H** is increased and vice versa until a suitable solution is attained, as the spreadsheet description to follow shows (linear approximation method). Graphical illustrations follow. The difference $\bar{\mathbf{n}} - \bar{\mathbf{n}}' = \Delta$ or d goes through zero at the solution. This sign change identifies the interval of the root. When the difference is negative, the trial **H** root is too small, when positive, **H** is too large, see Figs. III-7b and III-7e. In the bisection method of approximation, this shows which of a pair of bracketing trial roots to average with their mean for the next trial. This can be done easily on a computer one step at a time. But spreadsheet programming the whole process is possible since iterations converge rapidly. We summarize one simple process as follows. See also Appendix B and the exercises in spreadsheet calculations in Chapters I and II.

Solving pH Problems by Using a Spreadsheet

1. Make column headings for **pH**, **H**, α, $\bar{\mathbf{n}}$, $\bar{\mathbf{n}}'$, $\Delta = \bar{\mathbf{n}} - \bar{\mathbf{n}}'$, **pH** (for plotting sequences if wanted), and cells for entry of specific K_a, C_{AB}, and C_A values.
2. Enter trial values for **pH** and formulas for the other variables. (Economical programming forms for α and $\bar{\mathbf{n}}$ are shown in Chapter I and Appendix B.)
3. Step through trial **H**'s as described in the preceding and observe the signs of Δ until the desired precision is obtained in the equality of the $\bar{\mathbf{n}}$'s.
4. The interpolation (extrapolation) formula may be used for faster convergence. For a pair of **H** values $\mathbf{H}_1$, $\mathbf{H}_2$, which are close to or bracket the root, an improved root $\mathbf{H}_x$, calculated from the slopes of two straight lines (secants), is (between $\mathbf{H}_1$ and $\mathbf{H}_x$)

$$\mathbf{H}_x = \mathbf{H}_1 + \Delta_1(\mathbf{H}_2 - \mathbf{H}_1)/(\Delta_1 - \Delta_2)$$

or

$$\mathbf{H}_x = \mathbf{H}_1 + (\mathbf{H}_2 - \mathbf{H}_1)/(1 - \Delta_2/\Delta_1)$$

or between $\mathbf{H}_x$ and $\mathbf{H}_2$,

$$\mathbf{H}_x = \mathbf{H}_2 - (\mathbf{H}_2 - \mathbf{H}_1)/(1 - \Delta_1/\Delta_2)$$

whichever seems more convenient. This is the same as calculating the zero of the line of the differences in Δ (see Fig. III-7*e*). (See Fig. III-8 for the lines and points to use in derivation of these points slope formulas. These may be tested numerically in any of the spreadsheets that follow. Keeping signs correct is a problem in hand calculator work.)

We return to a spreadsheet example calculation after showing the nature of the convergence.

Graphical Illustrations of $\bar{\mathbf{n}}$ Convergence

It is convenient to choose some odd pK_a values to plot linear **H** curves clearly. First look at H_3A with all three pK_a values of 2.00.

The normal pH plot (**Fig. III-7*a***) is shown. To show the near linear behavior of the $\bar{\mathbf{n}}'$ plot, let us replot this versus **H** in **Fig. III-7*b*** and add the four $\bar{\mathbf{n}}'$ lines (dotted) for pure acid and three buffers (2.5, 1.5, and 0.5 buffer ratios) and total $C_A = 0.01$. This shows the intersections and that the $\bar{\mathbf{n}}'$ makes straight lines in this region ($\bar{\mathbf{n}}' = 3 - \mathbf{D}/0.01$ for the H_3A and $\bar{\mathbf{n}}' = 0.5 - 100\mathbf{D}$ for the last buffer, 0.005 *M* HA^{2-} and 0.005 *M* A^{3-}). Note that in the usual pH plots, **H** increases in the opposite direction so that all slopes are reversed.

Figure III-7*c* shows the $\bar{\mathbf{n}}'$ function vs. **H** on a linear scale; with $\mathbf{H} \times 10^7$. Here $\bar{\mathbf{n}}' = 1 - 10^6\mathbf{D}$ for very dilute solution (10^{-6} *M*) of HA^- and is a type of hyperbola of the form

$$xy - ax + bx^2 - c = 0$$

where x is **H** and y is $\bar{\mathbf{n}}'$ and the full **D** is used and multiplied through by **H**. Contrast with the usual pH plots in Figs. III-9 and III-14*b*. Note the basic region is entirely in the first segment of **H**, 0 to 10^{-7}. In this calculation pK_w 14.00 was used. We see why the $\bar{\mathbf{n}}'$ lines seem straight in Fig. III-7*b* on the acid side.

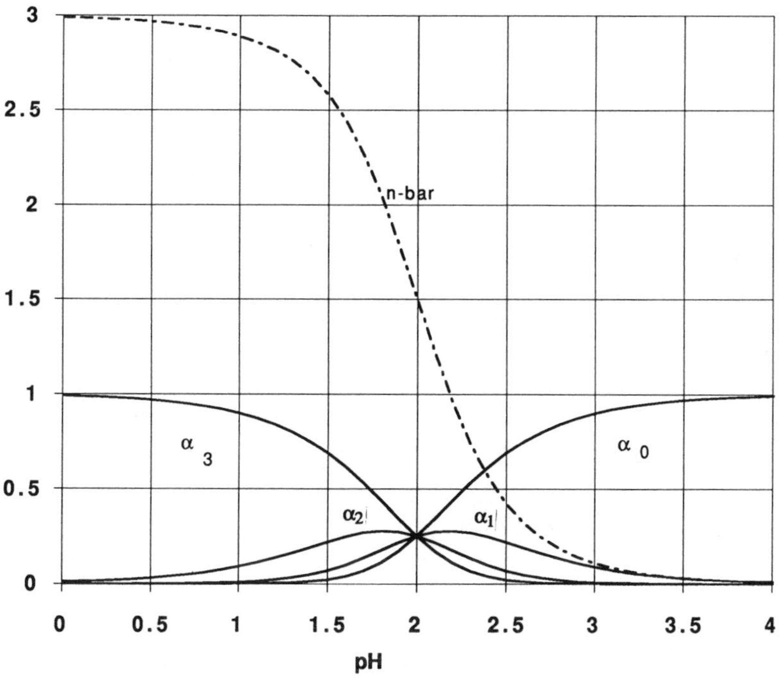

Fig. III-7*a*. The α, $\bar{\mathbf{n}}$ plot for H_3A with close-spaced pK_a's of 2.00.

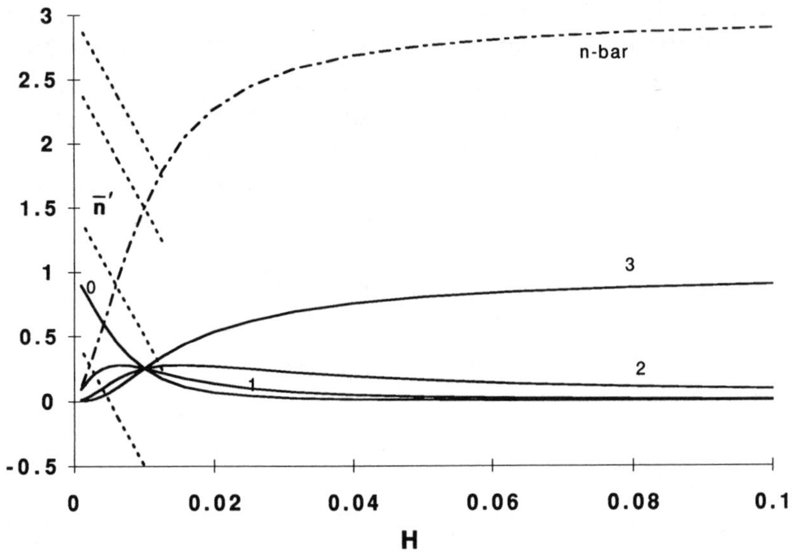

Fig. III-7*b*. The α, $\bar{\mathbf{n}}$ plot vs. linear **H** of H_3A with pK_a's of 2.00. The $\bar{\mathbf{n}}'$ for four mixtures are shown dotted.

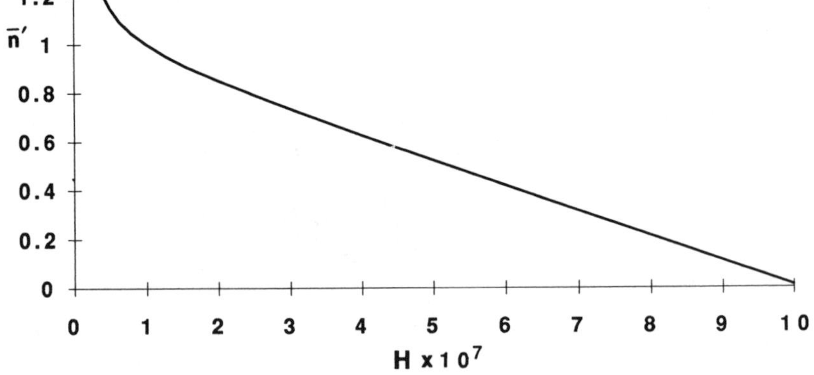

Fig. III-7*c*. The $\bar{\mathbf{n}}'$ function vs. **H** on a linear scale.

Away from the neutral point, the function approaches a straight line vs. **H** or **OH**. In acid solutions,

$$\bar{\mathbf{n}}' = a - \mathbf{bH} = 1 - 10^6\mathbf{H}$$

and in basic solutions

$$\bar{\mathbf{n}}' = 1 + 10^6\mathbf{OH}$$

Figures III-7*b*,*c* show how the acid side of this hyperbola seems to be straight. The basic portion cannot be seen with the **H** scale needed in that figure. The **OH** linear scale plot reverses the situation (**Fig. III-7*d***). In Fig. III-7*d* note the reverse directions from the previous plot. Here the whole acid portion is in the first segment.

The curvature in both $\bar{\mathbf{n}}$ and $\bar{\mathbf{n}}'$ shows up only in special regions. For an example, we chose an H_3A having pK_a's 6, 7, and 8. The $\bar{\mathbf{n}}'$ for 10^{-6} *M* acid alone is $3 - 10^6\mathbf{D}$. The plots of the $\bar{\mathbf{n}}$'s and their differences, *d* or Δ, are shown in **Fig. III-7*e***. The apparent answer is about 7×10^{-7}. The spreadsheet program gives 7.094×10^{-7}, p**H** 6.149.

The straight line approximation method (Norris, *Computational Chemistry*, pp. 72–73; see the Annotated Bibliography), can use the two $\bar{\mathbf{n}}$ curves (intersections of straight line segments) or the equivalent, the zero of their difference, the lower curve.

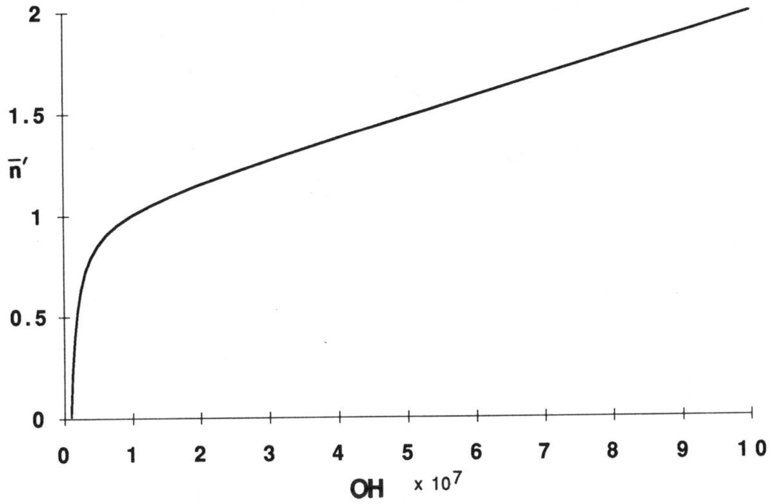

Fig. III-7*d*. The OH ($\times 10^7$) plot.

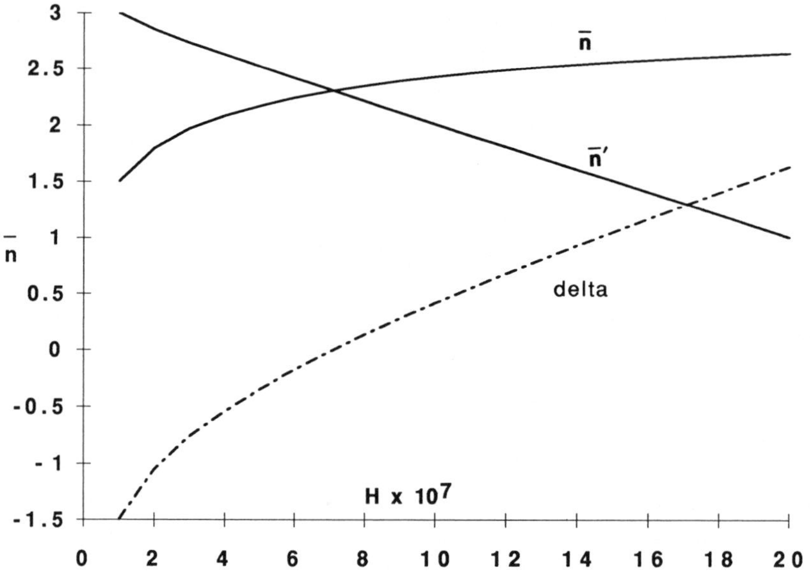

Fig. III-7*e*. Very dilute H_3A (p*K*'s 6, 7, and 8) solution described.

Numerical Solutions of Various Mixtures in H_3A Systems

Find the **pH** of a 1:1 buffer of H_3A and NaH_2A both at $C = 0.0100\ M$. The pK_a values are 2, 3, and 4. With such large, closely spaced *K* values (Fig. III-1), the approximate answer ($\mathbf{pH} = pK_1$) is far off. (This requires that 0.01 *M* **H** forms from the 0.01 *M* acid!) The region is suitable to begin calculation.

Here $C_A = 0.02\ M$ and C_{AB} (or C_H) $= 0.05\ M$ so that $\bar{\mathbf{n}}' = 2.5 - 50\mathbf{H}$. The fact that $\bar{\mathbf{n}} = 2.5$ while $\bar{\mathbf{n}}' = 2.0$ at **pH** 2.0 shows that proton balance is impossible for these conditions of **pH** and *C*'s. Examination of an α diagram (Fig. III-1) suggests that **H** could be around 0.005. First try a simple trial-and-error approach. Get $\bar{\mathbf{n}}$'s and plot them or take differences. We show the graphical method for visualization:

H	0.004	0.005	0.006
$\bar{\mathbf{n}}$	2.0830	2.1714	2.2418
$\bar{\mathbf{n}}'$	2.300	2.250	2.200
Δ	−0.217	−0.0786	+0.0418

Figure III-8 shows the plot of these values of $\bar{\mathbf{n}}$ vs. **H** crossing between 0.005 and 0.006 *M* H and $\bar{\mathbf{n}} = 2.5$ at 0.00563, $\bar{\mathbf{n}} = 2.218$ (**pH**

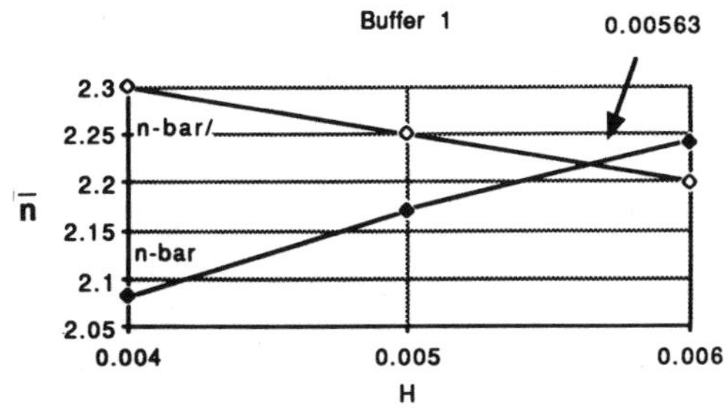

Fig. III-8. The **H** of a buffer, 0.0100 M each of H_3A and NaH_2A, in 2, 3, and 4 H_3A system.

2.249). Instead of α_3 and α_2 both at 0.5, we get 0.323 and 0.574 at $\bar{\mathbf{n}} = 2.2$, so that we have moved to the right on the α diagram. The system must adjust until the loss of protonated species provides the **H** of the abscissa.

Numerical interpolation to intersections treated as straight lines, as given before under step 4, is

$$\mathbf{H}_x = 0.005 + (-0.0786)(0.001)/(-0.0786 - 0.0418) = 0.00565$$

at p**H** 2.248. The agreement seems satisfactory, but this estimate can be combined with $\mathbf{H} = 0.006$ data for another estimate of $\mathbf{H}_x$. The final result is 0.00564 M for p**H** 2.2487 at $\bar{\mathbf{n}} = 2.2180$ obtained by the spreadsheet iteration shown in the next section.

It is helpful in all these examples to look at the computer printout of $\bar{\mathbf{n}}$ and α values as well as the plots. Note that $\bar{\mathbf{n}} = 1.5$ at p**H** 3 owing to symmetry but that $\bar{\mathbf{n}}$ is not exactly 2.5 at p**H** 2 or 0.5 at p**H** 4 owing to unsymmetrical overlap of species. Spreadsheets for several H_3A examples are described and shown in Appendix B.

Other Mixtures in the pK 2, 3, 4 Acid System

With an $\bar{\mathbf{n}}$ curve (vs. **H** or p**H**) one can find intersections of $\bar{\mathbf{n}}'$ for any mixture in the system (Figs. III-1 and III-2) (or program solutions in a variety of suitable iteration schemes for this simple relation). Table III-1 presents other possible cases for the 2, 3, 4 acid for which the first buffer (mixture 3) was previously detailed. This should clarify how $\bar{\mathbf{n}}'$ is written for any mixture. Note the $C_H/C_A = \bar{\mathbf{n}}'_{max}$ in acidic cases. The first approximations are described in what follows. Results are provided so that practice calculations by the reader can be checked. See also $\bar{\mathbf{n}}'$ values after Fig. III-2.

TABLE III-1. Weak H_3A system, pK_a's 2, 3, 4

Mixture	$\bar{n}' = C_{AB}/C_A - \mathbf{D}/C_A$	$\mathbf{H}_{est.}$ to Start	pH Result
1. 0.1 *M* H_3A, 0.01 *M* HCl	$3.1 - \mathbf{D}/0.1$	0.01–0.1	1.469
2. 0.01 *M* H_3A	$3 - \mathbf{D}/0.01$	$<\sqrt{K_1 \times 0.01}$	2.154
3. 0.01 *M* H_3A, 0.01 *M* H_2A^-	$2.5 - \mathbf{D}/0.02$	K_1	2.248
4. 0.01 *M* H_2A^-	$2 - \mathbf{D}/0.01$	$\sqrt{K_1K_2}$	2.696
5. 0.01 *M* H_2A^-, 0.01 *M* HA^{2-}	$1.5 - \mathbf{D}/0.02$	K_2	3.046
6. 0.01 *M* HA^{2-}	$1 - \mathbf{D}/0.01$	$\sqrt{K_2K_3}$	3.544
7. 0.01 *M* HA^{2-}, 0.01 *M* A^{3-}	$0.5 - \mathbf{D}/0.02$	K_3	4.10
8. 0.01 *M* Na_3A	$-\mathbf{D}/0.01$	$\sqrt{K_wK_3/C}$	7.90
9. 0.1 *M* Na_3A, 0.001 *M* NaOH	$(-0.001 - \mathbf{D})/0.1$	10^{-11}	10.8001

The first and last cases illustrate the use of C_{AB} to treat weak–strong mixtures (Eq. III-2). Note that $\bar{\mathbf{n}}'$ is positive for all cases including the last. Only the last two agree closely with the approximate answers. The closely spaced pK_a values and low C's were chosen to show such cases.

For mixture 2, use of the $\bar{\mathbf{n}}$ computer printout for the pK_a 2, 3, 4 acid will be illustrated with $C = 0.01$ *M*. The approximate p**H** 2.0 is a low limit estimate since it seems unlikely that the first step of a weak acid would be completely ionized. Here, $\bar{\mathbf{n}}' = 3 - \mathbf{H}/0.01$ (**OH** is about 10^{-12}, negligible in **D** in such an acid solution):

FROM COMPUTER PRINTOUT

pH	$\bar{n}$	$\bar{n}'$	Δ	H
2.0	2.4274	2.0	0.4274	0.01
2.1	2.3462	2.2057	0.1405	0.00794
2.2	2.2609	2.3690	−0.1081	0.00631

Interpolating graphically or numerically between the last two lines as before Fig. III-8, gives $\mathbf{H} = 0.00702$, p**H** 2.154. The student should note that on the computer printout and α diagram $[H_2A^-]$ is actually greater than $[H_3A]$ for this rather strong acid at this dilution. In the basic cases, the approximate solutions are close, and checking the $\bar{\mathbf{n}}$ values shows agreement. The pK_w of 13.80 was used for 0.1 *M* ionic strength. The p**H** plots are not useful, but linear ones (vs. **H** or **OH**) work for these cases. The last case (9) shows the great repression, in the presence of 0.001 *M* NaOH, of protonation of A^{3-}, which here contributes only a 0.0001 rise in p**H**. Compare with the CN^-, OH^- problem in the previous chapter. The cyanide ion is a much stronger base than the A^{3-} used here.

The ampholyte H_2A^- (4 in Table III-1) illustrates how bad the approximate approach can be. The expected halfway value of p**H** 2.5 is truly 2.7, over 30% off in **H**. Here, where **H** is near *C*, the equilibrium $[H_2A^-]$, 0.0058 *M*, is not near the analytical concentration, 0.01 *M*, and the amounts of H_3A and HA^{2-} formed are not equal, as shown by the α values in the computer printout for p**H** 2.7, $\alpha_3 = 0.116$ and $\alpha_1 = 0.290$. Texts usually use weaker acids and *C* near 0.1 *M*, which gives $\bar{\mathbf{n}}'$ close to its maximum (equilibrium concentrations close to the analytical *C*'s). This is apparent in the plots of Figs. III-2 and III-11 and in Section III-11.

Examination of the computer $\bar{\mathbf{n}}$ values and calculation of $\bar{\mathbf{n}}'$ for the examples in Table III-1 should give the reader a feeling for the method. Note that the round *C* values are used for clarity in the examples, but any can be used; e.g., a buffer of 0.0260 *M* H_3A and 0.0798 *M* H_2A^- has the $\bar{\mathbf{n}}'$ function $(0.2376 - \mathbf{D})/0.1058 = 2.246 - 9.452\mathbf{D}$.

Further practice in using $\bar{\mathbf{n}}'$ and C_{AB} in a real case, citric acid, is seen in the table of citrate mixtures in the exercises at the end of this chapter.

III-8. THE COMPLETE $\bar{\mathbf{n}}'$ FUNCTION: FURTHER EXAMPLES OF $\bar{\mathbf{n}}$, $\bar{\mathbf{n}}'$ INTERSECTIONS

Figures III-9*a,b* and **III-10** display the functions used for the preceding numerical computer solutions. First (Fig. III-9*a*) is a simple $\bar{\mathbf{n}}'$ plot of three dilutions of a 1:1 buffer of H_2A^-/HA^{2-} in any H_3A system (a 1:1 buffer refers to the ratio of analytical, not equilibrium, concentrations, $C_3 : C_2$), total C_A of (a), 0.1 *M*, (b), 0.001 *M*, and (c), 10^{-6} *M*. Note that $\bar{\mathbf{n}}'$ is independent of the *K*'s. It tells only the proton balance required with any chosen analytical concentrations and **H**. For example, to have p**H** 2 with this buffer ratio, $\bar{\mathbf{n}}'_{max} = 1.5$, for total $C_A = 0.1$ *M*, we must have

$$\bar{\mathbf{n}}' = (C_H - \mathbf{D})/C_A = (0.15 - 0.01)/0.1 = 1.5 - 0.1 = 1.4$$

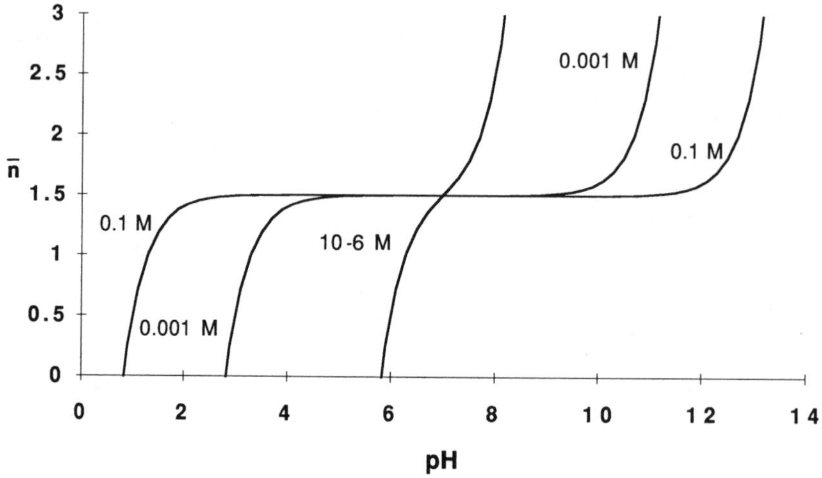

Fig. III-9*a*. Complete $\bar{\mathbf{n}}'$ curves for three 1:1 buffer mixtures of any H_3A system.

For $C_A = 0.001$ this becomes

$$\bar{\mathbf{n}}' = 1.5 - 10 = -8.5$$

This means that such a **pH** is impossible (it is more than the **H** available). To have **pH** 11, **pOH** 3, we must have, for the 0.1 *M* case,

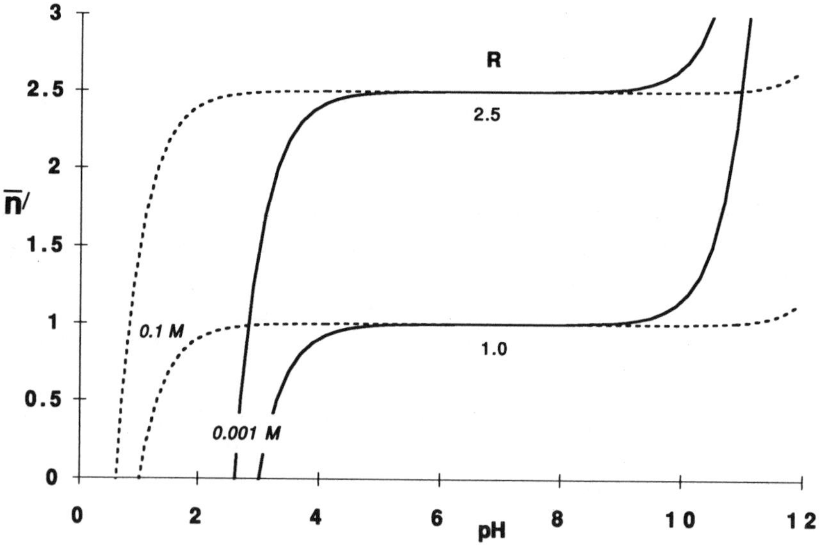

Fig. III-9*b*. The $\bar{\mathbf{n}}'$ curves for mixtures of different proportions, C_H/C_A ratios of 2.5 and 1.0.

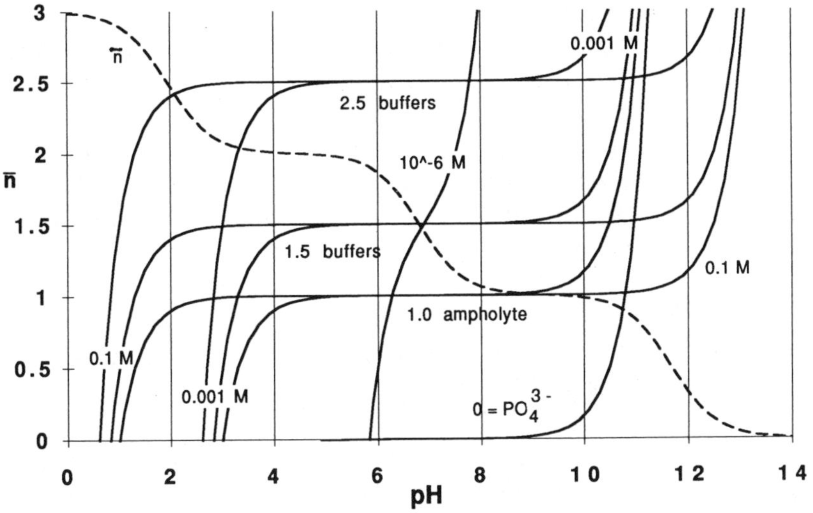

Fig. III-10. The $\bar{\mathbf{n}}$ (dashed line) and $\bar{\mathbf{n}}'$ (solid lines) for the phosphate cases: $\bar{\mathbf{n}}$ dashed, $\bar{\mathbf{n}}'$ solid lines.

$$\bar{\mathbf{n}}' = 1.5 - (10^{-11} - 10^{-3})/0.1 = 1.51$$

This signifies that 10% of species A must remove protons from water to make the 0.001 *M* OH^- and increase $\bar{\mathbf{n}}'$. For the 0.001 *M* case, $\bar{\mathbf{n}}'$ must be 2.5. All these conditions apply regardless of the acid chosen. Similarly, $\bar{\mathbf{n}}$ for a chosen acid is the same curve for any case of the *C*'s proposed. Figure III-9*a* shows three C_A molarities at one C_H/C_A ratio, 1.5 (equimolar H_2A^- and HA^{2-}).

Figure (III-9*b*) shows the effect of changing the C_H/C_A ratio (which is $\bar{\mathbf{n}}'_{max}$ or better, $\bar{\mathbf{n}}'$ at the neutral point). Only the 0.1 *M* and 0.001 *M* cases are shown. Note that the lines drop one unit below the maximum at $\log C_A = \mathbf{pH}$ on the acid side and rise one unit on the basic side at $\log C_A = \mathbf{pOH}$. This may be clearer in Fig. III-2.

Next we superimpose upon $\bar{\mathbf{n}}'$ lines the equilibrium $\bar{\mathbf{n}}$ curve for a specific acid (given K_a values), H_3PO_4, 0.1 *M* ionic strength, pK_a values 1.95, 6.80, and 11.67 (Fig. III-10). This shows the intersections, the only points where the equilibrium and material balance conditions are satisfied. These are solutions to the eight cases: the **pH** of the buffer and the HA^{2-} ampholyte having C_A of 0.1 or 0.001 *M* and one 10^{-6} *M*. This shows graphically the method of the computer solutions given before. The ampholytes intersect near $\bar{\mathbf{n}} = 1.0$ where the value is harder to read in this plot. Plotting (and numerical computing) vs. **H** instead of this log plot gives a clear intersection, as will be shown. Solutions of PO_4^{3-} alone have $\bar{\mathbf{n}}' = 0 - \mathbf{D}/C_A$, and thus it rises from the zero level in the range **pH** 10–14 to intersect at basic values for these molarities. Note in Fig. III-10 the very dilute buffer, 10^{-6} *M*, and the base alone, PO_4^{3-} (0.001 *M* Na_3PO_4), which is largely protonated to HPO_4^{2-} by water, reaching equilibrium $\bar{\mathbf{n}}$ near 1. One can easily sketch the $\bar{\mathbf{n}}$ curves for any other H_3A cases on Fig. III-10 to show the **pH** intersections for the eight mixtures in each H_3A case. Try citric, 2, 3, 4 and 6, 7, 8 acids.

III-9. CORRECT AMPHOLYTE RELATIONS USING THE COMPLETE APPROACH

Let us compare the traditional ampholyte equation with forms obtained from the $\bar{\mathbf{n}}$, $\bar{\mathbf{n}}'$ method to show a way around a required approximation in using the older method numerically. (The forms obtained in what follows are *not* needed for most calculations since the $\bar{\mathbf{n}}$ method is rigorous and simpler.)

When the K_a values are close together and large (rather acidic) and the solutions are dilute (0.01 *M* and below), the usual approximations give significant error. First consider the ampholyte H_2A^- in the H_3A system. Using Eqs. III-1 and III-2 with $\bar{\mathbf{n}}' = 2 - \mathbf{D}/C$ set equal to the full $\bar{\mathbf{n}}$ and collecting terms gives a correct relation of **H** to *C*,

$$2C - \mathbf{D} + (C - \mathbf{D})\mathbf{H}/K_3 - \mathbf{D}\mathbf{H}^2/K_2K_3 - (C + \mathbf{D})\mathbf{H}^3/K_1K_2K_3 = 0$$

This can be arranged to resemble the usual approximate text equation as

$$\mathbf{H}^2 = K_1K_2 \frac{C - \mathbf{D} + (2C - \mathbf{D})K_3/\mathbf{H}}{C + \mathbf{D} + \mathbf{D}K_1/\mathbf{H}} \qquad \text{(III-3)}$$

The fraction is the complete expression for the ratio $[H_3A]/[HA^{2-}]$ in the *C* molar ampholyte solution, H_2A^-. This is approximately but always less than 1 in acid cases and greater than 1 in basic. Examine the α diagrams of various acids to verify and clarify this. See the more detailed numerical discussion of ampholytes in Section III-11.

The common full equation contains the equilibrium concentration of the ampholyte species (Eq. III-4). It is usually solved approximately for **H** by inserting the analytical concentration of ampholyte and approximating **H** successively. The usual derivation for the $[H_2A]$ equation is from a proton gain–loss (see the diagrammatic example in Fig. III-3 for an alternative to charge balancing):

$$[H_3A] + \mathbf{H} = [HA] + 2[A] + \mathbf{OH}$$

Note that this contains only minor species and not the given C species. Next one converts each A term to $[H_2A]$ using equilibrium constant expressions:

$$\mathbf{H}[H_2A]/K_1 + (\mathbf{H} - \mathbf{OH}) = K_2[H_2A]/\mathbf{H} + 2K_2K_3[H_2A]/\mathbf{H}^2$$

$$\mathbf{H}^2 = \frac{K_w + K_2[H_2A] + 2K_2K_3[H_2A]/\mathbf{H}}{1 + [H_2A]/K_1} \qquad \text{(III-4)}$$

If one substitutes for $[H_2A] = C\alpha_2$ using the convenient form

$$\alpha_2 = (\mathbf{H}/K_1 + 1 + K_2/\mathbf{H} + K_2K_3/\mathbf{H}^2)^{-1}$$

one obtains the complete **H** equation in single variable, **H** (Eq. III-3).

For **pH** near 7 and C not too low, this relates to the approximate $\mathbf{H} = \sqrt{K_1K_2}$. For more acidic (or basic) cases, which we saw in the preceding examples for the 2, 3, 4 acid, the **pH** of the two ampholytes (cases 4 and 6 in Table III-1) were 2.7 and 3.54 for the 0.01 solutions instead of the approximate 2.5 and 3.5. For real cases (oxalic, tartaric, citric, and many other acids), results must be checked with the full equation if the rigorous method is not used. An iterative approach using trial **H** values on the right side of Eq. III-4 is possible, but convergence is slow for large, closely spaced K values. Ricci[1] regularly inserts $K_w/\mathbf{H}$ for **OH** in **D** and applies methods for numerical solution of high power equations. Some texts have used an approximate form omitting both the K_3 and the K_1 terms. This does not work for the first ampholyte (case 4) in Table III-1. The simultaneous $\bar{\mathbf{n}}$ method circumvents these problems, which arise from correct but poorly poised equations like the preceding (Eq. III-4).

Ampholyte Ilustrations

Calculation of dilution effects on ampholytes of a variety of systems shows effects of K values and of overlap in triprotics. Oxalic acid and the general diprotic HA^- behavior is discussed later (Figs. III-11 and III-12). Using the spreadsheet method (Section III-7), we obtain **pH** and $\bar{\mathbf{n}}$ values for the real and hypothetical systems that follow. The system pK_a's and the **pH** limit, p$\sqrt{K_1K_2}$, are given as headings for calculated **pH** for the C values of NaHA: 1, 0.1, and 0.01 M. The complete equation (III-3) is used.

Ampholytes of Three Diprotic Acids, $I = 0.1\ M$

In the following, (i) oxalic, p$K_a = 1.04$, 3.82, p$\sqrt{K_1K_2} = 2.43$; (ii) p$K_a = 1.0$, 2.0, p$\sqrt{K_1K_2} = 1.50$; (iii) p$K_a = 12.0$, 13.0, p$\sqrt{K_1K_2} =$ 12.50:

	(i)		(ii)		(iii)	
C	pH	$\bar{n}$	pH	$\bar{n}$	pH	$\bar{n}$
1	2.450	0.997	1.533	0.971	12.450	1.045
0.1	2.580	0.974	1.717	0.808	12.209	1.256
0.01	2.960	0.890	2.227	0.407	11.632	1.679

Notice the trends toward more neutral solution with dilution and that $\bar{n}$ decreases in acid and increases in base as required from examination of α diagrams like that for oxalic. (An alternative spreadsheet calculation and discussion of ampholytes was given by Parker and Breneman[2] using iteration in a set of conventional mass–charge–equilibrium balance equations.)

Ampholytes H_2A of Three Triprotic Acids, $I = 0.1\ M$

In the following, (i) citric, p$K_a = 2.93$, 4.36, 5.74, p$\sqrt{K_1K_2} = 3.645$; (ii) p$K_a = 1.0$, 2.0, 3.0, p$\sqrt{K_1K_2} = 1.50$; (iii) p$K_a = 2.0$, 2.0, 2.0, p$\sqrt{K_1K_2}$–2.00:

	(i)		(ii)		(iii)	
C	pH	$\bar{n}$	pH	$\bar{n}$	pH	$\bar{n}$
1	3.642	1.9998	1.519	1.970	1.824	1.985
0.1	3.645	1.9977	1.697	1.799	1.870	1.865
0.01	3.674	1.9788	2.173	1.329	2.102	1.210

The α diagram for citric acid is shown later (Fig. III-14a) and the 2, 2, 2 acid is shown in Fig. III-7, and no. 3 in the Exercises. The 1, 2, 3 acid diagram is like the 2, 3, 4 acid (Figs. III-1 and III-2) shifted down one log unit. Note that acidity may increase beyond the $\sqrt{K_1K_2}$ value by additional **H** from the third proton in H_3A, but $\bar{n}$ still cannot rise above 2 in acidic cases. (See the basic diprotic ampholyte case in the preceding diprotic iii.) In general, deviation from the approximate formula is greater in high and low pK_a systems and in higher dilutions.

III-10. MIXED SYSTEMS

$(NH_4)_2HPO_4$

Let us use C M diammonium hydrogen phosphate to illustrate calculation methods. This will show that seemingly different complete equations are identical and all such problems can be solved starting from the proton balance condition as expressed in the $\bar{\mathbf{n}}$, average bound proton number relation. Compare the complete approaches here with those in various earlier advanced texts. [See references in Chapters II, III and the Annotated Bibliography for these works: Butler (Ionic Equilibrium, p. 240), Ricci[1] (Hydrogen Ion Concentration, p. 310) and (Treatise on Analytical Chemistry, 1959, Part IB, Ch. 8, Sillen, p. 287; ibid, Ch. 12, Bruckenstein and Kolthoff, p. 443).] Ricci gets the full sixth power **H** equation for a general case of this type. Bruckenstein and Kolthoff get the full $\bar{\mathbf{n}}$ form equation, but state that is too complex to use. Butler sets up the full proton balance and uses it with a log concentration diagram of both phosphate and ammonium systems to get a logical approximation. Recently, Olivieri[3] used successive approximations to solve the set of equations. All of these make a vivid case for the directness and simplicity of the $\bar{\mathbf{n}}$ method following.

The general approach here is to set equal the bound protons expressed in material balance and equilibrium condition forms. The first is the total protonic acidity put in, C_H, minus that "net free" (on water) ($\mathbf{H} - \mathbf{OH} = \mathbf{D}$). The second is $\bar{\mathbf{n}}C$ for each weak acid system in a mixture. Here phosphate is C molar and NH_4^+ is $2C$ molar, which makes $C_H = 3C$. Using our master relation for mixed systems,

$$C_H - \mathbf{D} = \sum C_i \bar{\mathbf{n}}_i$$

the bound acid equals the sum of the equilibrium bound acid of each system (ammonium, α_{1a}, and phosphate, $\bar{\mathbf{n}}_p$), which yields three alternative interesting relations:

$$3C - \mathbf{D} = 2C\alpha_{1a} + C\bar{\mathbf{n}}_p \qquad \text{(III-5a)}$$

Then, using $\Sigma\,\alpha_n = 1$ and $\bar{\mathbf{n}}_p = \alpha_1 + 2\alpha_2 + 3\alpha_3$ (for the phosphate system, p) gives

$$3 - \bar{\mathbf{n}}_p = 2\alpha_{1a} + \mathbf{D}/C = 3\alpha_0 + 2\alpha_1 + \alpha_2 \qquad \text{(III-5b)}$$

or,

$$2\alpha_{0a} - \mathbf{D}/C = 2\alpha_3 - \alpha_0 + \alpha_2 \qquad \text{(III-5c)}$$

(Note that for a monoprotic system, $\alpha_1 = 1 - \alpha_0$.) Now, let us interprete the three. In Eq. III-5a, the total bound acidity is set equal to that on ammonia (a) plus that on phosphate (p). The other two are equivalent but reflect alternative starting species. In Eq. III-5b, we can postulate the original mixture as being C molar H_3PO_4 plus $2C$ molar NH_3. The protons lost by phosphate (going to equilibrium) are $(3 - \bar{\mathbf{n}})C$, which must equal those gained by ammonia, $\alpha_{1a}(2C)$, and water, **D**. (If basic, **D** is negative and tells proton loss by water and so it balances.) This is Eq. III-5b (times C). See Ricci[1] (pp. 301–310). In Eq. III-5c postulate a starting level: C molar HPO_4^- and $2C$ molar NH_4^+, the proton gainers equal the losers,

$$2[H_3PO_4] + [H_2PO_4^-] + \mathbf{H} = \mathbf{OH} + [PO_4^{3-}] + [NH_3]$$
$$2\alpha_3 C + \alpha_2 C = \alpha_0 C + \alpha_{0a}(2C) - \mathbf{D}$$

which is Eq. III-5c (times C). These illustrate the centrality of the proton condition as opposed to more artificial material–charge balance approaches that divert attention from the acid–base equilibria involved. Any of the Eq. III-5 set can be used in the spreadsheet method finding the intersection by the secant method.

With equilibrium K–**H** relations entered for all α expressions, these three equations are identical. Butler and Sillen use the third, approximating (as follows clearly from looking at the α diagram relations) α_3 very small, **D** small, p**H** about 8, and α_0 small. They conclude that the mixture is at the crossing of the log C of the remaining species, NH_3 and $H_2PO_4^-$. (This assumes that every proton lost by NH_4^+ is gained by HPO_4^-.) The result is p**H** 8.07 using K^0 values for the acids and $C = 0.1\ M$. At this ionic strength ($0.3\ M$), α_0 is not so small and the diagram is shifted to the left (pK_3 changes from 12.3 to about 11.5). The method is clear but the result is not numerically accurate. This is the general problem with the log C diagram method. The diagrams do give a clear picture of the species relations and can lead to a correct choice of the species that must be considered. After achieving a solution, it can be checked in a complete equation with newly corrected K values for the final ionic strength. Olivieri[3] solves the problem for $C = 0.0001\ M$, also using K^0 values, to get **pH** 8.04. We get **pH** 7.88 below for $\boldsymbol{I}$ corrected K_a's.

The method proposed here is to use computer-calculated $\bar{\mathbf{n}}$ and α_{1a} values in Eq. III-5b with $\mathbf{D}/C$ negligible to find the **pH** that satisfies them. This is close to Butler's method, but it does not omit any term in $\bar{\mathbf{n}}$.

From computer results at 0.1 **pH** unit intervals we have

pH	7.8	7.9	8.0	At 7.88
$3 - \bar{\mathbf{n}}$	1.9092	1.9266	1.9408	1.9245
$2\alpha_{1a}$	1.9374	1.9217	1.9024	1.9251

This used K values for $\boldsymbol{I} = 0.1\ M$. Omitting the $\mathbf{D}/C$ term affects the fifth figure in the preceding values for $0.01\ M$ solution. So below $0.001\ M$, C becomes significant. See Fig. III-12 for dilution effects on an ampholyte. By omitting the $\mathbf{D}/C$ term, we have obtained the high C limiting **pH** value for a given $\boldsymbol{I}$.

To reiterate our theme, all correct methods of approaching the proton balance achieve the same K–$\mathbf{H}$–C relationships. Therefore, one might as well use a single $\bar{\mathbf{n}}$ relation that works as well for all cases and keeps attention on the essential proton balance, which is the point of acid–base equilibrium. Combined with α diagrams, this offers a more coherent view of the subject. The $\bar{\mathbf{n}} = \bar{\mathbf{n}}'$ relation (for single polyprotic systems) provides a mathematically well-conditioned pair of iteration functions for the solution of all the problems asking for **pH** given K_a's and C's. They are always of opposite slope (vs. **H**) and have but one intersection, one positive root. Put together, they give the full $n + 2$-power **H** equation, which multiplied out to show the **H** polynomial could be solved by standard, long methods (instead of the simple and direct $\bar{\mathbf{n}}$-intersection, computer or graphical solution of this book). Others used the full equation method, as mentioned at the start of Section III-10. In mixtures of systems, as in the example just presented, a summation of $\bar{\mathbf{n}}$'s gives the rigorous equation.

The Citrate–Oxalate Mixed System

Find the **pH** and species in a mixture having analytical C's of $0.0300\ M$ sodium oxalate and $0.0200\ M$ citric acid at ionic strength $0.1\ M$, 25°C. We see that $C_{AB} = 0.0600\ M$. The total H is, as above, the bound H in each system plus **D**,

$$\bar{\mathbf{n}}_{ox}(0.03) + \bar{\mathbf{n}}_{cit}(0.02) + \mathbf{D} = 0.06$$

The pK_a's of these systems are between 1 and 6, so there is enough H to protonate each system and make the **pH** acidic. We run a spreadsheet from, say, **pH** 2–5 and find that the left crosses 0.06 between **pH** 3.5 and 3.6. Then finer divisions are run near 3.6, showing that the **pH** is near 3.585 (spreadsheet portion that follows). The α diagrams (Figs. III-11 and III-14)-show the major species and $\bar{\mathbf{n}}$ at this **pH**. The HOx^-, Ox^{2-}, H_2Cit^-, H_3Cit, and $HCit^{2-}$ are all appreciable. In the preceding equation we get H bound to oxalate 0.0190, to citrate 0.0407, **D** = 0.00026 for a total of 0.0600 *M* to three significant figures.

pH	H	$\bar{\mathbf{n}}_{ox}$	$\bar{\mathbf{n}}_{cit}$	D	Total H
3.5	0.0003162	0.679335	2.091994	0.0003162	0.062536
3.6	0.0002512	0.626294	2.027000	0.0002512	0.059580
3.58	0.0002630	0.637171	2.039984	0.0002630	0.0601778
3.585	0.0002600	0.634463	2.036738	0.0002600	0.0600287
3.59	0.0002570	0.631748	2.033492	0.0002570	0.0598793
3.595	0.0002541	0.629025	2.030246	0.0002541	0.0597298

I have not seen a problem of this type in any text or treatise. I do not know of any other numerical or graphical method of doing it. An extension to a weak acid–weak base polyprotic titration is detailed in Chapter IV for citric acid and ethylenediamine. A similar situation is the preparation of universal buffer mixtures in Chapter IV where adjustment of ionic strength to the needed value is described. These two applications are simplified by a reverse solution: Choose the **pH** and solve for the C_{AB} required to obtain it in a given volume, ionic strength, and *C* values of the systems. A single point illustration follows.

How much NaOH (C_{sb}) is required to prepare 1 l of buffer with **pH** 4.000 with the *C*'s for the system just treated?

The new $C_{AB}(C_H)$ will be 0.06000 – C_{sb}. In the equation for these systems, we get, using $\bar{\mathbf{n}}$ values from oxalate and citrate system tables,

$$\bar{\mathbf{n}}_{ox}(0.03) + \bar{\mathbf{n}}_{cit}(0.02) + \mathbf{D} = 0.06000 - C_{sb}$$

$$0.3985(0.03) + 1.760(0.02) + 0.0001 = 0.06000 - C_{sb}$$

for which $C_{sb} = 0.01275$ mol NaOH/l to raise the **pH** to 4.000. Similarly, lowering the **pH** to 3.000 requires 0.0159 mol HCl/l, the $+C_{sa}$ (in place of $-C_{sb}$).

III-11. CONCENTRATION DEPENDENCE OF pH IN POLYPROTIC SYSTEMS

Dilution effects on polyprotic systems can be calculated simply from the master $\bar{\mathbf{n}}$ relations in the form

$$C_A = \frac{\mathbf{D}}{C_H/C_A - \bar{\mathbf{n}}} \tag{III-6}$$

This is the $\bar{\mathbf{n}}'$ relation, but if the equilibrium condition $\bar{\mathbf{n}}$ value at the chosen **H** is used, the true C_A for the specific substance is obtained. For example, solutions of H_3A alone are summarized by $C_A = \mathbf{D}/(3 - \bar{\mathbf{n}})$. So, if a table of $\bar{\mathbf{n}}$ vs. **pH** is available for a particular acid like citric, one can calculate the concentrations of the acid that will give chosen **pH** values. This is another instance of reversal of variables for easier computing: the reverse (finding **H** for chosen C is longer). For other types of mixtures, the total C_A can be found by inserting the specific C_H/C_A ratio of the mixture. For example, an ampholyte C molar H_2A^- has ratio 2 so that $C = \mathbf{D}/(2 - \bar{\mathbf{n}})$. A 1:3 buffer of H_3A and H_2A^- has ratio $\frac{9}{4}$, etc. In this section we conceive of starting with a mixture of the given ratio at high analytical concentrations, say, 1 M. Then water is added to effect dilution. Throughout $\boldsymbol{I} = 0.1$ M is used. (The mathematical form is the point here so that the conditional constants are not adjusted to the high starting ionic strengths.)

Let us look first at the oxalic acid system α's. **Figure III-11** shows the α and $\bar{\mathbf{n}}$ lines for oxalic acid.

Figure III-12 shows the dilution curves for three mixtures: the acid, H_2Ox (or H_2A); the ampholyte, HA^-; and the buffer, equimolar HA^- and A^2.

The C_H/C_A ratios (Eq. III-6) are 2, 1, and 0.5 for three cases in Fig. III-12. As an example calculation, take **pH** 1 for the pure oxalic acid case. Using pK_a's 1.04 and 3.82 yields $\bar{\mathbf{n}} = 1.522$, which in Eq. III-6 gives

$$C = 0.1/(2 - 1.522) = 0.208\ M \qquad \log C = -0.68$$

Note that all solutions approach neutrality at infinite dilution. However, some solutions have a high C–**pH** limit while others do

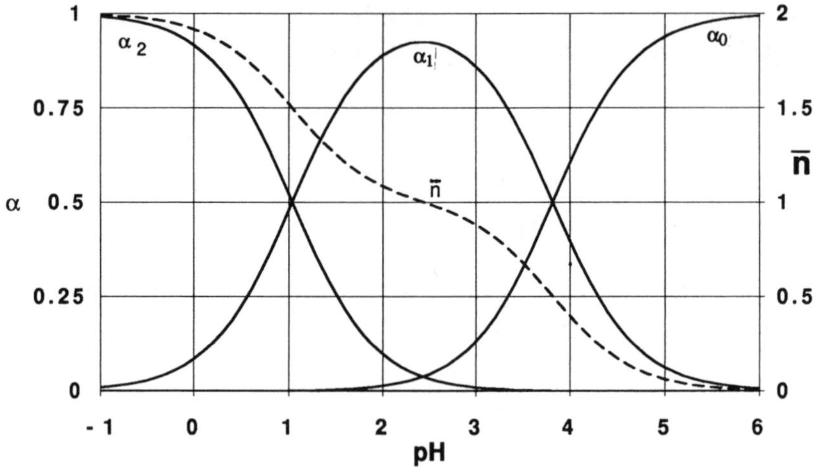

Fig. III-11. Alpha and $\bar{\mathbf{n}}$ lines for oxalic acid, $\boldsymbol{I} = 0.1\ M$, 25°C.

not. The α diagrams and the chemistry of competing equilibria help explain this (see Fig. III-11).

As C increases, a smaller fraction of the added material has to ionize to form the **H**. For H_2A (and A^{2-} basic, not shown), this means that α_2 (or α_0) may continue to increase. These are asymptotic to 1 but have no maximum, while α_1 has a maximum. Proton balance

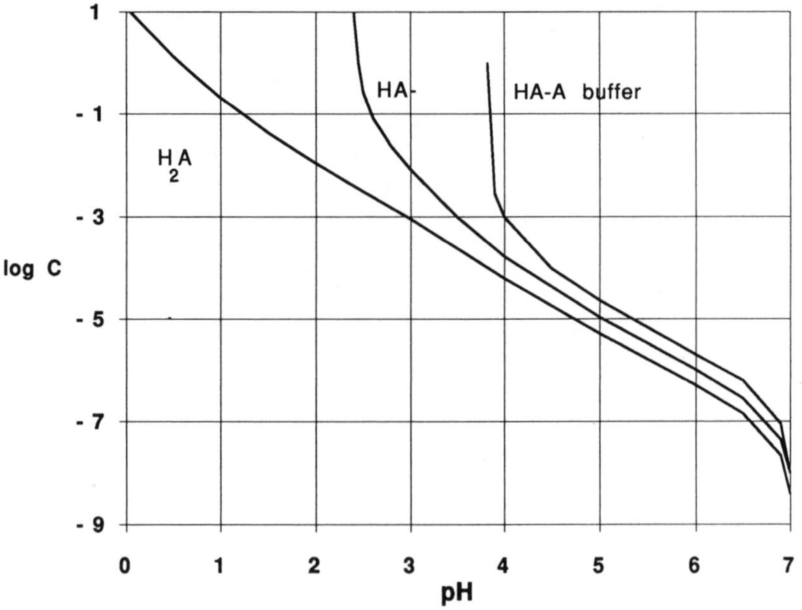

Fig. III-12. Dilutions in the oxalate system.

shows that no concentration of NaHOx (alone) can move left of $\alpha_{1(\max)}$, where there would be more H_2Ox than Ox^{2-} formed, making $\bar{\mathbf{n}}$ greater than 1. This cannot happen in this solution, as it would increase C_H. (It can happen in triprotic systems as shown in what follows.) (Bases can increase in $\bar{\mathbf{n}}$ by removing H^+ from water, producing OH^-.) Any protons lost from the original HOx^- must form H_2Ox and net **D** of H^+. The proton gain–loss balance for this pure ampholyte solution is

$$[H_3O^+] + [H_2Ox] = [OH^-] + [Ox^{2-}] \quad \text{or} \quad \mathbf{D} = [Ox^{2-}] - [H_2Ox]$$

That is, α_0 must be greater than α_2 in acidic solution, and $\bar{\mathbf{n}}$ must decrease. Thus, increasing the concentration of original ampholyte can only bring α_0 and α_2 closer to their crossing from the higher pH side (bring $\bar{\mathbf{n}}$ up toward 1). The H^+ freed, **D**, is the increasingly small difference between H_2Ox and Ox^{2-} as C increases.

For oxalic acid, a species plot of α vs. log C would show all the features of the α diagram up to **pH** 7. However, other acids with weaker constants can be quite different. Looking at phosphoric acid with pK_a values 1.95, 6.80, and 11.67 at ionic strength 0.1 M and 25°C (see Fig. III-6), we see that the α curves must remain at the **pH** 7 values at high dilutions (**Fig. III-13**). Here (left end) HPO_4^{2-} and $H_2PO_4^-$ are the major species. The species PO_4^{3-} is not appreciable at any dilution of H_3PO_4 alone. Recall similar distinctions (Chapter II)

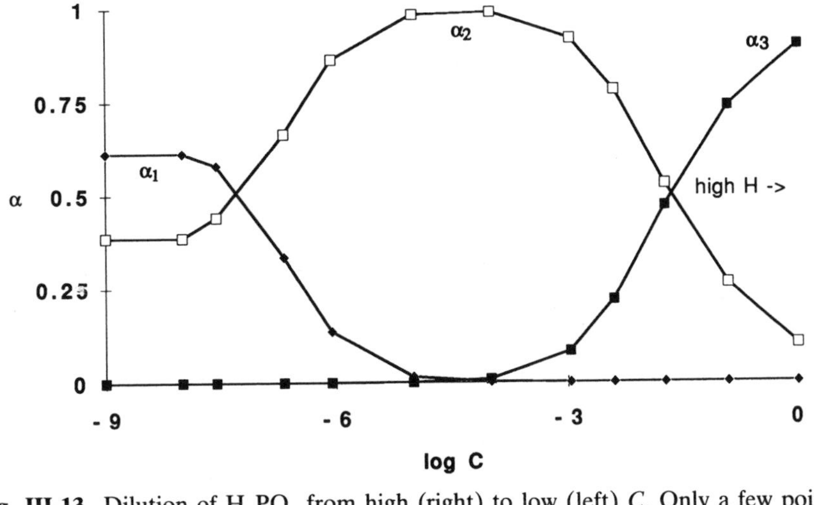

Fig. III-13. Dilution of H_3PO_4 from high (right) to low (left) C. Only a few points were calculated. The curves are, of course, smooth.

among monoprotic acids. Fairly strong ones are highly ionized at high dilution, but those with pK_a's above 7 are not.

III-12. FURTHER EXAMPLE CALCULATIONS

Citric Acid. Citric acid has pK_a's 2.93, 4.36, and 5.74 at 25°C and 0.1 *M* ionic strength. Its α, $\bar{\mathbf{n}}$ diagrams are given in **Figs. III-14*a*,*b***. Contrast these with H_3PO_4 in Fig. III-6. Examination of this diagram for various specified concentrations allows rapid choice of the **pH** region involved. The maximum value of $\bar{\mathbf{n}}'$ is C_H/C_A in acidic solutions. For example, equimolar buffers of H_3A and H_2A^- can approach $\bar{\mathbf{n}} = 2.5$ (if no proton loss occurs) near **pH** 3. In fact, some **H** must be formed from the acids so that the equilibrium $\bar{\mathbf{n}}$ will be less than 2.5. The full equations (III-1 and III-2) can be used. Similarly, we can decide that other common mixtures of citrates will be near **pH** 3.5 (pure H_2A^- solution), 4.3 (1:1 buffer H_2A, HA^{2-}), 5.0 (pure HA^{2-} solution), and 5.7 (1:1 buffer HA^{2-}, A^{3-}). The $\bar{\mathbf{n}}'$ lines are similar to the previous (2, 3, 4) case (Fig. III-2).

We show three methods of plot and/or spreadsheet solutions: semilog, linear $\bar{\mathbf{n}}$ vs. **H** (or **OH**), and log-log (Appendix B). The spreadsheet for the linear solution may be made first using the secant

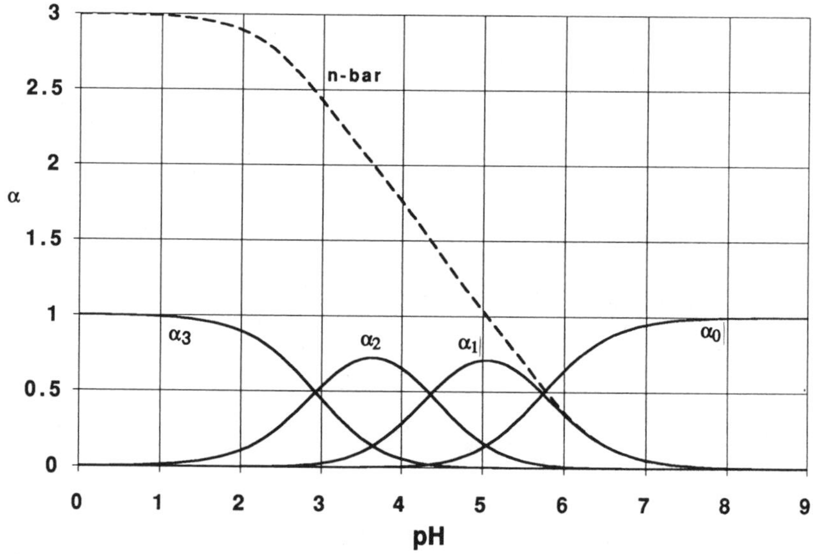

Fig. III-14*a*. Alpha, $\bar{\mathbf{n}}$ lines for citric acid on equal ordinate to show addition of α's to form $\bar{\mathbf{n}} = 3\alpha_3 + 2\alpha_2 + \alpha_1$.

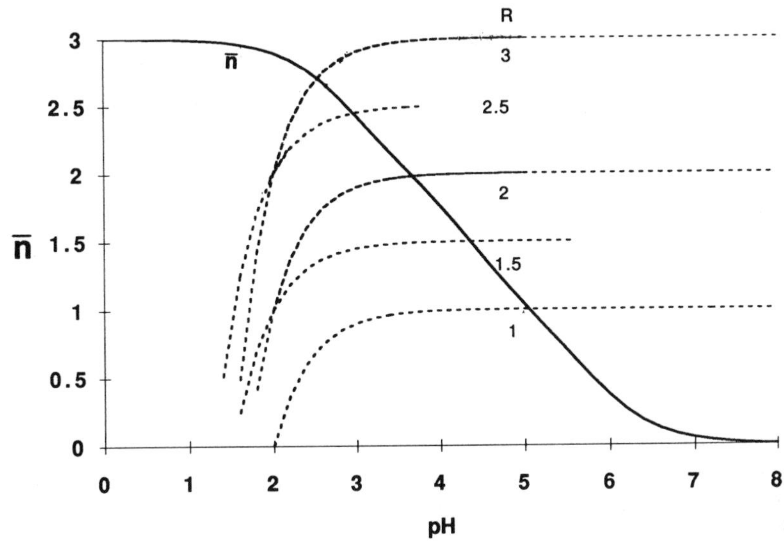

Fig. III-14*b*. Citric acid system ($\bar{\mathbf{n}}$, solid lines, $\bar{\mathbf{n}}'$, dotted lines): 0.01 *M* acid, buffers, and ampholytes; **pH**, from top, 2.5, 3.0, 3.7, 4.3, 5.0.

intersection method. The other two methods can be tried as a check simply by changing the spreadsheet formulas to log 10( ) for $\bar{\mathbf{n}}$ and then computing the log of both $\bar{\mathbf{n}}$ and **H** again. The secant formulas will do as well for references to the linear values as to the log values.

The intersections at curved portions of $\bar{\mathbf{n}}'$ in Fig. III-14*b* indicate significant differences from the results given by approximate methods.

Example. The pure 0.01 *M* H_3Cit is approximately $\sqrt{K_1 C} = 2.47$ (true value is 2.533). Each species $C = 0.01$ *M* Fig. III-14*b*, i.e., total A is 0.02 *M* for the buffers and 0.01 for the ampholytes. $\bar{\mathbf{n}}'$ lines can cross each other if their total C_A's are not the same.

***Basic* 0.01 *M* $\mathbf{Na_3}$ Citrate Case.** This case would be in the flat portion of the log curve at the right bottom, so another graph (**Fig. III-15**) is constructed with $\bar{\mathbf{n}}$ and $\bar{\mathbf{n}}'$ vs. **OH**. Although **H** could as well be used, it reverses the slopes. The nonlogarithmic plot is more nearly linear to better show the intersection that occurs at $\bar{\mathbf{n}} = 0.000933$. A portion of the computer spreadsheet for this plot is shown (pK_w 13.80 is used to agree with the 0.1 *M* ionic strength *K*'s used for the citrate system; thus, **pH** 8.8, rather than 9.0, corresponds to **pOH** 5.00):

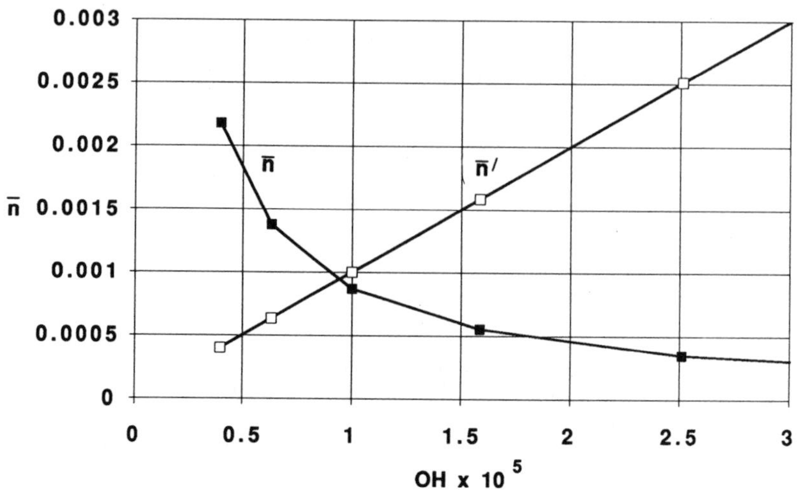

Fig. III-15. A 0.01 *M* sodium citrate intersection plot: **pH** 8.78, **pOH** 5.02 (from the intersection, $[OH^-] \cong 9.5 \times 10^{-6}$). By interpolation or computer iteration, $\mathbf{OH} = 9.38 \times 10^{-6}$, **pH** 8.772. The approximate $\mathbf{H} = \sqrt{K_3 K_w / C_A}$ gives **pH** 8.77. For stronger base, PO_4^{3-}, polyamines, etc., the deviation would be larger.

pH	$[OH^-] \times 10^5$	$\bar{\mathbf{n}}$	$\bar{\mathbf{n}}' \cong 100[OH^-]$	$\Delta \times 10^3$
8.4	0.398	0.00218	0.000398	1.78
8.6	0.631	0.001379	0.000631	0.748
8.8	1	0.00087	0.001	−0.13
9	1.585	0.00054926	0.001584893	−1.04
9.2	2.512	0.00034663	0.002511886	−2.16
9.4	3.98	0.00021873	0.003981072	−3.76

Logarithmic Scale Plot. The logs of $\bar{\mathbf{n}}$ and $\bar{\mathbf{n}}'$ plotted vs. **pH** often yield more linear intersections, especially in basic cases. Seven cases are shown in **Fig. III-16**. Here all cases can be shown on one plot, in contrast with those in the preceding. In addition to the plots shown in Fig. III-14*b*, in Fig. III-16 a buffer of 0.1 ratio is shown. It contains *C*'s of 0.0182 *M* for citrate and 0.00182 *M* for $HCit^{2-}$, **pH** 6.70.

For further practice, the ampholyte near $\bar{\mathbf{n}} = 2.0$ (0.01 *M* H_2Cit^-) gives **pH** 3.674 at $\bar{\mathbf{n}} = 1.979$. A table of a citrate system and answers for practice are given in the exercises the end of this chapter.

Linearity of *C* Terms: Buffer Design

Calculation of C's, the Inverse Problem. A simplification the master relation allows is that any mixture is linear in the analytical concen-

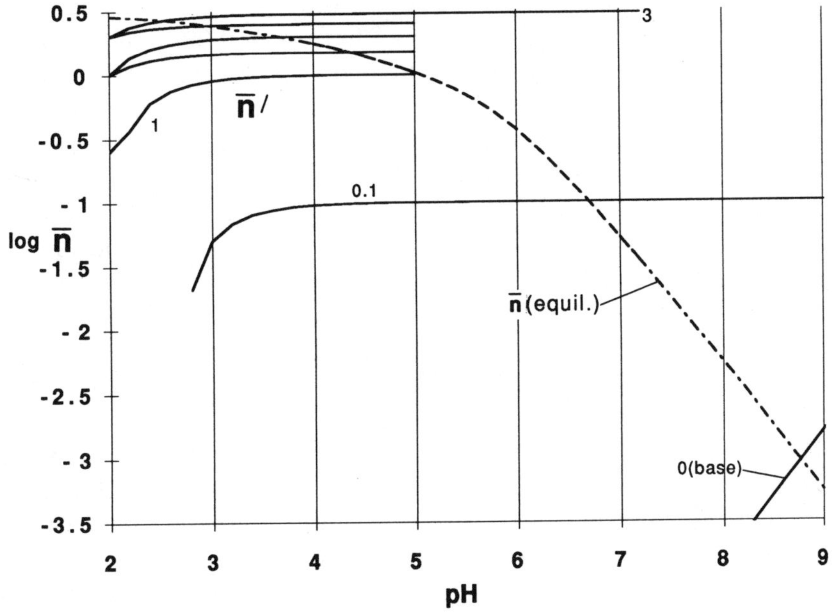

Fig. III-16. Log $\bar{\mathbf{n}}$–p**H** plots of ampholytes, buffers, and basic (citrate) (as in III-14*b*), and a buffer of 0.1 ratio; $\bar{\mathbf{n}}$, dashed line; $\bar{\mathbf{n}}'$, solid lines.

trations, *C*'s. Put $\bar{\mathbf{n}}$ into the material balance equation III-1, multiply out, and collect terms to get (one to three of the *C*'s may be zero for specific mixtures as in previous examples; more than one set of *C*'s can represent identical solutions as explained in Section III-6, "Equivalent Solutions"):

$$(3-\bar{\mathbf{n}})C_3 + (2-\bar{n})C_2 + (1-\bar{\mathbf{n}})C_1 - \bar{\mathbf{n}}C_0 = \mathbf{H} - \mathbf{OH} \quad \text{(III-7)}$$

Examples

1. What composition citrate buffer of C_3 and C_2 produce a p**H** of 3.301 with total citrates 0.04 *M*?

Using ionic strength 0.1 *M* and pK_a values 2.93, 4.36, and 5.74, using Eqs. III-1 and III-2 first calculate that the equilibrium $\bar{\mathbf{n}}$ = 2.2231 (α, $\bar{\mathbf{n}}$ tables for citric and other systems are found in Appendix D). Then take $C_3 = x$ molar, $C_2 = 0.04 - x$ molar. From Eq. III-7 we get

$$(3 - 2.2231)x + (2 - 2.2231)(0.04 - x) = \mathbf{H} - \mathbf{OH}$$

With $\mathbf{H} = 5.00 \times 10^{-4}$ M, we get $x = 0.00942$ M (C_3) and $C_2 = 0.0306$ M. From the $\bar{\mathbf{n}}$ calculation, we have $\alpha_0 = 0.0002$, $\alpha_1 = 0.0575$, $\alpha_2 = 0.6613$, and $\alpha_3 = 0.2810$, which yield ionic strength 0.0334 M, so that adding 0.0666 more will give the 0.1 M required for the constants used. A mixture of the C_3 and C_2 found (H_3Cit and NaH_2Cit) plus 66.6 mmol of, say, KCl per liter should give the **pH** required. Note everything is in molarity units here. For the pa_H meter reading expected, we add 0.083 to get pa_H 3.384 assuming Kielland table f_+ for hydronium ions. Remember that the C's are the correct amounts, C_n's, needed to prepare this solution. They are not the equilibrium species concentrations (α_n C's). At equilibrium, $[H_3A] = \alpha_3(0.04) = 0.0112$ M, and $[H_2A^-] = 0.0264$ M. About 6% of the 0.04 M C_A is in HA^{2-} and A^{3-}.

Assuming that K_1 can be used for a simple buffer solution, the approximate method makes an interesting comparison. (This is the monoprotonic approximation, pretending that only the first proton is lost and C's are the equilibrium species concentrations.)

Here $K_1 = \mathbf{H}[H_2A^-]/[H_3A] = 0.0005(0.04 - x)/x$ gives us $x = 0.012$ and $C_2 = 0.0280$. This is wrong because we omitted the amount forming **H**, HA^{2-} and A^{3-}. The overlap with the K_2 species is appreciable. This mixture would turn out over 0.1 pH unit too acidic. Even though **H** is quite small for correction of C_3 and C_2 terms here, the answer is still off because of the α's. Appreciable HA^{2-} forms (Fig. III-14*a*).

2. How could 1 l of this buffer be prepared starting with 0.100 M solutions of citric acid and NaOH? From the $\bar{\mathbf{n}}$–$\bar{\mathbf{n}}'$ relation and $\mathbf{D} = 0.000500$ M, $\bar{\mathbf{n}} = 2.2231 = (C_H - \mathbf{D})/0.0400$, $C_H = 0.08942$ is required. The starting amount of 1 l (400 ml of 0.100 M acid) had 3(0.04 mol) = 0.1200 M acidity, so 0.03058 must be neutralized to obtain the equilibrium amount 0.08942 M. Thus, 305.8 ml of 0.100 M NaOH is required. Note once again that identical solutions can be made in more than one way and that two C terms suffice to specify any single system.

Overlap Effects on the pH of Citrate Buffers

Example. Look at a buffer 0.001 M in citric acid and 0.050 M in NaH_2 citrate (the point of greatest error in **Fig. III-17).** The 1:50 ratio of conjugates put in the approximate (He–Ha) K_1 equation gives **pH** 4.629. Thus (from $C_H = 0.103$ M and $C_A = 0.051$ M) $\bar{\mathbf{n}}' = 2.019$. This is incompatible with equilibrium values: $\bar{\mathbf{n}} = 1.3$ or 0.037 M of the acidity "free" making the **pH** 1.4 and α_1 ($Hcit^{2-}$) the

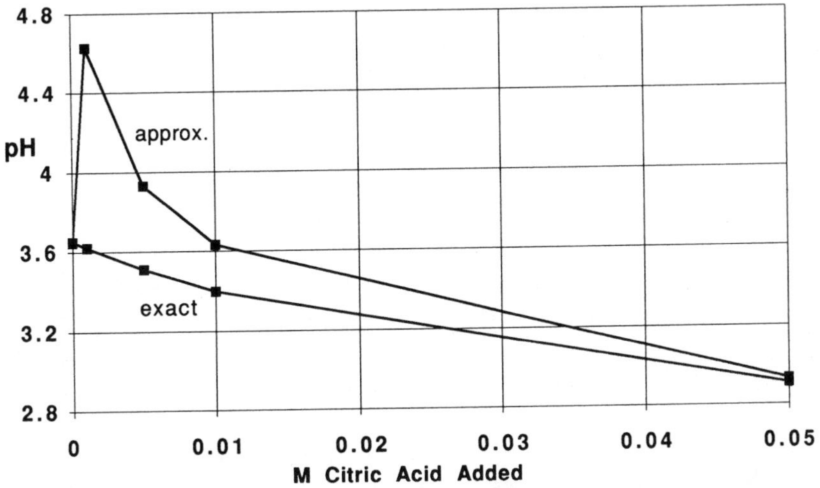

Fig. III-17. Left (at 0 *M* citric acid added): pure 0.05 *M* NaH_2 citrate solution having approximate **pH** 3.645 and exact equation **pH** 3.649. Buffers with added H_3 citrate are not well treated by the approximate buffer (He–Ha) equation at buffer ratios far from maximal, 1:1 buffers.

major citrate species! Searching out the **pH** that gives $\bar{\mathbf{n}} = \bar{\mathbf{n}}' = 2.015$ at **pH** 3.618, we have a buffer for which the approximate equation is over 10-fold off! The $\bar{\mathbf{n}}$–α table in Appendix D) (or diagram, Fig. III-14*a*) lets the student see this clearly.

SUMMARY

1. Deducing $\bar{\mathbf{n}}'$ and C_{AB} for mixtures in one or more polyprotic systems is described in Sections III-1 and III-6 and in the citrate table in the exercises.
2. Chemical argument to produce proton balances for polyprotic mixtures is aided by bar diagrams of species shifts in Section III-2.
3. Regularities of α functions are related to K_a spacings. These lead to the sketching of approximate α curves and estimation of **pH** of mixtures (Section III-4).
4. Rationalization of $\bar{\mathbf{n}}$, $\bar{\mathbf{n}}'$ functions are used for design of methods for solving complete **H** equations (Sections III-7 and III-8).
5. In the ampholyte problem, rigorous treatment by $\bar{\mathbf{n}}$ methods and the numerical consequences are contrasted with approximate methods (Section III-9).

6. Multiple systems like $(NH_4)_2HPO_4$ and mixtures of oxalate and citrates are treated in Section III-10.
7. Some mixtures have a high C limiting **pH** and some do not (Section III-11).

EXERCISES

The tables of α and $\bar{\mathbf{n}}$ values for several systems in Appendix D are given to higher precision than can be read off small graphs. These can be used in considering these problems especially if calculator solving is required rather than microcomputer use.

1. a. A simple acid for round-value mental calculations is the diprotic acid having pK_{a1} 1.5 and pK_{a2} 2.5. Consider $\bar{\mathbf{n}}$ and $\bar{\mathbf{n}}'$ values for a 0.0100 M solution of the acid, H_2A, alone. Sketch, without calculations, the approximate α and $\bar{\mathbf{n}}$ curves. Rationalize the α, $\bar{\mathbf{n}}$, and $\bar{\mathbf{n}}'$ values at **pH** 2.00. (The correct **pH** is 2.00, rather than 1.8, the approximate $\sqrt{K_aC}$.)

b. Consider the same conditions as in **a** for a 0.0100 M NaHA solution. (The correct **pH** is about 2.4, not the approximate 2.0 from $\sqrt{K_{a1}K_{a2}}$.) Some calculation of the $\bar{\mathbf{n}}$'s will be needed to establish an accurate **pH** in this case.

c. From the sketch in part **a** estimate the **pH** of buffers that have ratios of 1:1, 1:10, and 10:1 in this system.

d. D-Tartaric acid (pK_a's 2.82 and 3.97 for $I = 0.1\ M$) is a real case having about one log unit K spacing. Investigate the effect of this on the pure acid solution, on the ampholyte, and on some buffers in the range from 1 to 10^{-4} M total tartrate. First make a rough estimate of **pH** and $\bar{\mathbf{n}}$ and then use a spreadsheet for the tartrate system, just changing the $\bar{\mathbf{n}}'$ term for each new case tested.

[*Ans.*: **Ampholyte:** limit **pH** 3.395; 1 M, **pH** 3.396 ($\bar{\mathbf{n}} = 0.9996$); 0.001 M, **pH** 3.666 ($\bar{\mathbf{n}} = 0.784$). These results are printed to one more significant figure than justified by the uncertainties in the constants so that differences in the calculation process may be detected. **First buffers** (K_{a1}): 0.1 M H_2T, 0.01 M HT^-, **pH** 2.099, $\bar{\mathbf{n}} = 1.837$; 0.01 H_2T, 0.1 M HT^-, **pH** 3.287, $\bar{\mathbf{n}} = 1.086$.]

2. The mixtures specified in the following table by analytical C's of weak and strong acids and bases added in the first six columns were entered into the **H**, $\bar{\mathbf{n}}$ calculation spreadsheet to obtain the results in the last two columns. The reader might use this sheet to practice computation methods as well as deduction of C and $\bar{\mathbf{n}}'$. The citric acid pK_a values for 25°C and $\boldsymbol{I} = 0.1\ M$ used were 2.93, 4.36, and 5.74 and pK_w = 13.80. See examples in the chapter (such as Figs. III-14 and III-15).

The need for complete calculations can be judged by doing the approximate methods, for example, the first $\sqrt{K_a C}$ is 2.465, or 19% error in **H** (not in p**H**!). For the rest, approximate equation p**H** values are 3.76, 5.36 (3.0, strong alone approximation), 4.32, 3.645, 5.050, 8.77 (10.8, strong alone), and 6.694.

CITRATE SYSTEM EXAMPLE SHEET

Put into Mixture						Deductions			Results	
H_3A	H_2A^-	HA^{2-}	A^{3-}	H^+(s)	OH^-(s)	$\bar{\mathbf{n}}'$	C_{AB}	C_A	p**H**	$\bar{\mathbf{n}}_{eq}$
0.01						3 − 100**D**	0.03	0.01	2.533	2.707
0.01					0.012	1.8 − 100**D**	0.018	0.01	3.958	1.789
0.01			0.024			0.706 − 29.4**D**	0.03	0.034	5.484	0.706
0.01				0.001		3.1 − 100**D**	0.031	0.01	2.451	2.746
0.01	0.01	0.01	0.01	0.002	0.003	1.475 − 25**D**	0.059	0.04	4.390	1.474
	0.01					2 − 100**D**	0.02	0.01	3.674	1.979
		0.01				1 − 100**D**	0.01	0.01	5.055	0.999
			0.01			0 − 100**D**	0	0.01	8.770	9.33E − 4
			0.01	0.002	0.001	0.1 − 100**D**	0.001	0.01	6.699	0.09999
			0.01	0.001	0.002	−0.1 − 100**D**	−0.001	0.01	10.8000_4	8.7E − 6

One can try other mixtures and also the dilution effects on them. Compare the results with the $\bar{\mathbf{n}}$ and α diagrams to reason the differences between actual results and approximations. For example, why does citric acid contribute so much **H** in the presence of strong acid (p**H** 2.451 case) while citrate hardly affects the same excess strong base (p**H** 10.8 case)?

3. Compare the species distributions in three H_3A acid systems for the several analytical mixtures given: (a) phosphoric (Fig. III-6); (b) citric (Fig. III-14); (c) acid having all three pK_a's 2.000 (Fig. III-7).

Mixtures: The 1:1 buffers of $\bar{\mathbf{n}}_{max}$ of 2.5, 1.5, and 0.5 at C total 0.2, 0.02, and 0.002. Pure H_3A and pure Na_3A at $C = 0.1$, 0.01, 0.001. Pure ampholytes, H_2A^-, HA^{2-}, at the same three C's. Many of these are shown in the chapter, so focus on the pK 2 cases to explain why the approximate or expected species are so far off in some of the mixtures. Sketch the $\bar{\mathbf{n}}'$ on an $\bar{\mathbf{n}}$-diagram copy.

4. (Refer to the exercise 3.) All of the following statements *appear* true for the phosphoric acid diagram in **3**. The last two are fase and are not required by the equilibrium equations. Explain in terms of the species shifts required to reach equilibrium starting from the analytical C values. These may be deduced in helpful ways from the bar diagrams, $\bar{\mathbf{n}}$–$\bar{\mathbf{n}}'$ intersection plots, and α–$\bar{\mathbf{n}}$ diagrams.

- **a.** Adjacent α curves cross at $\mathbf{pH} = pK_a$, the K_a step involving the two α species.
- **b.** $\bar{\mathbf{n}}$ is exactly the sum of $\alpha_1 + 2\alpha_2 + 3\alpha_3$.
- **c.** The 1:1 buffers (those having equal analytical concentrations of two conjugates, such as H_2A^- and HA^{2-}) have $\mathbf{pH} = pK_a$.
- **d.** Ampholytes have $\mathbf{pH}$ halfway between two pK_a values.

5. The most celebrated metal ion chelating agent is EDTA (ethylenediaminetetraacetic acid, H_4Y). It is usually treated as a tetraprotic acid for uses above $\mathbf{pH}$ 3. However, it has six base positions, four carboxyl and two amino, with pK_a values 0, 1.5, 2.00, 2.68, 6.11, and 10.17 at 25°C and ionic strength 0.1 M. The first two are rough estimates. These give the diagram in **Fig. III-18**.

For which ampholytes and 1:1 buffers at $C = 0.010\ M$ will the $\mathbf{pH}$ be at the crossings (buffers) and maxima (ampholytes)? Explain the possibility that the abscissa $\mathbf{pH}$, or $\mathbf{pOH}$, is reasonable for those cases and also for 0.0100 M pure H_6Y^{2+} (say, the dihydrochloride) and pure Na_4Y solutions. Use rough mental

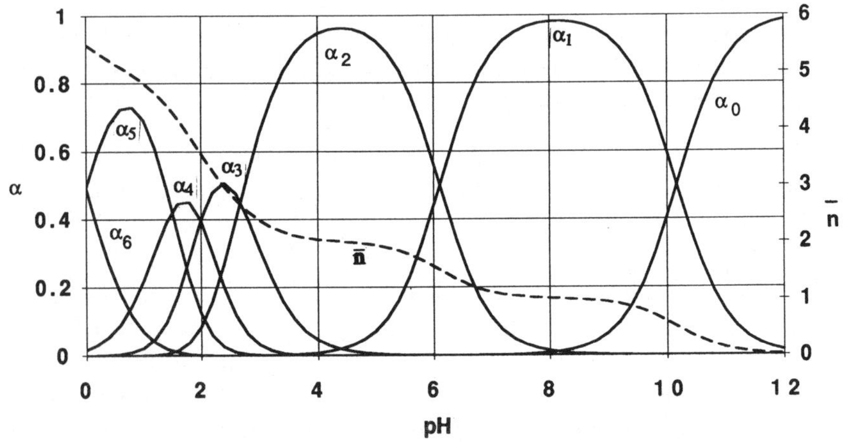

Fig. III-18. The α, $\bar{\mathbf{n}}$ diagram for EDTA showing all seven species, $\alpha_6, \ldots, \alpha_0$.

calculations of $\bar{\mathbf{n}}'$ to estimate the **pH** of all of these cases. *Hint*: Start from assumptions that lead to contradictions such as the following: (i) 0.0100 *M* EDTA dihydrochloride could produce **pH** 2.0 if it lost all of the first proton but the species would have $\bar{\mathbf{n}} = 5.0$, which is seen to occur at **pH** 1.0; (ii) pure 0.0100 *M* Na_4Y has to be near **pH** 12 to make α_0 close to 1. But, to have **pH** near 12, 0.01 *M* **OH** must be formed (and 0.01 *M* HY^{3-}), which requires **pH** near 8! Then continue estimates to reach compatible $\bar{\mathbf{n}}$, $\bar{\mathbf{n}}'$ values by eye. This is a visual walk through the numerical computing methods of this chapter.

REFERENCES

1. Ricci, J. E., *Hydrogen Ion Concentration*, Princeton University Press, Princeton, NJ, 1952.
2. Parker, O. J. and Breneman, G. L., *J. Chem. Educ.*, **67**, A5 (1990).
3. Olivieri, A. C., *J. Chem. Educ.*, **67**, 229 (1990).

IV

BUFFERS, ACID–BASE TITRATIONS, LABORATORY STRATEGIES, AND K_a DETERMINATIONS

IV-1. TITRATION EQUATIONS

The master relation (Eq. I-5 or III-1) leads simply to the complete titration curves for weak acids, H_nA, where $n = 1, 2, 3, \ldots$. Consider the titration of v_0 milliliters of C_0 molar H_nA by C_b molar NaOH. The C_{AB} (acid left to titrate) at any point is $nC - C_{sb}$. Each C must be calculated in mmoles divided by the new volume in milliliters at each point, $V = v_0 + v_{sb}$, $C_{sb} = C_b v_{sb}/V$, and $C = C_0 v_0/V$. This greatly reduces the complexity of the formulas to be derived. In the master relation, the acid balance at any point is

$$C_{AB} = nC - C_{sb} = \bar{\mathbf{n}}C + \mathbf{D} \tag{IV-1}$$

Define the degree of titration $\phi = C_{sb}/C$, the mmoles of base added per mmoles of acid. This runs from 0 to 1 for a monoprotic acid and is just the fraction titrated. It runs from 0 to n for others, conveniently designating the segments of the curve as in the three waves of H_3PO_4. Divide through Eq. IV-1 by C to get

$$n - \phi = \bar{\mathbf{n}} + \mathbf{D}/C \quad \text{or} \quad \phi = n - \bar{\mathbf{n}} - \mathbf{D}/C \tag{IV-2a}$$

To see the significance, think of a portion of a curve between p**H** 5 and 9 where the last term is very small. For H_3A, after one equivalent of acid is neutralized, $\bar{\mathbf{n}} \cong 2.0$, so $\phi = 1.0$ (segment titrated) and so on to 3.0 at the final equivalence point where all three equivalents of base have been added per mole of acid taken. The $\mathbf{D}/C$ term makes it exact and is large at the start and end of most titrations. Curves for two strengths of H_3A acids are shown in **Fig. IV-1**. These are easily calculated from a computer sheet of $\bar{\mathbf{n}}$ values since a titration curve (Eq. IV-2a) is basically a slight modification of the $\bar{\mathbf{n}}$ curve. One chooses **pH** and calculates $\bar{\mathbf{n}}$ and thus ϕ. The reverse is much harder. One may iterate v_b between the ϕ of Eq. IV-2a with the balance,

$$\phi = C_b v_b / C_0 v_0 \quad \text{using} \quad C = C_0 v_0 / (v_0 + v_b)$$

However, v_b can be solved if it is preferred to ϕ as the plotting variable. Using the preceding relations in Eq. IV-2a gives

$$v_b = \frac{(n - \bar{\mathbf{n}}) C_0 v_0 - \mathbf{D} v_0}{C_b + \mathbf{D}} \tag{IV-2b}$$

For a monoprotic acid, where $1 - \bar{\mathbf{n}} = 1 - \alpha_1 = \alpha_0$, $\phi_1 = \alpha_0 - \mathbf{D}/C$, or

$$v_b = \frac{(1 - \alpha_1) C_0 v_0 - \mathbf{D} v_0}{C_b + \mathbf{D}} \tag{IV-3}$$

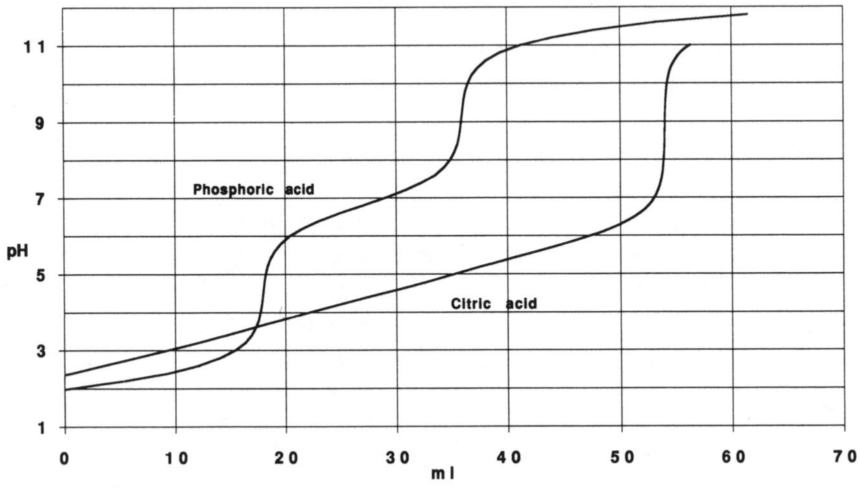

Fig. IV-1. Plots of complete titration curve equations for phosphoric and citric acids. Both are for 90 ml of 0.0200 *M* acid titrated by 0.100 *M* NaOH. There are 18 ml base per segment and 54 ml to the final equivalence point.

The significance here is clear: α_0 $(=1-\alpha_1)$ is the fraction of HA deprotonated.

Equations I-1 and I-2 (the α and $\bar{\mathbf{n}}$) can be combined in any proportions to express $\bar{\mathbf{n}}$ in useful forms. For example, the H_3A case is $\phi = 3 - \bar{\mathbf{n}} - \mathbf{D}/C$. Subtracting Eq. I-2 from 3 times Eq. I-1 gives

$$\phi = 3\alpha_0 + 2\alpha_1 + \alpha_2 - \mathbf{D}/C$$

Again the α terms give the number of protons *removed* from H_3A.

Program the spreadsheet columns for **H**, $\bar{\mathbf{n}}$, v_b, and **pH**. The last two terms are then in the proper sequence for plotting in the form of Fig. IV-1 using the PASTE SPECIAL command in Excel and choosing the scatter diagram option. The point markers are deleted for these line plots. To understand the negative v_b values obtained from the entry of low **pH** values into the preceding formulas, see the next section.

An $\bar{\mathbf{n}}, \bar{\mathbf{n}}'$ calculation as in Chapter III is needed to determine the precise **pH** of the starting 0.02 *M* acids. For H_3PO_4 this gives **pH** 1.983 ($\bar{\mathbf{n}} = 2.481$) and for citric acid, **pH** 2.363 ($\bar{\mathbf{n}} = 2.783$). For such strong acids the curves are near straight to start, while weaker cases jump into the buffering region as NaOH is added. (See Fig. IV-2*e*.)

Use of C_{AB} in Full pH Range Titration Curves

In formulas like IV-2b, **pH** values below the true start of the titration curve produce negative v_b values. This signified negative base, which we define as strong acid in our C_{AB} function. That is, some HCl, say, of C_{sa} at the same value as C_{sb}, must be added to attain the **pH** demanded. The C_{AB} function allows us to explain this and to derive a better formula for titration of mixtures of a weak and a strong acid by a strong base. (Multiple weak acids are treated later.) The strong acid and base are taken as monoprotic: a multiplier for their polyprotism will serve if, say, H_2SO_4 and $Ba(OH)_2$ were used.

Putting in the definition of C_{AB} shows, for bound acidity per A group,

$$\bar{\mathbf{n}} = \frac{C_{wa} + C_{sa} - C_{sb} - \mathbf{D}}{C_A}$$

For the preceding cases, assuming only weak acid to start, we took $C_{sa} = 0$, which produces the undesirable negative values of v_0. [The curves are plotted starting with the first positive v_b (v_{sb}) value.] Here we show a more complete process. We multiply the right side up and down by total V, which equals v_0 (of weak acid taken) plus $v_{sa} + v_{sb}$. This converts all the C terms to millimoles (for A) and milliequivalents for the others. Here $C_{wa} = n(C_0)v_0$ milliequivalents of protons. These can be expressed in v terms or in the v of each originally taken since v increases as C's decrease. At equilibrium, **D** must be given in the actual equilibrium volume, and we use $C_A V = C_0 v_0$ to give

$$\bar{\mathbf{n}} = \frac{nC_0v_0 + C_{sa}v_{sa} - C_{sb}v_{sb} - \mathbf{D}(v_0 + v_{sa} + v_{sb})}{C_0v_0}$$

Now we see that at low **pH** we need to add some strong acid and no strong base. Solving for v_{sa} gives

$$-v_{sa} = \frac{(n - \bar{\mathbf{n}})C_0v_0 - \mathbf{D}v_0}{C_{sa} - \mathbf{D}}$$

Compare with Eq. IV-2b to see why we can get a negative value for v_{sb}, which we can now interpret correctly as the strong acid v_{sa}. Clearly, we can start with given amounts of weak and strong acid and titrate with strong base using the formula with all the terms in it. (To start any calculation exactly at $v_{sb} = 0$, one might first do a calculation of the true **pH** of the starting mixture. There is usually no need for this since we can make the spreadsheet for **pH** 0–12, say, and plot starting with the first positive value of v_{sb}.

IV-2. EFFECT OF p*K* SPACING AND CONCENTRATIONS

It requires a space of about 1.6 log units between K_a values to produce a noticeable bump in **pH** titration curves. (Citric acid in the preceding has only about a 1.4 log unit spacing.) Concentration also plays a role as shown in the oxalic acid curves in the next section (Fig. IV-2*b*. For a six-proton acid **Fig. IV-2*a*** shows spacings 0.5, 0.5, 2, 3, and 3 units. There were 25 ml of 0.06 *M* H_6A titrated by 0.2 *M* NaOH: 7.5 ml base per segment. The lack of a clear final equivalence point at 45 ml occurs as with phosphoric acid, which also has a very basic final species.

The slopes are simple Δ calculated quotients (times 10) for 0.1 **pH** unit segments of the calculated titration curve. [EDTA would have a similar curve if the H_4Y form were soluble enough to titrate at these conditions. It is H_6A^{2+} owing to the two nitrilo groups (Fig. III-18).]

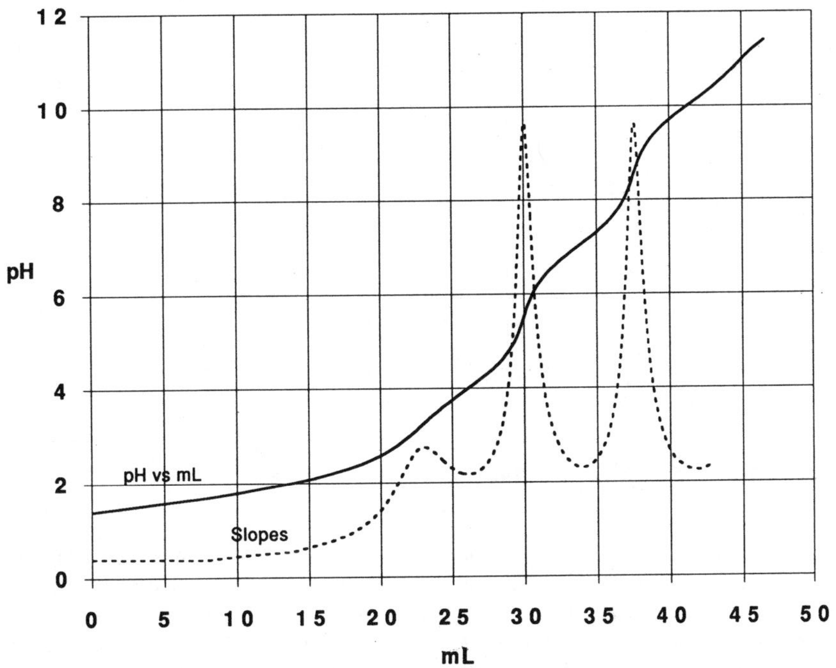

Fig. IV-2*a*. Fifty milliliters of 0.03 *M* H_6A titrated with 0.2 *M* NaOH shows the effect of K_a spacing on the slope of titration curves. This is a six-*K* acid having pK_a values 1, 1.5, 2, 4, 7, 10, which shows spacings 0.5, 2, and 3 in several **pH** regions.

Concentration Effects

The titration curves shift upward as C is lowered and the height of any intermediate equivalence point rise is lowered. **Figure IV-2*b*** shows this for oxalic acid.

EXERCISES

The reader might investigate plots for H_2A acids with varied spacings and C's to see when the jumps observed in the preceding arise.

IV-3. WEAK ACID–WEAK BASE MIXTURES

Although titration of a weak acid by a weak base is not often needed, the curve also represents a series of buffers of the mixed system that could be of use. Acetic acid by ammonia is often given to show that

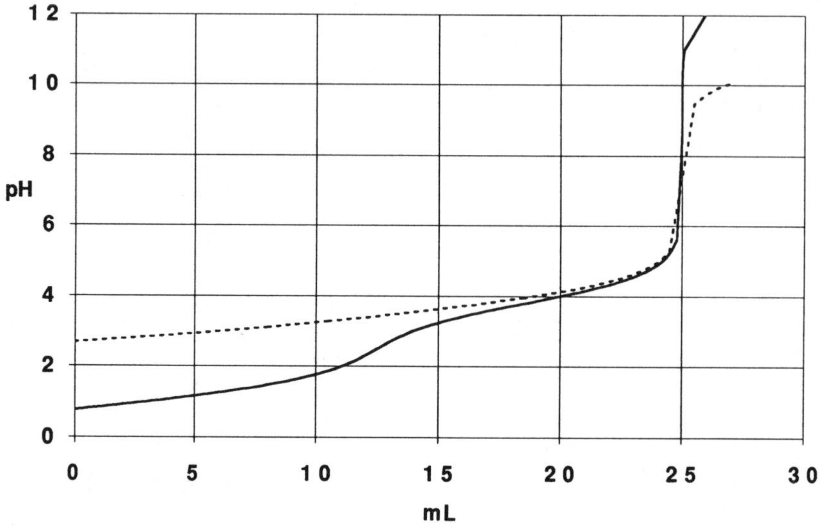

Fig. IV-2*b*. Titration curves for oxalic acid: bold line, 25.00 ml of 0.500 M H_2Ox by 1.000 M NaOH; dotted line, 25.00 ml of 0.00200 M acid by 0.004 M NaOH. The pK_a's are 1.04 and 3.82 ($I = 0.1$ M, 25°C). Note that the first equivalence rise is visible only at higher C for this separation of K's. (These illustrate the C effect but are not truly valid at the high I of the first. The pK_a values at $I = 1$ M are actually 1.04 and 3.55.)

the equivalence point is poorly defined and the buffer capacity of ammonium acetate solution is quite low. Note that the total number of acid protons is unaffected in this type of titration. During the first segment during which ammonia removes protons from acetic acid (or water or H_3O^+; equilibrium mathematics does not distinguish), the sequence of $HA–A^-$ buffers is like that formed by NaOH titration. However, near and beyond the equivalence point, the new ammonia added forms buffers with the NH_4^+, and acetate becomes effectively inert. The acidity that began as HA is now in ammonium and net hydronium. Not until two equivalents of ammonia have been added do we attain the 1:1 ammonia system buffer. Unbuffered weak base is never reached in this type of titration. (The equivalence point salt for monoprotics is treated in Section II-5.)

Let us show a more interesting case that is easily handled by the $\bar{\mathbf{n}}$ method: citric acid and 1,2-ethanediamine, "ethylenediamine,"en. The log K_a constants are 2.93, 4.36, and 5.74 for citric acid and 7.08 and 9.89 for enH_2^{2+}. All are at 25°C and $I = 0.1$ M. Let us take 25 ml (v_{0c}) of 0.04000 M (C_{0c}) citric acid and titrate it with 0.06000 M (C_{0e}) diamine (**Fig. IV-2*c***).

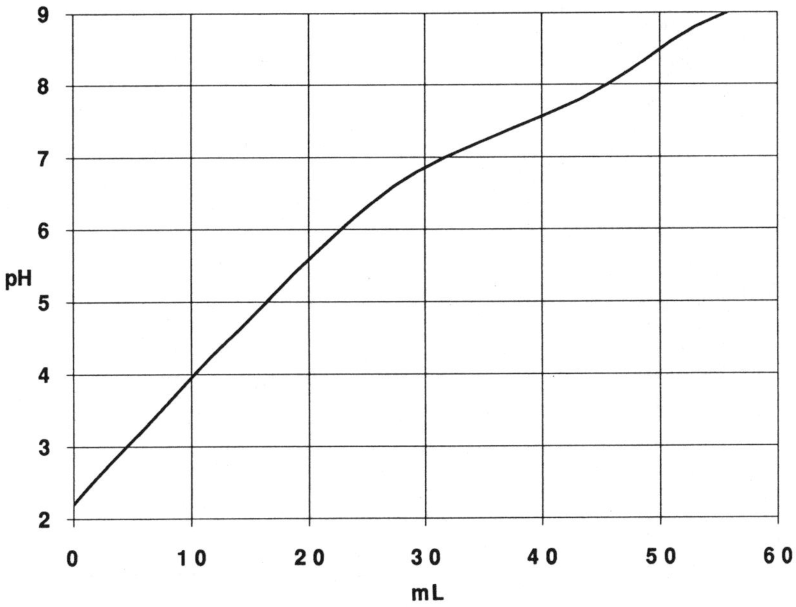

Fig. IV-2*c*. Titration curve of 25 ml 0.04 *M* citric acid 0.06 *M* ethanediamine.

Use the general acid balance (Eqs. IV-1 and IV-2) with the sum of the two systems, with total volume at any point $V = v_{0c} + v_{en}$. The mmoles of total acidity, $C_{AB}V$, holds constant since no strong base is added to reduce it. It is diluted:

$$C_{AB} = (3C_{0c}v_{0c})/V$$

This equals the bound **H**, $(\bar{\mathbf{n}}_c C_{0c}v_{0c}/V) + (\bar{\mathbf{n}}_{en} C_{0en}v_{en}/V)$ plus **D**. Solving for v_{en} gives the explicit titration curve:

$$v_{en} = [(3 - \bar{\mathbf{n}}_c)C_{0c}v_{0c} - \mathbf{D}v_{0c}]/[\bar{\mathbf{n}}_{en}C_{0en} + \mathbf{D}]$$

We see that buffering obtains over the whole range shown. A quantity of diamine equivalent to the citric acid acidity has been added at 25 ml. After this point a sequence of en buffers form, all having the same total acidity equivalents. Note that this is not the same curve that would be obtained by titration by NaOH of a totally protonated mixture of citric acid and enH_2^{2+} in which protons are removed to form water.

Next we plot the $\bar{\mathbf{n}}$ values obtained in the course of calculating the titration curve spreadsheet. **Figure IV-2*d*** shows variations of $\bar{\mathbf{n}}$ during

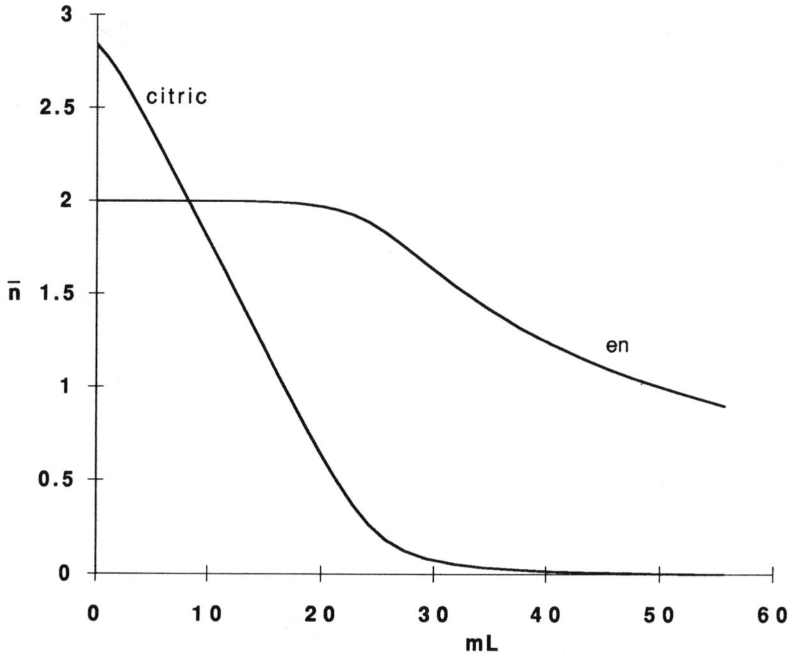

Fig. IV-2*d*. Variations of n̄'s during the citric acid–en titration.

the citric acid–en titration. As suggested by the pK_a values, the diamine can remove most of the citrate protons before it forms its own $\bar{\mathbf{n}}$ less than 2 in buffers. At the start, the diamine forms mainly enH_2^{2+} when added to excess citric acid. Check that we reach $\bar{\mathbf{n}}_{en}$ of about 1.5 at 37.5 ml and 1.0 at about 50 ml.

Titration Error Caused by Carbonate in NaOH Solutions Used to Titrate Acids

The system just discussed leads us to a more difficult problem that can be solved by the same approach. All titrations of acids by standard NaOH solution face the problem of the weak base CO_3^{2-} present. With great care, this can be kept below 2 parts per thousand of the NaOH so that the end point error is negligible if very pure NaOH (50% solution) is used to prepare the titrant. However, air contact increases the carbonate:

$$CO_2 + 2OH^- \rightarrow CO_3^{2-} + H_2O$$

The use of the solid reagent introduces 1–2% Na_2CO_3. The total of these bases (weak plus strong) can be determined by titration with strong acid. However, that total base normality can cause error in titration if the same end point **pH** is not used in both standardization of the base and in titration of an unknown acid. The protons placed on carbonate in the starting acid solution are not all removed at the usual weak acid end point, about **pH** 8–9, where HCO_3^- remains. Thus, more strong base is required to raise the **pH**, and this larger volume overestimates the acid *if* the total base normality is used in calculation. The possibility of partial loss of CO_2 during the titration increases the uncertainty of the acting normality of the mixed base being used.

To clarify this, let us calculate a titration curve for an exaggerated case: Titrate 50 ml of 0.1 M acetic acid by a base solution that has 0.9 M NaOH (the C_{sb0}) and 0.1 M CO_3^{2-}, the C_{C0}. (This situation would occur if 0.1 mmol of CO_2 were absorbed per ml by 1.1 M NaOH removing 0.2 M of the OH^- and producing 0.1 M CO_3^{2-}. This might be possible if the base were quite old with long air contact.) The total base is 1.1 N. This system is like the previous weak–weak case but with the addition of strong base. The total bound proton level balance becomes, using $C_{AB} = C_{A0}v_0/V - C_{sb0}v_b/V$

Bound protons = HA put in − strong base added − free H^+

$$\bar{\mathbf{n}}_A C_A + \bar{\mathbf{n}}_C C_C = C_{A0}v_0/V - C_{sb0}v_0/V - \mathbf{D}$$

where $C_A = C_{A0}v_0/V$, the total acetate at any point, that calculated to the new total volume $V = v_0 + v_b$. Here, C_C is the new total carbonates C:

$$\bar{\mathbf{n}}_A C_{A0}v_0/V + \bar{\mathbf{n}}_C C_{C0}v_b/V = C_{A0}v_0/V - C_{sb0}v_b/V - \mathbf{D}$$

As before, this is linear in v_b (the volume of the mixed base being added), so we can solve for it in this specific case,

$$v_b = \frac{C_{A0}v_0(1 - \bar{\mathbf{n}}_A) - \mathbf{D}v_0}{\bar{\mathbf{n}}_C C_{C0} + C_{sb0} + \mathbf{D}}$$

Compare with the monoprotic case (Eq. IV-3), which this becomes if C_{C0} is zero, if the base is pure NaOH. Putting in the C values and using $\bar{\mathbf{n}}_A = \alpha_{1A}$ and $\alpha_0 = 1 - \alpha_1$, we get

$$v_b = \frac{5\alpha_{0A} - 50\mathbf{D}}{\bar{\mathbf{n}}_C 0.1 + 0.9 + \mathbf{D}}$$

On a spreadsheet we calculate the v_b for a series of **pH** values, 2–12, and repeat for pure NaOH of 1.1 *N* to get the curve expected for pure NaOH of the same total available base as the strong base (**Fig. IV-2*e***). Assuming no CO_2 is lost can be roughly valid for fast titration of cold dilute acid solution.

Carbonate forms buffers at lower **pH** so that more of the mixed-base solution is required to shift the carbonate buffer to **pH** 8. Compare the protonation shifts for the similar case above: citric acid plus a diamine. [A similar curve would result if the diamine were substituted by pure carbonate and no CO_2 was lost.] Addition of a weak base does not change the *available* acidity while a strong base does, by forming water. Boiled acidic carbonate also acts as a strong base, removing available **H** as water, the CO_2 not being available in the solution for reaction:

$$H_2CO_3 \rightarrow CO_{2(g)} + H_2O$$

Thus, we need to consider titrant possibilities: (1) pure NaOH, (2) pure Na_2CO_3 solutions, or (3) NaOH + Na_2CO_3; where each has the same total number of equivalents of base. We also must decipher the effect of the **pH** chosen for the end point: what proton level is left.

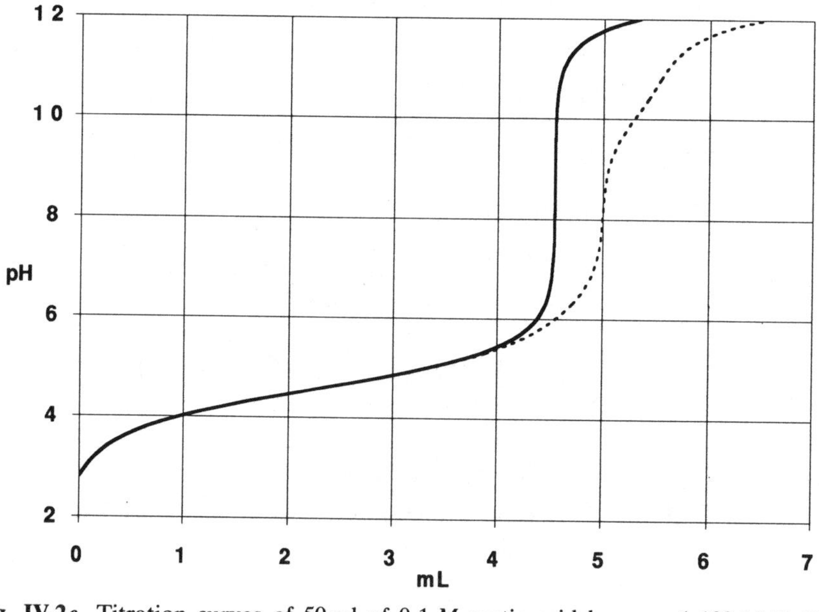

Fig. IV-2*e*. Titration curves of 50 ml of 0.1 *M* acetic acid by pure 1.100 *M* NaOH (solid line) and by base having 0.900 *M* NaOH and 0.100 *M* Na_2CO_3 (dotted).

For 1, the correct **pH** is neutral for strong acids, and for weak acids it is the **pH** of the specific pure NaA solution. For 2, it is the **pH** of pure H_2CO_3 for strong acids (or neutral for boilout). For weak acids the titration curve may be too gradual for end point determination, as in Fig. IV-2*c*. If the acid is stronger than carbonic, boilout or stirring under an air jet might shift the neutralization close to completion. For 3 it is seldom practical to use the **pH** of pure H_2CO_3 (about 4) or of pure deprotonated CO_3^{2-} (above 12) so that the fairly good rise around pure HCO_3^- is used (about **pH** 8). Case 1 is seen clearly in the acetic acid curve in Fig. IV-2*e*. Case 2 is mainly limited to determination of carbonates by titration with strong acid.

Case 3 requires use of the N (equivalents per liter) of base protonated when the end **pH** is reached. At **pH** just above 8, all the original OH^- is protonated to water and the carbonate is protonated to HCO_3^-. Thus, the effective base N of our example solution is not 1.100 but 1.000 meq/ml.

Note the removal of the second proton (from HCO_3^-) in the dotted line at **pH** above 9. At **pH** about 8.8, the pure base equivalence point occurs at 4.55 ml of 1.1 M NaOH. But with the carbonated base we get **pH** 6.0 when 4.55 ml is added. This requires that about 4% of the acetate remains as HAc, while roughly half the carbon is H_2CO_3 and half HCO_3^- (α diagram, Fig. IV-7). The protons do balance, as can be seen in the 5.00 mmol of acetate (and original H^+) and 0.455 mmol of carbon, which leads to 0.2 mmol HAc and 0.23 mmol each of H_2CO_3 and HCO_3^-, adding to the 0.89 mmol acidity remaining after we add the 4.1 mmol of strong base. If one omits the carbonate and takes the usual acetate end point at about **pH** 8^+ at 5.00 ml, the acetate is quantitatively deprotonated, but the carbonate is still half protonated so that 10% of the original C_H is left. A calculation with 1.1 as the base N gives 5.50 mmol HAc for a 10% positive error.

If the base were standardized vs. strong acid or vs. a weak acid like acetic or KHP to this 8^+ **pH** end point, it would be 1.000 M (called NaOH) and the calculation would give correct results. Remember that this N will change as the bases ages and absorbs more CO_2.

Note that the slope of the carbonate–base curve is lower than that of the pure NaOH, so that the end point will be less well defined. This and the loss of variable portions of CO_2 make use of a base with more than a few parts per thousand of its basicity as carbonate an invalid titrant for high precision use.

One can investigate further the results of varied K_a's and carbonate content. An early article by McAlpine[1] gives the results of lab tests on the speed of loss of CO_2 and validity of titration results. A recent article by Michalowski[2] shows many curves for varied proportions of carbonate and strengths of acids. (The latter uses $\bar{\mathbf{n}}$ as Z and the Eq. 5 for titration is identical to the one derived here from the master relation.)

IV-4. STATISTICAL EFFECTS IN TITRATIONS

Let us test the initially plausible hypothesis that 0.1 M HA of pK_a 3.000 should have the same titration curve as an equal volume of 0.03333 M H_3A, all pK_a's 3.000. Program the spreadsheet as before for the triprotic case and also for the monoprotic case (**Fig. IV-3**).

Details of the differences that occur for statistical reasons are provided by King.[3] Briefly, our taking of all constants 10^{-3} does not make the acid positions the same. The polyprotic has three positions from which to lose protons. In all, there are 12 isomeric steps if we identify the different positions in all the species of H_3A, H_2A^-,

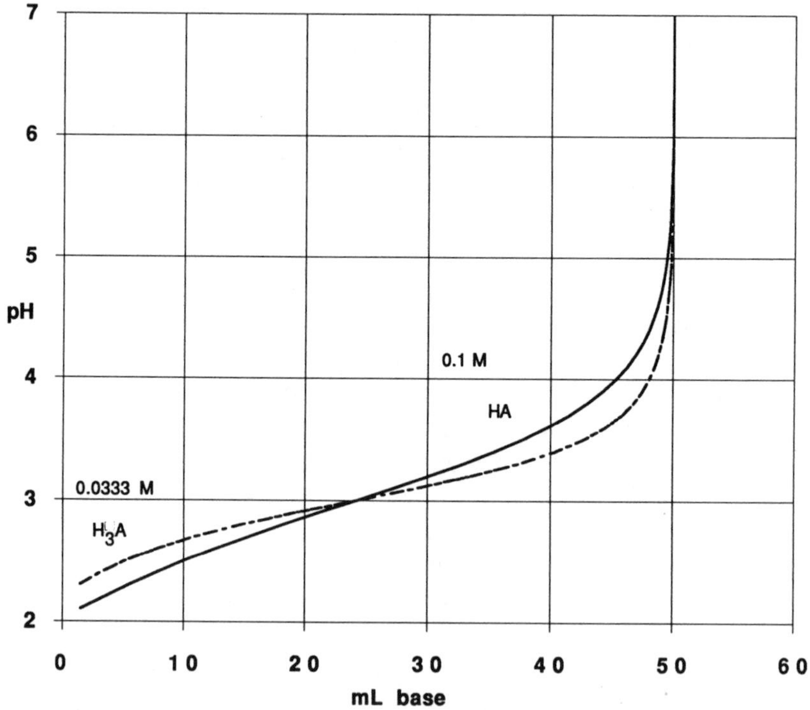

Fig. IV-3. Complete equation titration curves for H_3A and HA having all pK_a values 3.00 and each a total titratable **H** of 5 meq in 50 ml. This shows where differences occur for statistical reasons. The NaOH is 0.100 M.

HA^{2-}, and A^{3-}. When these are correctly accounted for by taking these step K_a's (microscopic constants) as 0.001, we find the macro pK_a's are 2.523, 3, and 3.477 instead of all 3. (Instead of 1000, the three *formation* constants are 3000, 1000, and 333.3.) These then give a titration curve falling exactly upon the monoprotic of pK_a 3.000. These exactly adjust the $\bar{\mathbf{n}}$ to equal 3 times α_1 of the monoprotic. Note that β_3, $K_1K_2K_3$, is still 10^9 in formation.

IV-5. SLOPE, BUFFER CAPACITY, AND METHODS FOR DETERMINATION OF EQUIVALENCE POINT BY LINEAR FUNCTIONS (GRAN PLOTS)

Closely related to the titration curve is its slope and the system buffer index **B**. The latter is the increment in strong base giving one **pH** unit increase. The titration curve slope is a volume-adjusted inverse of **B**, the change in **pH** per milliliter of base increment. It can be calculated by analytical differentiation or easily estimated by the delta fraction on the titration curve calculation table. This is shown in **Figs. IV-4**

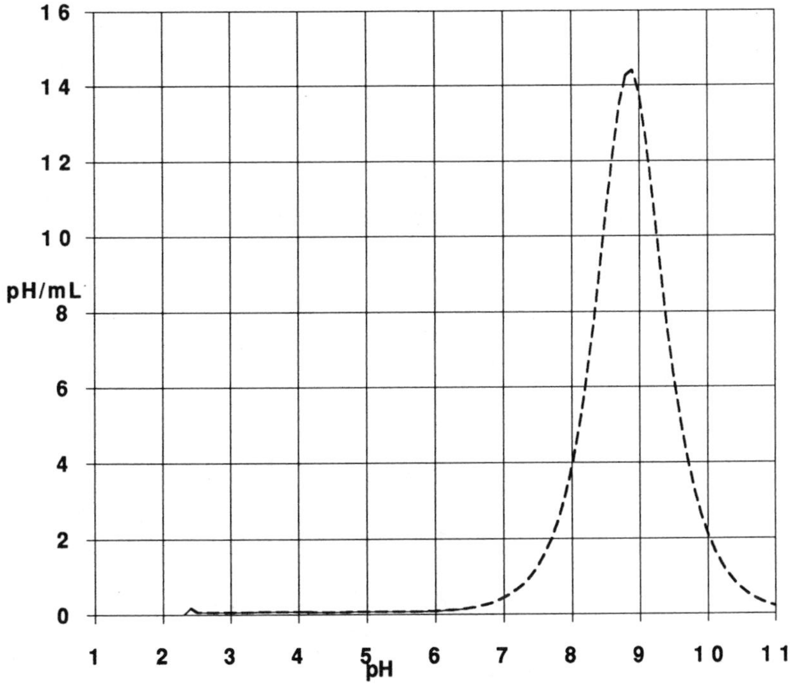

Fig. IV-4. Slopes calculated from 0.1 **pH** unit intervals on the citric acid titration curve of Fig. IV-1.

and **IV-5*a*,*b***. The maximum is quite sensitive to the **pH** increment taken for the delta method. Figure IV-4 shows slopes calculated from 0.1 **pH** unit intervals on the citric acid titration curve of Fig. IV-1. The maxima of these slopes are the minima of the buffer capacity. The best buffering occurs at the minima, about **pH** 2, 7, and 11 in the phosphate system. Figure IV-5*a* shows the slope calculated by the delta method at 0.2 **pH** intervals for the phosphoric acid titration shown in Fig. IV-1. This emphasizes the lack of a visible third end point for the very basic PO_4^{3-}.

In 1950, Gran[4] discovered the near linearity of the inverse of the slopes in Fig. IV-5*a* if plotted vs v_b. **Figure IV-5*b*** shows the slope and its inverse, $\Delta v_b/\Delta$**pH**, calculated from the same data used for Fig. IV-5*a*. This slope is a slight variant of the buffer index, the Δ molarity of added base required to change the **pH** by one unit. Such plots can be used to estimate equivalence points, but the more linear method detailed in Gran's 1952 paper[4] and later in this section is now favored. Let us call the inverse Gran's first method. Both methods have the attraction that a few points not too far from the end point can be used to extrapolate to a fairly accurate result. Both of Gran's

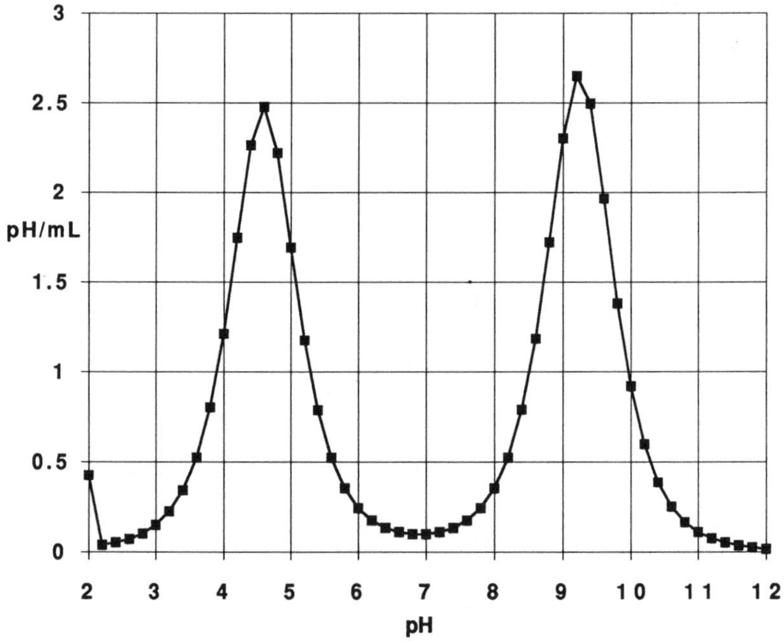

Fig. IV-5*a*. Slope calculated by the delta method at 0.2 **pH** intervals for the phosphoric acid titration shown in Fig. IV-1. This emphasizes the lack of a visible third end point for the very basic PO_4^{3-}.

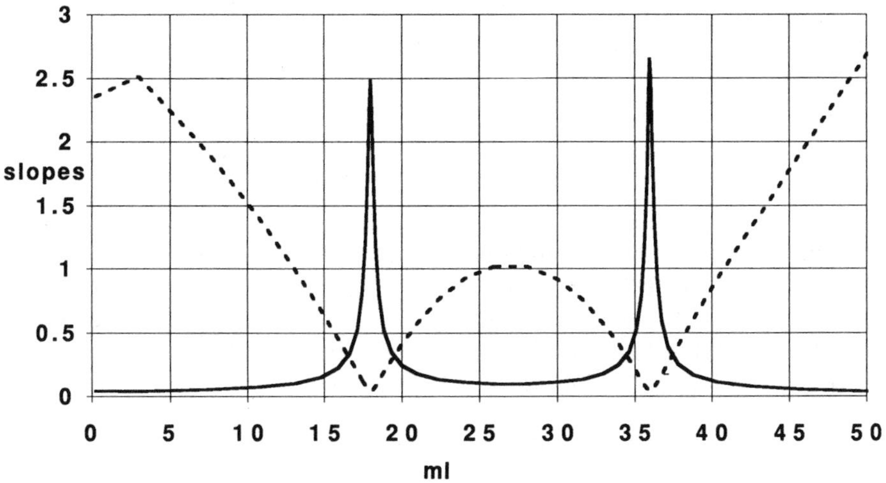

Fig. IV-5*b*. Gran's first method plot. The slope times 0.1 and the inverse slope, dotted, of the titration curve for phosphoric acid is plotted vs. volumes of base used.

papers are in English and give extensive mathematical treatments of these linear plot methods. They are not exactly linear and have serious deviations under some conditions to be described in what follows.

Titrations of Mixtures of Acids

Equation IV-1 is simply expanded to the sum of bound plus "free" protons for mixtures similar to the mixture problems in Chapter III and in Section IV-8:

$$C_{\mathrm{AB}} = \mathbf{D} + \sum_{i=1}^{n} \bar{\mathbf{n}}_{\mathrm{a},i} C_{\mathrm{a},i}$$

The last term sums protons bound to each base group i to n, say, acetic, succinic, and citric acid in a mixture.

Linear Titration Functions

Titration curves themselves (**pH** vs. amount of titrant added), their slopes (first derivative), and even the second derivatives (inflections) have been used to determine the end point and estimate the equivalence point. Various problems can make these tedious or even impossible. A linear function that could be extrapolated to the equivalence point would be fine. Gran[4] devised several methods. A complete one will be described. (An approximate one is given in the exercises at the end of the chapter.) First let us examine **pH** measurement on the weak acid solution alone to see why **pH** is a poor measure of the molarity of a weak acid.

Example. Estimate the concentration of acetic acid if its solution in 0.1 *M* NaCl reads pa_H 2.625 ± 0.005. Use the 0.1 *M* ionic strength pK_a, 4.56 ± 0.01. Using the Kielland table value for f_+ (0.83 ± 0.01) for **H** gives **pH** 2.706 ± 0.011. In the full Charlot equation for a weak acid alone (Eq. II-6), we get (**OH** negligible)

$$C_1 = \mathbf{D}/\alpha_0 = \boldsymbol{H}(1 + \boldsymbol{H}/K_a) = 0.299 \pm 0.008\ M$$

This 3% uncertainty probably represents more favorable precision than can be expected in **pH** measurement. Pure acid, known ionic strength, and a linear glass electrode response are all required. So this is the answer to the question, why don't we just read the **pH** if we want to know how much acid is present. (One might consider how precise results might be if the meter were calibrated for molarity **pH** with 0.0100 *M* HCl and then used in an unknown solution of nearly the same C_{sa}.)

Gran's Linear Method

Strong acid (initially v_a of C_a^0 molar acid) is titrated by a base (C_b). The volume at any point is $v_a + v_b = V$. The millimoles of acid left at any point is **D***V*. Since this must decrease in direct proportion to the strong base added, the linear function is simple. This can be found (to the two to three significant figures just shown) from meter readings:

$$\mathbf{D}V \cong \mathbf{H}V = \frac{10^{-pa_H}}{f_+} V = C_a^0 v_a - C_b v_b \qquad \text{(IV-4a)}$$

The only variable on the right is v_b. Therefore, the left millimole function, which can be evaluated at any point from the volumes and meter reading, must be a straight line function of the volume of base added (holding ***I*** constant). At the equivalence point,

$$C_a^0 v_a = C_b v_b$$

Therefore, the line of Eq. IV-4a hits zero at the equivalence point. The precision of the intersection can be better than that of each **H** value since we make the best line through points approaching the end, and the slope C_b and v_b are known to higher precision. After equivalence, added base increases **OH** directly in proportion to the v_b added beyond that at equivalence, v_e:

$$10^{\mathrm{pa_H}-14}V/f_- = -\mathbf{D}V = C_b(v_b - v_e) \qquad \text{(IV-4b)}$$

This (left) function can be plotted on the same graph to give another straight line intersecting zero at v_e.

This might be useful for very dilute solutions, but generally, the strong titrations are not in need of aid for end point determination. The weak cases are more interesting since they may exhibit small or unobservable end point jumps.

Weak Monoprotic Titrations

In the complete equation (II-7) it is convenient to calculate the C's of the conjugates at each point by using

$$C_{\mathrm{AB}} = (C_1^0 v_a - C_b^0 v_b)/V \qquad C_{\mathrm{wb}} = C_0 = C_b^0 v_0/V \quad (\text{for } v_b < v_e)$$

$$v_e = C_1^0 v_a / C_b^0 \quad (\text{for } v_b \text{ at equivalence})$$

This produces (in Eq. II-7)

$$\mathbf{H} = K_a \frac{C_1^0 v_a - C_b^0 v_b - \mathbf{D}V}{C_b^0 v_b + \mathbf{D}V}$$

Multiplying out and dividing by C_b^0 yields

$$\mathbf{H}v_b + \left[\frac{\mathbf{H}V\mathbf{D}(1 + K_a/\mathbf{H})}{C_b^0}\right] = K_a(v_e - v_b) \qquad \text{(IV-5)}$$

Again the right side decreases directly with v_b, the only variable, but the left presents problems. However, while the titration curve is near neutral **pH**, **D** makes the bracketed term negligible compared to the first term, and

$$\mathbf{H}v_b = \frac{10^{-\mathrm{pa_H}}}{f_+} v_b \cong K_a(v_e - v_b) \qquad \text{(IV-6)}$$

This is the approximate linear function usually used for a monoprotic weak acid or for one step of a polyprotic acid assuming other steps are well separated. Thus one holds ionic strength constant (so that f_+ can be put into the constant), takes several **pH** readings at v_b steps during the titration, and calculates the left Gran function, which should be a straight line going to zero at equivalence (where $v_e = v_b$). Note the difference from the strong case (Eq. IV-4) where total volume V is required. For not very low C values and pK_a about 4–8, good results may be obtained. However, for more dilute solutions and especially for stronger acids, p$K_a < 5$ (or much weaker acids, p$K_a > 9$), erroneous v_e are found. The bracketed term in Eq. IV-5 is not negligible. If K_a is known, the correction (the bracketed term) can be estimated and added. The same strong excess base function in the preceding can be added to check the v_e intersection.

These observations and extensions to very complex (polyprotic) titrations are treated by Schwartz[5] and by the author.[6]

Mixture of Strong and Weak Acids

An interesting application occurs for certain favorable mixtures of a strong and a weak acid. It has been derived from a charge balance formulation,[9] but we show how it flows from the master equation (II-7b) using the C_{AB} form, which makes clear how it works. This shows the advantage of the complete approach in Eq. II-7b: this case has a weak system in the presence of excess strong acid (at the start) and with excess strong base after the second equivalence point.

Use the preceding C terms and $C_{AB} = C_{wa} + C_{sa} - C_b$. (In these titrations we are using C_b for the changing C of the C_{sb} added.) Let us take, for example, a 50-ml mixture that is 0.1 M each in HCl and acetic acid and is titrated by 0.1 M (C_b^0) NaOH. Initial C_0 (like sodium acetate) is zero. Putting these into Eq. II-7b and rearranging, we get

$$C_b - C_{sa} + \mathbf{D} = C_{wa}/(1 + \mathbf{H}/K_a)$$

Use the preceding sliding C terms (adjusted for the new volume at each point, $V = v_0 + v_b$) and v_e terms for the net equivalence volumes of the base for the strong and the weak acids, all multiplied by V to yield, in millimoles,

$$v_b C_b^0 = v_{es} C_b^0 + v_{ewa} C_b^0/(1 + \mathbf{H}/K_a) - \mathbf{D}V \qquad \text{(IV-7)}$$

The meaning of this is in millimoles of the materials: that of base added, of strong acid, of weak acid neutralized, and of net **H** or **OH** free at the point. To illustrate, take the case given, 5 mmol of each acid and thus 50 ml NaOH to each equivalence point: (solve Eq. IV-7 for **D**V)

$$\mathbf{D}V = 5 + 5\alpha_0 - v_b(0.1)$$

Thus, only the strong acid and the excess strong base Gran functions are used in this method. At the start, the p**H** is near 1, $\mathbf{D} \cong 0.1$, α_0 of acetic acid is near zero, and $v_b = 0$. Thus, $\mathbf{D}V = 5$. At 5 ml of NaOH, α_0 of acetic acid is still small (strong acid still present) and the final term is 0.5, making $\mathbf{D}V = 4.5$. In this way one sees that the α_0 term stays small while most of the strong acid is titrated, and the left term (the strong acid Gran function shown before) decreases approximately linearly to zero so that a straight line through most of the points will hit zero at v_{es}. Continuing through the buffer region of the weak acid shows that **D**V remains close to zero until excess base is added after 100 ml titration. (Try the midpoint buffer of acetic acid to confirm, $\alpha_0 = 0.5$ and $v_b = 75$ ml base, which makes the right side zero.) After 100 ml, we have the preceding strong base Gran function and the **D**V line falls off, marking the second end point, as in a similar case (different quantities) in **Fig. IV-6**. For a stronger weak acid like HSO_4^-, this would not work so well. A related equation could be derived for two weak acids if the K_a values are small and well separated.

Unknowns. The preceding explains why the Gran approach to a mixture works. For an actual unknown mixture, one adds NaCl to make high constant ionic strength* and takes pa_H readings at, say, intervals of 10 ml of NaOH. Calculate **D**V and plot vs. milliliters of base to find the two v_e points. Another example follows.

The example was programmed (Eq. IV-7) in the following order:

$$\text{p}\mathbf{H} \quad \mathbf{D} \quad \alpha_0 \quad v_b \quad \mathbf{D}V = (50 + v_b)\mathbf{D}$$

Thus, to plot the expected titration curve, choose a list of p**H** (from 1 to 12 by increments of 0.2 was used) and calculate the rest on a

*$\boldsymbol{I} = 0.4\ M$ is a good choice as it lies at the center of a rather flat region of the Davies equation curve for f. The f is 0.73 for $\boldsymbol{I}$ of 0.3–0.5 M (see Appendix A).

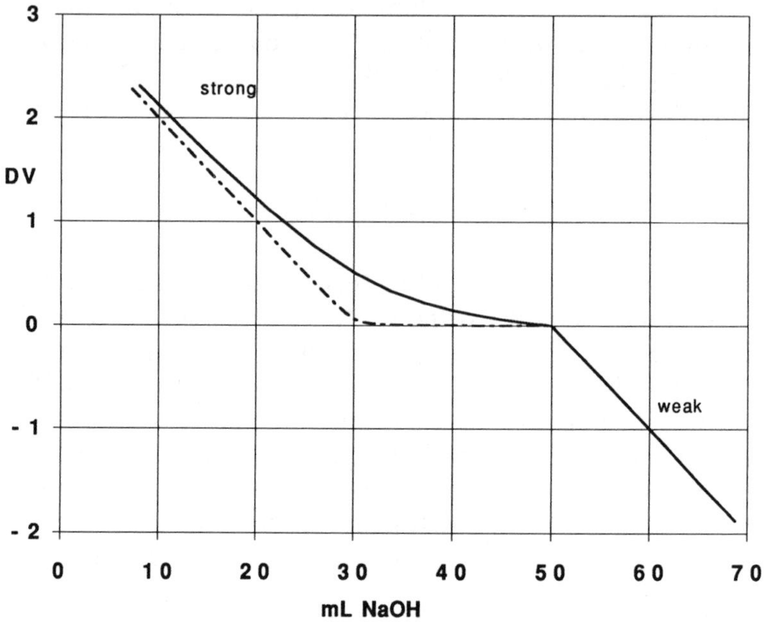

Fig. IV-6. Titration of a 50.00-ml mixture of acid that is 0.0600 M in HCl and 0.0400 M in HAc (dashed line). Plot is of **D***V* vs. milliliters of 0.1000 M NaOH. Equivalence volumes are 30 ml for strong, 50–30 ml for the weak. As in the first example, $\mathbf{D}V = 3 + 2\alpha_0 - 0.1v_b$. The solid line shows the effect if chloroacetic acid is the weak acid (pK_a 2.66).

spreadsheet. Plot the last two columns. Again we see the advantage of inverting variables: from the **D***V* equation calculate the v_b needed to give the chosen **pH** rather than the reverse ($v_b = (3 + 2\alpha_0 - 50\mathbf{D})/(\mathbf{D} + 0.1)$). Note the curvature of lines near the intersections. These are the results expected from a laboratory titration *if* the pK_a (4.56) and pK_w (13.80) assumed are correct ($\boldsymbol{I} = 0.1\ M$). For an actual experiment, K_w is required (for **D** in the basic portion), f_+ is required to convert pa$_H$ to **pH**, and a guess is needed about the amount of curvature to ignore in extrapolating the lines. The Gran plot has value in such cases since the usual **pH** vs. milliliters of base titration curve would not show clearly where the first equivalence point is. Unless the second acid is rather weak, there can be large uncertainty in the line drawn. In the chloroacetic case one would probably overestimate the strong and underestimate the weak by about 5 ml. Moving the α_0 term to the **D***V* side would correct this curve. That requires estimation of v_{ewa} and iteration.

Single Titration Analysis of Buffer Mixtures and Very Weak Bases

Examination of the preceding method and its plot will show how an unknown mixture of HAc (x millimoles) and NaAc (y millimoles) can be analyzed by a titration with NaOH if a known excess (s millimoles) of HCl is first added. If the acid is weak enough (conjugate base strong enough), the Ac^- will be nearly all protonated and a plot can be made similar to Fig. IV-6. The remaining strong acid segment found (d) will correspond to $s - y$. Thus, $y = s - d$. The weak segment found (w) is then $x + y$, so that $x = w - s + d$. This result would be difficult to obtain for acetic acid without the Gran method since the **pH** rise between the strong and weak acid would be indistinct. For an acid of pK_a 6 or higher the titration curve itself would give a jump.

A solution of a weak base alone can be treated similarly since w will be the result. The excess of HCl need not be known since x is zero. A weak base could be titrated directly by a strong acid applying an analogous **OH** Gran derivation for the linear function.

Polyprotic Acids. Strong base titrations of polyprotic acids can be plotted so that straight lines extrapolating to each successive equivalence point occur, but only if the K_a values are so well separated that appreciable **pH** regions between the equivalence points can be well approximated by the relation of a single K_a. That is the case for H_3PO_4 and H_2CO_3. (Details are given in Ch. 7 of the 1975 book by the author.[6]) Complete relations can be derived for any polyprotic from the general $\bar{\mathbf{n}}$ titration relation (Eq. IV-2b). Devising linear functions is quite complex. The reader can try it for a diprotic, H_2A, with pK_a's only 2 units apart. The article by Schwartz[5] treats this. It is formidable. For H_2A we can easily get

$$\bar{\mathbf{n}} v_e/2 + \mathbf{D}V/C_b = v_e - v_b$$

Again the right side decreases linearly to zero at the final equivalence point. The problem is to calculate the left from **pH** data and the usual volumes. This also requires the K_a's (for $\bar{\mathbf{n}}$) and successive estimates of v_e. A computer treatment seems quite feasible. Using K_a's in the Gran function would require controlled temperature and ionic strength so that conditional K_a values will be constant.

IV-6. PRACTICAL ASPECTS: IONIC STRENGTH CONTROL

Constant ionic strength, I, during titrations is desirable for linearized methods and for the determination of equilibrium constants. A sliding set of values of K_a's can be handled by using K_a^0F with the appropriate activity coefficients quotient term F throughout. Besides complexity, this has the fault that reliable estimates of f's are not available except at very low I, introducing possibly large uncertainties. As briefly outlined in Chapter I and detailed further in Appendix A, thermodynamic equilibrium constants can be obtained if activities of all species in the K expression can be obtained or if a series of K_a's at decreasing I can be extrapolated to zero ionic strength. Conditional K_a's for a fixed I can be found if equilibrium molarities of every species can be determined. But how do we find the $[H^+]$? The pH meter responds to the activity of H^+_{aq} so that an estimate of its f_+ is required. This is often the choice in work of moderate precision at low I, 0.1 M and below. Our following example with citric acid uses this approach. The glass electrode–reference half-cell method can give reliable results if the standardizing solutions and the unknowns are quite similar, in I and type of solutes, so that both activity coefficients and liquid junction potentials may be assumed constant.

An alternative often used recently* is to calibrate the pH meter in terms of **H** *concentration* at one I, which must then be part of the medium of the K measurements. For example, pH measurements of buffers of the system of unknown K's done in a medium of 1 M KCl can be taken as p**H** (molarity) if the meter is calibrated with known concentrations of strong acid like HCl (0.01–0.001 M as p**H** 2 and 3) in 1 M KCl or of buffers of some well-studied system, like acetic acid, for which the K_a and thus p**H** in 1 M KCl is known. The book by R. G. Bates (Ch. 9), and that of Harned and Owen (see the Annotated Bibliography) treat such methods in detail. They also describe the best research methods using a hydrogen electrode and a reference electrode without liquid junction. Such methods still require one activity coefficient, but a series of measurements can be extrapolated to get K_a^0 to good accuracy as in the citric acid work of Bates and Pinching at NBS.[10]

*A recent example of this method is given in the "Experimental" section of the paper by C.M. Gerritsen and D.W. Margerum, *Inorg. Chem.* 1990, 29, 2757. The pH meter was set with known $HClO_4$ mixtures in a medium of 1 M $NaClO_4$ and used to determine the conditional molarity K_a of a very weak acid, HOCl, in this medium.

The liquid junction potential refers to the potential difference which is set up across a region of large difference of concentration, as we have where the calomel or AgCl electrode with its 1–4 *M* KCl meets the test solution containing the glass electrode. This potential may be large, but it may be nearly the same in the standardizing and unknown solutions if I and concentration of buffers are closely similar (Ch. 3 in Bates' book). This has been found valid in many investigations compared with research methods using cells without a liquid junction. Consult the references for details. This is by way of explaining the need for the following discussion of I and K_a determination experiments. The inherent uncertainties in activity and junction potential effects require special arrangements in order to achieve even moderate accuracy in K_a determinations.

Constant ionic strength I is simply achieved for uncharged weak *monoprotic acids* being titrated by NaOH. One needs to add neutral salt (NaCl) to the HA to make the molarity NaCl equal to the molarity of NaOH. This allows one to choose any desired ionic strength. For example, to get 100 ml of 0.02500 *M* HA with 0.050 *M* as the desired I, make the NaOH 0.0500 *M*. Add 5.0 mmol NaCl (0.29 g) to the 100 ml of acid. (Note that HA makes negligible contribution to I.) One can check that each milliliter of NaOH produces 0.05 mmol of NaA (strong electrolyte), keeping 0.05 mmol of ionic material/ml. At the equivalence point, we have 2.5 mmol NaA and 5 mmol NaCl in a volume of 150 ml. Alternatively, the NaOH can be more dilute if its 0.05 *M* ionic strength is made up with NaCl. It could be 0.04 *M* NaOH with 0.01 *M* NaCl present in it. The reader might check that this works and gives $I = 0.05\ M$ at every point up to equivalence.

For different charge types, NH_4Cl (HA has 1+ charge) and $NaHSO_4$ (HA is 1-), arrangements can be made to keep I constant. For C molar NH_4Cl, the NaOH must be the same concentration, C, with added C molar NaCl present in the NaOH only. This makes up for the loss of two ions in the reaction making NH_3.

For NaHA, the NaOH must have half the molarity of the starting acid. The dilution then balances the increase of ionic strength caused by the $\frac{1}{2}Cz^2$ term of A^{2-} being formed. These ideas depend on weak HA so that its ion contribution is small. This would not apply to HSO_4^- (pK_a 2.0).

For an uncharged base vs. a strong acid (NH_3 or THAM vs. HCl), one needs salt in the amine at the same C of the acid used. For example, 0.02 *M* THAM containing 0.05 *M* KCl titrated by 0.05 *M* HCl would hold a constant I of 0.05 *M*.

For *polyprotic* cases, constant low $\boldsymbol{I}$ is not simply achievable unless the pK_a values are well separated (3^+ units). Flooding can give nearly constant high strength: 0.02 M H_3A with 2 M NaCl titrated with 0.02 M NaOH, which is also 2 M in NaCl. At the three equivalence points, this would have approximate $\boldsymbol{I}$ of 2.01, 2.02, and 2.03. Since high $\boldsymbol{I}$ precludes use of calculated f values, one may wish to make some compromises to obtain low $\boldsymbol{I}$. The following scheme works if the K_a values are well separated so that three clearly defined charge-type regions obtain. When this is not true, $\boldsymbol{I}$ is still close to the desired value (e.g., citric acid) so that small corrections can be made (using $\alpha_n C_A$ for the actual equilibrium species, as in the universal buffer example at the end of Section IV-8).

Approximate Constant $\boldsymbol{I}$ Scheme for Polyprotic Cases, H_3A

1. First Segment. H_2A^- is formed. This is exactly analogous to the preceding HA cases. The starting $\boldsymbol{I}$ of the H_3A and of the NaOH (if necessary) must be adjusted to the same value with KCl, KNO_3, etc. A simple arrangement is used for this example: acid, 0.05 M H_3A, 0.10 M KCl; base, 0.05 M NaOH, 0.05 M KCl. (Other combinations may affect the water volumes needed in 2 and 3.) The use of KCl is desirable to reduce the junction potential error in pH work aimed at determination of K_a values when a calomel reference electrode is used. The $\boldsymbol{I}$ will be held at 0.1 M and the volume becomes $2V_0$ at the end of this segment, mainly NaH_2A solution (with KCl).

2. Second Segment. Buffers of H_2A^- and HA^{2-} are formed. Continue the titration with the same NaOH (as in step 1) and add a volume of water equal to half the new NaOH solution used to reach each point being measured. Alternatively, dilute the first NaOH with half its volume of water, for this segment of titration.

3. Third Segment. Buffers of HA^{2-} and A^{3-} are formed. Continue as before but add water equal to the new NaOH volume added.

Carrying through the calculations for a few buffers and the ampholytes will clarify the process and suggest other schemes and extension to H_4A or basic species titrated by HCl, e.g., determination of the very acid lower constants of EDTA or other amino acids can be done starting with unprotonated forms and adding HCl.

IV-7. K_a DETERMINATIONS

The K_a values can be calculated after experimental determination of $\bar{n}'$ values, usually by potentiometric methods for **pH** in mixtures of known C's. The experimental $\bar{\mathbf{n}}'$ is taken as $\bar{\mathbf{n}}$ in the equations to follow. Bjerrum's arrangement of the $\bar{\mathbf{n}}$ equations I-4 or III-1 is illuminating:

$$K_1 = \frac{3-\bar{\mathbf{n}}}{\bar{\mathbf{n}}-2}\mathbf{H}\left[1 - \frac{(1-\bar{\mathbf{n}})K_2}{(\bar{\mathbf{n}}-2)\mathbf{H}} + \frac{\bar{\mathbf{n}}K_2K_3}{(\bar{\mathbf{n}}-2)\mathbf{H}^2}\right]^{-1} \tag{IV-8}$$

When K_2 and K_3 are much smaller than K_1 and $\mathbf{H}$ is near K_1 ($\bar{\mathbf{n}}$ near 2.5), the first term is just the approximate expression for K_1. (At $\bar{\mathbf{n}} = 2.5$, Eq. IV-8 is just $K_1 = \mathbf{H}$, as it must be.) The other terms are then corrections for overlap of the other species. Approximate values for the other K's can serve for approximate calculation of K_1. Similar arrangements are made for calculations of K_2 and K_3, and the first approximations are iterated to constant values. It is worth the trouble of devising a series of solutions of one ionic strength. Although insertion of activity coefficient values can give K^0 estimates, the K values at one ionic strength are accurate experimental values (for the specific mixture) requiring no assumptions about activity coefficients except for conversion of pa_H to $\mathbf{pH}$. There are calibration (vs. $\mathbf{H}$) schemes and extrapolation to zero $\boldsymbol{I}$ that can be used. See the discussion in the previous section.

The advantage of obtaining enough experimental $\bar{\mathbf{n}}'$ values at constant $\boldsymbol{I}$ to plot a curve of $\bar{\mathbf{n}}$ vs. $\mathbf{pH}$ is seen at certain rounded values read from the experimental curve. For example, when $\bar{\mathbf{n}} = 2.5$ (call $\mathbf{H} = \mathbf{H}_1$), Eq. IV-8 gives

$$K_1 = \mathbf{H}_1[1 + 3K_2/\mathbf{H}_1 + 5K_2K_3/\mathbf{H}_1^2]^{-1}$$

Similar checks on the other K's are possible at $\bar{\mathbf{n}} = 1.5$, 0.5. Then we may check products at these two points on the curve.

$$K_1K_2 = \mathbf{H}^2/(1 + 2K_3/\mathbf{H}) \quad (\text{at } \bar{\mathbf{n}} = 2.000 \text{ only})$$

$$K_2K_3 = \mathbf{H}^2(1 + 2\mathbf{H}/K_1) \quad (\text{at } \bar{\mathbf{n}} = 1.000 \text{ only})$$

Plotting Methods

Equation IV-8 can be arranged as

$$\frac{3-\bar{\mathbf{n}}}{\bar{\mathbf{n}}-2}\mathbf{H} = K_1K_2[(\bar{\mathbf{n}}-1)/(\bar{\mathbf{n}}-2)\mathbf{H}] + K_1 + (K_1K_2K_3/\mathbf{H}^2)\bar{\mathbf{n}}/(\bar{\mathbf{n}}-2) \tag{IV-9}$$

In the acid region, where the last term is small ($\mathbf{pH} \ll pK_3$, a plot of the left side vs. the bracketed part of the first term on the right should be a straight line of slope K_1K_2 and intercept K_1. A similar rearrangement shows that in the higher **pH** region, the left term is small and a plot of $\mathbf{H}(2-\bar{\mathbf{n}})/(1-\bar{\mathbf{n}})$ vs. $\bar{\mathbf{n}}/(1-\bar{\mathbf{n}})\mathbf{H}$ has slope K_2K_3 and intercept K_2. Thus, constant $\boldsymbol{I}$ data for $\bar{\mathbf{n}}, \mathbf{H}$ lead easily to the conditional constants of the acid. Drifting $\boldsymbol{I}$ during the titration will produce inconsistent data since the K_a values shift. The wide variations of older published "equilibrium constants" are partly the result of this experimental problem.

A Laboratory Example: Citric Acid

Data for citric acid obtained at close to constant ionic strength 0.1 *M* will illustrate the problems and precision expected. To 49.96 ml of 0.05005 *M* citric acid (0.100 *M* in KCl) was added successive 12.50-ml pipet portions of 0.0511 *M* NaOH that was 0.0500 *M* in KCl. The pa_H was read using a glass–calomel electrode pair calibrated vs. four standards [0.01 *M* HCl (0.09 *M* KCl) 2.098, KHP 4.008, 0.1 *M* acetate 4.652, and phosphate 6.865. See Buffer Standards in Appendix C]. All were at 25°C. The **pH** was taken to be pa_H −0.083, the $\log f_+$ value was from the Kielland table. The $\bar{\mathbf{n}}$ was calculated from the **pH** and the known analytical concentrations as $\bar{\mathbf{n}}' = (C_H - \mathbf{D})/C_A$. The C_H is (3 mmol citric acid taken − mmol NaOH added)/V. In these acidic solutions $\mathbf{D} = \mathbf{H}$. Here, $C_A = 2.5005$ mmol taken/V. For the first point, $V = 62.46$ ml. See Table IV-1.

Calculations: Single-*K* Method

First estimate approximate pK_2 and pK_3 as the **pH** at $\bar{\mathbf{n}} = 1.5$, 0.5 from a plot of the data that gives about 4.3 and 5.8. This would

TABLE IV-1. Citric Acid Neutralization Data

V (ml)	pa_H	pH	$\mathbf{H} \times 10^3$	$\bar{\mathbf{n}}$
62.5	2.641	2.588	2.77	2.675
75.0	3.003	2.920	1.20	2.453
87.5	3.353	3.270	0.537	2.215
100.0	3.728	3.645	0.2265	1.969
112.5	4.095	4.012	0.0973	1.718
[a]	5.910	5.827	1.49×10^{-6}	0.4999

[a]A fresh preparation: 0.01111 *M* in each Na_3Cit and Na_2HCit, calculated to have $\boldsymbol{I}$ at 0.100 *M*.

require more data if the constants were not so close together, making the $\bar{\mathbf{n}}$, **pH** line almost straight for citric acid (Fig. III-14).] Then, from Eq. IV-8,

$$K_1 = \frac{2.77 \times 10^{-3}(3 - 2.675)/(2.675 - 2)}{1 + 0.045 + 0.0004}$$

$$= 1.28 \times 10^{-3} \qquad \mathrm{p}K_1\ 2.89$$

Note that in the denominator, the K_2 term affects the result by 4.5%, and the K_3 term affects the result by 0.04%. Next, at $\bar{\mathbf{n}} = 1.718$,

$$K_2 = \mathbf{H}\frac{(3 - \bar{\mathbf{n}})\mathbf{H}/K_1 - (\bar{\mathbf{n}} - 2)}{\bar{\mathbf{n}}K_3/\mathbf{H} + (\bar{\mathbf{n}} - 1)} = 4.95 \times 10^{-5} \qquad \mathrm{p}K_2\ 4.30_5$$

Note that the $\bar{\mathbf{n}} = 1.969$ point will give a loss of significant figures in the important $(\bar{\mathbf{n}} - 2)$ term so it is not a good choice for a single-point calculation, but it is valid for the plotting method that follows. Next, in the mixture at $\bar{\mathbf{n}} = 0.5$,

$$K_3 = \mathbf{H}/\bar{\mathbf{n}}K_1K_2[(3 - \bar{\mathbf{n}})\mathbf{H}^2 + (2 - \bar{\mathbf{n}})\mathbf{H}/K_2 + (1 - \bar{\mathbf{n}})]$$

$$K_3 = 1.62_5 \times 10^{-6} \qquad \mathrm{p}K_3\ 5.79$$

Finally, we plotted the terms in Eq. IV-9 for the first four points to get slope 6.40×10^{-8}, or $\mathrm{p}K_1K_2$ 7.19_4, and intercept 1.28×10^{-3}, or $\mathrm{p}K_1$ 2.89. The agreement of the calculations means only that the data are fairly consistent, not that they are right! A remeasurement of the last solution several days later with a different glass electrode gave $\mathrm{pa_H}$ 5.905 and $\mathrm{p}K_3$ 5.78. The actual difference before rounding to two significant figures was 0.012 log units, or about 3% of the K value.

Note: Do not take percentages of logarithmic values or **pH**. Significant figures in logs start roughly after the decimal point, but use the antilog for accuracy.

One may question whether more closely spaced data, say, every 4 ml, would yield a better defined $\bar{\mathbf{n}}$, **pH** curve and more reliable results. The longer time required risks drift of the electrode system and cumulative buret errors. Preparation of pipetted mixtures near $\bar{\mathbf{n}}$ values favorable for calculation (half integers as well as 2 and 1 for the products shown before) seems preferable. Use of larger volumes of mixtures, when the substances are available, also reduces error from evaporation and temperature changes. These data were obtained as described, before the better scheme for maintaining ionic strength described in the previous section was deduced. Thus, data at $\bar{\mathbf{n}}$ below 2 increased in $\boldsymbol{I}$. For this reason the mixture for $\bar{\mathbf{n}} = 0.5$ was synthesized fresh. Starting afresh at several points during the mixture preparation and working in a closed vessel would be desirable for the best work. Without such precautions, K results from simple titration of an acid are valid to no more than one decimal of the log value. Advanced student results have been good following the controlled $\boldsymbol{I}$ method described. It is instructive to consult the research methods used by Bates and Pinching,[10] who report pK^0 values of 3.128, 4.761, and 6.396. They used a cell without the problems of the (unknown) liquid junction potential:

$$\mathrm{Pt;\ H_2,\ \mathbf{H}_x^+\ (citrate\ mixtures),\ KCl_{aq},\ AgCl;\ Ag}$$

They varied the KCl, extrapolated to zero KCl, and then extrapolated these K values to zero ionic strength. Thus, no activity coefficients are used in getting K^0's.

Examination of some published values for citric acid at 25°C in four ionic media illustrates the $\boldsymbol{I}$ effect[7,8]:

$\boldsymbol{I}$	0	0.1 M	1 M	3 M
pK_1	3.128	2.87	2.80	2.40
pK_2	4.761	4.35	4.08	3.83
pK_3	6.396	5.83	5.33	4.88

The specific electrolytes can greatly effect the values, especially for the higher C's and charges. The 3 M medium was $NaClO_4$. Consult the references for details of the media, methods, and uncertainties. (See the abbreviated example in Appendix A.)

IV-8. FURTHER BUFFER TOPICS

Universal Buffer Design

A mixture of acids for making buffers over a wide **pH** range is sometimes convenient. The acidity preserves them longer, whereas buffers themselves degrade from CO_2 or bacterial growth. Calculation of proportions of acid and base for complex mixtures is not seen in texts, although researchers have published them using complete equations.[11] With small computers or even a programmable calculator, α and $\bar{\mathbf{n}}$ are readily found for chosen **pH**'s and the problem solved. We demonstrate with a mixture similar to those of ref. 11. [A single-point calculation was described in Section III-9 for citrate–oxalate systems.]

The acids are phosphoric, succinic, and boric (**p**, **s**, **b**) covering the range of pH 2.5–11. The 0.1 **M** ionic strength, 25°C, pK_a values are **p** = 1.95, 6.80, 11.67, **s** = 4.00, 5.24, and **b** = 9.00. Stock solutions of standardized acids can be made and stored separately or mixed so that precisely 0.01000 mol of each acid can be measured for each liter of final buffer mixture. The available acidity is thus 0.06 *N*. The calculation falls into two parts:

(i) Find the amount of NaOH needed to reduce the 0.06 to the needed C_H.
(ii) Find the ionic strength contribution and thus how much KCl is needed to make it 0.1 *M*.

We shall work in **pH** (molarity), so that the buffer will give a pa_H reading (NBS scale) of 0.083 units higher. Alternatively, this could be subtracted at the start to obtain round values of pa_H (and slightly different C_H results).

(i) *For pH 9.00.* From Eq. I-6, for these 0.01 *M* acids we have total H,

$$C_H = \sum C_i \bar{\mathbf{n}}_i + \mathbf{D} = 0.01(\bar{\mathbf{n}}_p + \bar{\mathbf{n}}_s + \bar{\mathbf{n}}_b) + \mathbf{D} \qquad \text{(III-8)}$$

Calculate α's and $\bar{\mathbf{n}}$ at **pH** 9.00 for each acid:

	Species Charge of				
	3−	2−	1−	0	$\bar{\mathbf{n}}$
HBO_2, α_0, α_1			0.5	0.5	0.5
H_2Succ, α_0, α_1, α_2		0.9998	0.0002	10^{-9}	0.0002
H_3PO_4, α_0, α_1, α_2, α_3	0.0021	0.991	0.0062	10^{-9}	1.004

The $\bar{\mathbf{n}}$'s add to 1.5042 to give $C_H = 0.015042 - 1.6 \times 10^{-5} = 0.015026$. (We used p$K_w$ 13.80 to obtain $\mathbf{D} = -\mathbf{OH} = -1.6 \times 10^{-5}$.) Subtracting this from 0.06 yields 0.044974 *M* of the acid to be removed by NaOH, or 44.97 ml of the 0.1000 *M* base/l final solution.

(ii) *Ionic Strength.* For example, the contribution of Na_3PO_4 is $C_0 = \alpha_0(0.01)$, and from the ***I*** form, $0.5(Cz^2)$ for $3Na^+$ and

PO_4^{3-} ions, we get $0.5(3\alpha_0 + 9\alpha_0)C = 6\alpha_0 C = 0.00013$. In this way we get the ionic strength contributions of the equilibrium species:

$$C(\alpha_{0b} + [6\alpha_{0p} + 3\alpha_{1p} + \alpha_{2p}] + [3\alpha_{0s} + \alpha_{1s}]) + \mathbf{OH} = 0.06492\ M$$

using the α values in the preceding table. So, to obtain $\boldsymbol{I} = 0.1\ M$, we need to add 0.0351 mol KCl/l. This must be done to make the K_a's used valid and to achieve the **p**H chosen, 9.00. Laboratory tests of such mixtures have usually agreed with predicted pa_H to about ±0.01 unit. (One can compare the usual exact approach calculation from scratch by Rosenthal and Zuman.[12] Further practice can be checked with these results for selected **p**H points (3–9):

	3	4	5	6	7	8	9
mmol NaOH/l	9.10	15.25	22.98	29.87	36.06	40.30	44.97
mmol KCl/l	89.0	84.4	73.4	60.3	48.0	40.4	35.1

A plot of **p**H vs. millimoles of NaOH constitutes a constant volume–constant ***I*** titration curve. The author[13] has published rapid volumetric methods for making precise standard buffers for laboratory use, and checked their **p**H meter measurements vs. NBS standards. Brief directions for these standards are given in Appendix C.

Shifts in Buffer pH Caused by CO_2

1. Consider the NBS standard buffer 0.0250 *m* in Na_2CO_3 and in $NaHCO_3$. The standard pa_H, called pH_s, given is 10.012 at 25°C. This has ionic strength 0.1 *M*, and the conditional pK_2 for H_2CO_3 is 9.93. Note that in the following equation **D** is negligible to 1/250, $M = m$ to the precision needed, and the pure buffer pH_s is correct (letting $x = 0$). How many moles (x) of CO_2 per liter of buffer will change the **p**H (and pH_s or pa_H) by 0.1 unit? Assume the reaction $xCO_2 + xCO_3^{2-} \rightarrow 2xHCO_3^-$ causes the shift, as justified by the α in Fig. IV. 7 (α_2 is negligible at this **p**H. The pK_1 at 6.16 has no effect on this buffer to the precision used).

Using the Charlot equation (II-7) and the new pa_H as 9.829 gives

$$10^{-9.83} = 10^{-9.93}\ \frac{0.0250 + 2x + 0.00011}{0.0250 - x - 0.00011}$$

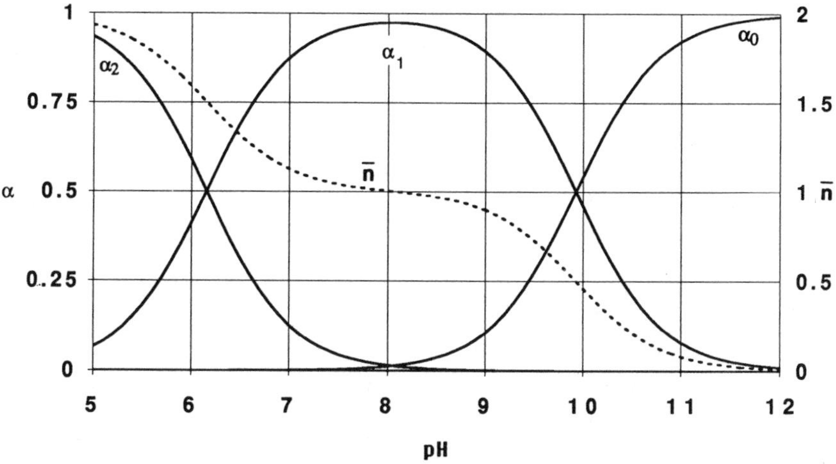

Fig. IV-7. An α, $\bar{\mathbf{n}}$ carbonate system diagram: pK_1 6.16, pK_2 9.93, $\boldsymbol{I} = 0.1\ M$.

A pK_w of 13.80 was used to calculate **OH** for **D**. The equilibrium ratio on the right equals $10^{0.1} = 1.26$. It was 1.00 before CO_2 addition. Thus, $x = 0.00191$ mol/l, or 1.91 mmol. This corresponds to about 48 ml of pure CO_2 at room conditions or 145 l dry air that has 0.033 vol% CO_2. It is unlikely that a few openings of a buffer container would account for such a shift in **pH**. Note that the wide spacing of the pK_a values in both problems allows precise use of the monoprotic approximations. Use of $\bar{\mathbf{n}}$ (α_1 in these cases) and bound proton concepts simplifies handling of complex problems.

Continued CO_2 Absorption by Initially Pure Na_2CO_3. A complete equation for the carbonate system requires the full $\bar{\mathbf{n}}$ function as **pH** goes lower:

$$\bar{\mathbf{n}} = (C_H - \mathbf{D})/C = (2x - \mathbf{D})/(0.0250 + x) = \alpha_1 + 2\alpha_2$$

where x is the added CO_2 in moles per liter. Let us start with 0.0250 M Na_2CO_3 alone and calculate the **pH** as x CO_2 is dissolved in it. With computer-calculated $\bar{\mathbf{n}}$ and **D**, we solve for x:

$$x = (0.025 + \mathbf{D})/(2 - \bar{\mathbf{n}})$$

Figure IV-8 results. It is similar to a titration plot for carbonate by strong acid. Here, however, total carbonates is increasing as CO_2 is added. Above **pH** 8 (pure bicarbonate, Fig. IV-7), buffers of carbo-

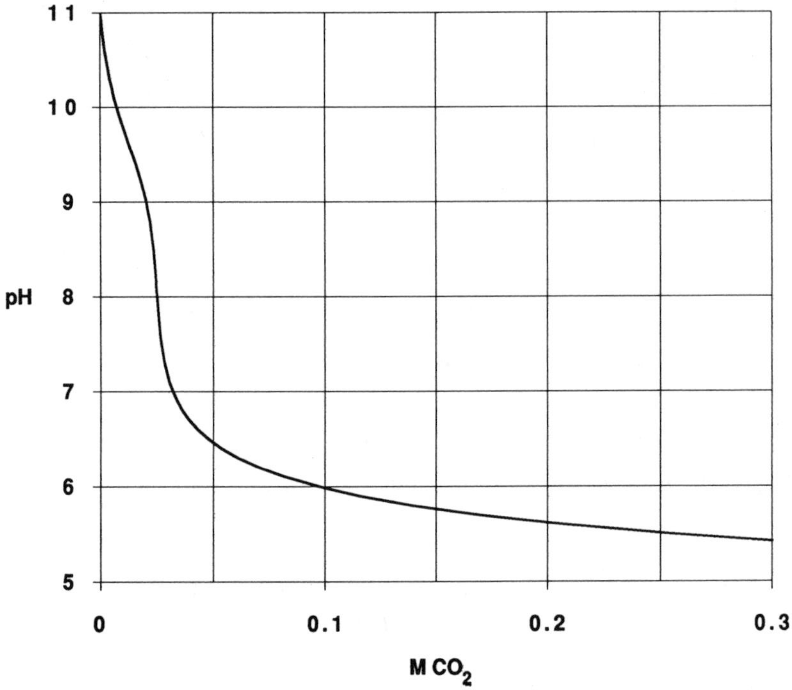

Fig. IV-8. The **pH** of 0.0250 *M* sodium carbonate as CO_2 is added.

nate and bicarbonate form. As more CO_2 (H_2CO_3) is added, buffers of bicarbonate and carbonic acid form below **pH** 8. Note that a 1:1 buffer of HCO_3^- and H_2CO_3 occurs when 0.075 *M* CO_2 has been added, **pH** about 6.16, giving 0.05 *M* of each species. (The abscissa is just the added CO_2, not the total carbon species, which is 0.1 *M* at this buffer point.)

2. Examine a *phosphate buffer* originally at pH 11.000 for the same shift in **pH**. Use pK_3 11.67 at $\boldsymbol{I} = 0.1$ *M* (25°C) and take HPO_4^{2-} and PO_4^{3-} as the only acting species. First, we must calculate the analytical *C*'s originally taken given that 0.0200 *M* total phosphate, *C*, is present. At **pH** 11.000, $\bar{\mathbf{n}} = 0.8246$; $\mathbf{D} = -\mathbf{OH} = -0.00159$ *M*. From $\bar{\mathbf{n}} = (C_H - \mathbf{D})/C$, $C_H = 0.0149$ *M* and from $\bar{\mathbf{n}} \cong \alpha_1$ and $\alpha_1 + \alpha_0 \cong 1$, we get $C_1 = 0.0149$ *M* and $C_0 = 0.0051$ *M* (0.00159 *M* PO_4^{3-} removed protons from water to make OH^- and HPO_4^{2-}). Then the equilibrium concentrations are 0.0165 *M* HPO_4^{2-} and 0.0035 *M* PO_4^{3-}, a ratio of 4.71. As before, this must change by a factor of 1.26, to 5.93, to shift **pH** by 0.1 unit. The chemical changes required are

$$xCO_2 + 2xPO_4^{3-} \rightarrow 2xHPO_4^{2-} + xCO_3^{2-}$$

$$yCO_2 + yCO_3^{2-} \rightarrow 2yHCO_3^-$$

$$(x + y)CO_2 + 2xPO_4^{3-} \rightarrow 2xHPO_4^{2-} + (x - y)CO_3^{2-} + 2yHCO_3^{2-}$$

Thus, we must satisfy the equilibrium ratios of the phosphate and of the carbonate conjugates at the new pH, 10.90; $\mathbf{D} = -\mathbf{OH} = -0.00126$:

$$PO_4^{3-}: \quad \mathbf{H}/K_3 = 5.93 = (C_1 + 2x - \mathbf{D})/(C_0 - 2x + \mathbf{D})$$

$$CO_2: \quad \mathbf{H}/K_2 = 0.107 = 2y(x - y)$$

Since y is not involved in the phosphate ratio, we find x first, 0.000477 M. Inserting this into the carbonate equation, we get $y = 2.42 \times 10^{-5}$ M. The total $x + y$ is 0.000501, or 0.501, mmol CO_2/l

SUMMARY

The general relations among α, $\bar{\mathbf{n}}$, C_{AB}, and $\mathbf{H}$ are especially helpful in treating a series of mixtures as found in titrations. Strong, weak, and multiple acid–base titrations have simple and exact formulations. Effects of carbonate in base titrant solutions are shown. Traditional **pH** vs. volume of titrant methods as well as linear (Gran) plots are described. Means of controlling ionic strength to prevent shifting of effective K_a's are illustrated in calculation of the three K_a's of citric acid from laboratory **pH** data (Section IV-7).

EXERCISES

1. An approximate Gran method is to plot the inverse slope near the equivalence point. Show that this is approximately a straight line (vs. milliliters of strong base) for 0.1 M weak acid HA of pK_a 5 titrated by 0.1 M NaOH. The inverse slope is $\Delta v_b/\Delta \mathbf{pH}$.

2. Show that titration of 10.00 ml of 0.1000 M sodium acetate by 0.1000 M HCl gives a poor curve of **pH** vs. milliliters of the acid but a clear straight line for extrapolation if $\mathbf{D}V$ vs. milliliters is plotted.

3. Derive a relation for **H** of a 0.02 M solution of sodium acetate as a function of moles per liter of CO_2 absorbed from zero to 0.02 M.

4. Compare the titration curves of H_3A, pK_a's 2, 3, and 4, and equivalent equal amounts of three monoprotics, HA, with pK_a's 2, 3, and 4. Relate to Fig. IV-3.

REFERENCES

1. McAlpine, R. K., *J. Chem. Educ.*, **21**, 589 (1944).
2. Michalowski, T., *J. Chem. Educ.*, **65**, 181 (1988).
3. King, E. J., *Acid-Base Equilibria*, Macmillan, New York, 1965, p. 222.
4. Gran, G., *The Analyst*, **77**, 661 (1952). *Acta Chem. Scand.* **4**, 559 (1950).
5. Schwartz, L. M., *J. Chem. Educ.*, **64**, 947 (1987).
6. Guenther, W. B., *Chemical Equilibrium*, Plenum Press, New York, 1975, Ch. 7.
7. Sillen, L. G. and Martell, A. E., Eds., *Stability Constants of Metal–Ion Complexes*; Special Publications, No. 17, 25, The Chemical Society, London, 1964, 1971.
8. Martell, A. E. and Smith, R. M. *Critical Stability Constants*, Plenum, New York (6 vols.), 1974–1989.
9. Castillo, C. and Jaramillo, A., *J. Chem. Educ.*, **66**, 341 (1989).
10. Bates, R. G. and Pinching, G. D., *J. Am. Chem. Soc.* **71**, 1274 (1949).
11. Avdeef, A. and Bucher, J., *Anal. Chem.*, **50**, 2137 (1978). The α diagrams and titration curves for a universal buffer mixture are shown.
12. Rosenthal, D. and Zuman, P., in *Treatise on Analytical Chemistry*; Kolthoff, I. M. and Elving, P. J., Eds., Wiley-Interscience, New York, 2nd ed., 1979; Part I, Vol. 2, p. 203.
13. Guenther, W. B., *The Analyst*, **113**, 683 (1988).

V

METAL–LIGAND EQUILIBRIA

INTRODUCTION

As with much chemical symbolism, the abbreviated formulation of metal ion–ligand complexing is deceptively oversimplified. This Lewis acid–base association is written as

$$\text{(a)} \qquad M + L \rightarrow ML\,, \quad ML + L \rightarrow ML_2\,, \quad \text{etc., to } ML_n$$

The metal ion M (for $[M^{m+}]$) is a Lewis acid able to accept one or more electron pairs. The ligand L (of any charge or neutral) is a Lewis base with one or more electron pairs to share. In aqueous solution, water strongly hydrates metal ions through the base position on oxygen, and the ligand less strongly through H, or the positive end of the water dipole. Therefore, the complex formation is in reality a competitive reaction of the ligands H_2O and L for the acid positions on the metal ion:

$$\text{(b)} \qquad Ni(H_2O)_6^{2+} + NH_3 \rightarrow Ni(H_2O)_5(NH_3)^{2+} + H_2O\,, \quad \text{etc.}$$

Five successive steps each subtract one H_2O and add one NH_3 to reach $Ni(NH_3)_6^{2+}$. The stepwise substitution of single ligands proceeds (shown in Chapter I and Appendix B). In aqueous solution, ligands compete with up to 55.4 *M* water. The K_{f1} for reaction (b) is 631 (in 2 *M* NH_4NO_3 used for buffering the **pH** and for ionic strength control). This represents a strong complexing in competition with so much higher water molarity. Remember, however, the convention of taking water activity as 1 in these equilibrium expressions so that it can be absorbed in the K. Both (a) and (b) are correctly described by a K_{f1} if this unit activity is valid or if the conditional constant is used for some lower water activity that is held constant, as in the flooding method. That is, many constants are given for such media as 1 M $NaClO_4$, and we assume that the water activity, say, 0.95, is part of the constant:

$$K_{f1} = \frac{[ML]}{[M][L]}$$

Thus, 1 M ammonia with trace [Ni(II)] has the ratio NiL/Ni 631 in a medium of 2 M NH_4NO_3 rather than $631/(H_2O)$, where the water activity might be about 52. The point to grasp is that (a) may seem to describe the affinity of L for Ni^{2+} (as a gaseous reaction of two particles!), but it really describes the tendency of NH_3 to replace a coordinated water molecule on Ni^{2+}_{aq} in water solution. Changing the activity of water by adding alcohol or $NaClO_4$ may greatly change the conditional constant. The K_1^0 is commonly unavailable for such cases for reasons that will become clear.

The range of affinities given by ML formation constants is large, weak complexing being much more prevalent than in the case of proton complexing of bases. Therefore, calculation methods taking overlap into consideration will be even more important here. One example, the first step in chloride complexing will suffice. Approximate ML constants (as log K_{f1}) for metal ions with one Cl^- are

M ion	Ni^{2+}	Co^{2+}	Fe^{3+}	Cu^{2+}	Pb^{2+}	Bi^{3+}	Cd^{2+}	Ag^+	Hg^{2+}
log K_1	−0.5	−0.3	1	1	1.6	2	2	3	6.7

The significance of $K_1 = 10^1$ for Cu^{2+} is that [for low concentration of Cu(II)] the ratio, R_{1-0}, $[CuCl^+]/[Cu^{2+}]$ in 1 M HCl is 10:1 and in 0.1 M HCl is 1:1. In 1 M HCl, the ratio for the Hg(II) system is $10^{6.7}$. Just as in K_a cases, the equilibrium constant gives a measure of the ratio of conjugate species in the presence of 1 M complexing agent, **H** or **L**.

V-1. FUNDAMENTAL RELATIONS

Mononuclear ML complexing is mathematically much like acid–base complexing except that there is a simplification because the **H–OH** mirror variable is absent. Where proton and OH^- competition is not significant (to be treated later), the master relation for bound ligand number is

$$\bar{\mathbf{n}}'_{ML} = \frac{C_L - [L]}{C_M} = \bar{\mathbf{n}} \qquad \text{(V-1)}$$

where the equilibrium $\bar{\mathbf{n}}$ in terms of step formation constants (using the usual species ratio forms) is*

$$\bar{\mathbf{n}} = \frac{K_1\mathbf{L} + 2K_1K_2\mathbf{L}^2 + 3K_1K_2K_3\mathbf{L}^3 + \cdots}{1 + K_1\mathbf{L} + K_1K_2\mathbf{L}^2 + K_1K_2K_3\mathbf{L}^3 + \cdots}$$

$$= \frac{\beta_1\mathbf{L} + 2\beta_2\mathbf{L}^2 + 3\beta_3\mathbf{L}^3 + \cdots}{1 + \beta_1\mathbf{L} + \beta_2\mathbf{L}^2 + \beta_3\mathbf{L}^3 + \cdots} \qquad \text{(V-2a)}$$

where $\mathbf{L}$ is [L], the unbound ligand concentration, with the charge not shown:

$$\beta_1 = K_1 = \frac{[ML]}{[M][L]} \qquad K_2 = \frac{[ML_2]}{[ML][L]}, \qquad \text{etc.}$$

$$\beta_2 = K_1K_2 \qquad \beta_3 = K_1K_2K_3$$

Extension to systems that have up to n ligands per metal is analogous (to $n\beta_nL^n$ in the $\bar{\mathbf{n}}$ expression and $\beta_n\mathbf{L}^n$ for the ratio ML_n/M).

The relation to species fractions is like that seen with acid–bases:

$$\bar{\mathbf{n}} = \alpha_1 + 2\alpha_2 + 3\alpha_3 + 4\alpha_4 + \cdots$$

*Sometimes the ratio form of Eq. I-8 is more convenient to calculate with a small calculator or to program for computers. Let us show this for six steps in formation direction. This covers ML, as well as acid–base in the formation direction ($K_{f1} = 1/K_{a,n}$, here $n = 6$). Let

$$b = R_{1-0} = [ML]/[M] = K_1\mathbf{L} \qquad c = K_2\mathbf{L} \qquad d = K_3\mathbf{L}$$

$$e = K_4\mathbf{L} \qquad f = K_5\mathbf{L} \qquad g = K_6\mathbf{L}$$

Then V-2 can be arranged to give

$$\bar{\mathbf{n}} = \frac{b[1 + c(2 + d\{3 + e[4 + f(5 + 6g)]\})]}{1 + b[1 + c(1 + d\{1 + e[1 + f(1 + g)]\})]} \qquad \text{(V-2b)}$$

The denominator is just that of α_0 (Eq. V-3). For any number of complexes fewer than six, let $g, f, \ldots$ be zero. See also Appendix B. On most calculating instruments all the parentheses are the same, not as shown here for clarity.

The α fractions are derived as before:

$$\alpha_0 = (1 + K_1\mathbf{L} + K_1K_2\mathbf{L}^2 + K_1K_2K_3\mathbf{L}^3 + \cdots)^{-1}$$

$$\alpha_1 = \frac{K_1\mathbf{L}}{1 + K_1\mathbf{L} + K_1K_2\mathbf{L}^2 + K_1K_2K_3\mathbf{L}^3 + \cdots} = \alpha_0K_1\mathbf{L} \qquad \text{(V-3)}$$

$$\alpha_2 = \alpha_0K_1K_2\mathbf{L}^2 = \alpha_1K_2\mathbf{L}$$

$$\vdots$$

(See Eq. I-8. Mathematically, these constants are the inverse of the *dissociation* constants used in elementary texts for ML and acid–base systems. IUPAC and the major reference compilations are now all in terms of formation constants as explained in Chapter I and Appendix B.)

The following examination of speciation in systems of known constants will ease later treatment of the scientifically more interesting problem of the determination of species present and the unknown constants.

V-2. TYPICAL SYSTEMS: MONODENTATE LIGANDS

A typical system showing the rather even close spacing of *d*-metal ion complexes is the Ni(II)–NH_3 shown in **Fig. V-1*a***.

Note the addition of α's to make $\bar{\mathbf{n}}$, which has no sharp rise for cases of such weak complexing. For example, 0.01 *M* Ni(II) in 1 *M* NH_3 has under half of the metal in the ML_6 species. Appreciable amounts of ML_5 and ML_4 are present even with the 100-fold excess of ligand present.

Note in **Fig. V-1*b*** that each log α is related to log α_0 by a constant (log β) plus n log([NH_3]) (see Eq. V-3):

$$\log \alpha_1 = \log \alpha_0 + 2.80 + \log \mathrm{A}$$

$$\log \alpha_2 = \log \alpha_0 + 5.04 + 2 \log \mathrm{A}$$

$$\log \alpha_3 = \log \alpha_0 + 6.77 + 3 \log \mathrm{A}$$

$$\log \alpha_4 = \log \alpha_0 + 7.96 + 3 \log \mathrm{A}$$

$$\vdots$$

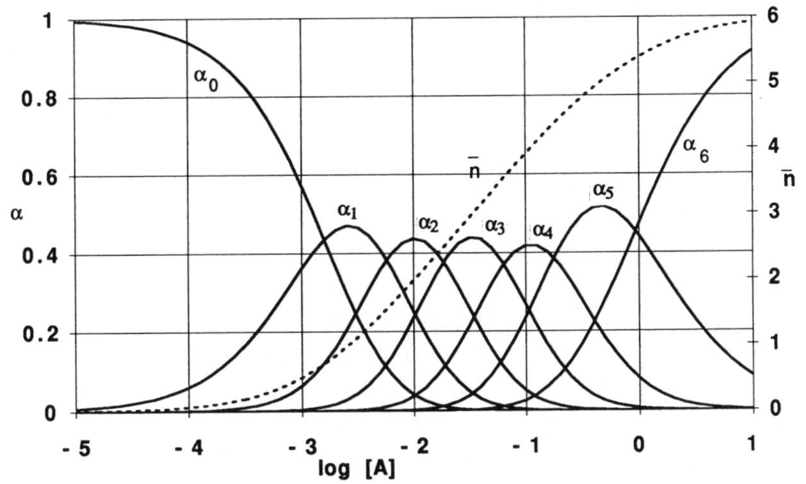

Fig. V-1*a*. Ni(II)–NH_3 α, $\bar{\mathbf{n}}$ diagram: log K_f 2.80, 2.24, 1.73, 1.19, 0.75, 0.03 for 30°C and $\boldsymbol{I} = 2\ M$ (NH_4NO_3). (See Section V-5.)

This has the feature that any measurement of one α curve and its log A values allows determination of all of the other α's. Differentiation shows that all are related to $\bar{\mathbf{n}}$:

$$\frac{\delta \log \alpha_n}{\delta \log \mathrm{A}} = n - \bar{\mathbf{n}}$$

e.g., the slope of log α_3 at a given log A is $3 - \bar{\mathbf{n}}$ at that log A point.

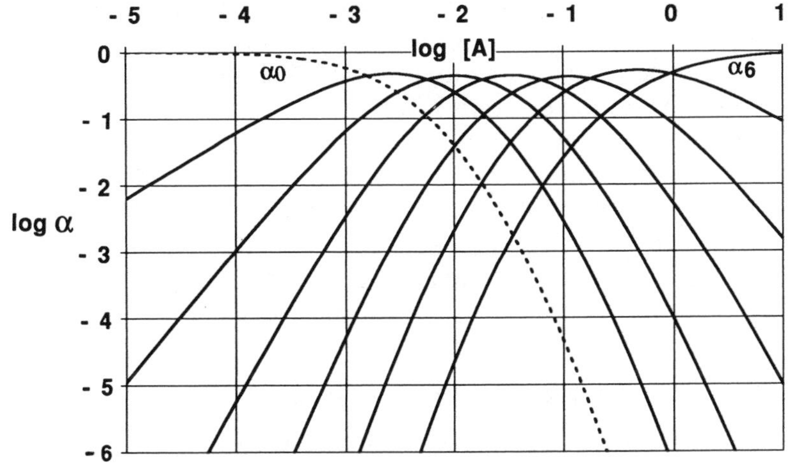

Fig. V-1*b*. Plots of log α_0 to log α_6 vs. log[NH_3] for the Ni(II)–NH_3 system: α_0 shown dotted; compare its inverse, $\mathbf{F}_0$, in Fig. V-6 (**A** = ammonia).

At log A = −2.5, we see that the $\bar{\mathbf{n}}$ on the previous figure is about 1.0, the slopes of log α_n for $n = 0, \ldots, 6$ are −1, 0, 1, 2, 3, 4, 5.

Species Calculations for Given Mixtures

Consider the problem of estimating species in mixtures. Since high ligand concentrations are required for this rather weak complexing system, the amount of NH_3 left free will not be negligible, as **H** or **OH** often are in protonic equilibria. Take 0.01 *M* Ni(II) with 0.100 *M* total NH_3. Try assuming a large fraction of Ni(II) is $Ni(NH_3)_6^{2+}$. This would leave 0.04 *M* NH_3 free. Take log 0.04 = −1.4. At that value of log **L** we see that $\bar{\mathbf{n}}$ is about 3 and the six-complex is not abundant. So we must move to our correct method, calculating $\bar{\mathbf{n}}' = 10 - 100\mathbf{L}$ (Eq. V-1) for a range of trial **L** values to find the compatible $\bar{\mathbf{n}}$. A quick rough estimate is made:

L	0.04	0.05	0.06	0.07	0.08	0.09	0.10
$\bar{\mathbf{n}}'$	6	5	4	3	2	1	0
log **L**	−1.4	−1.3	−1.22	−1.15	−1.1	−1.05	−1.00

Looking at Fig. V-1*a*, we see that this $\bar{\mathbf{n}}'$ line crosses $\bar{\mathbf{n}}$ between 3 and 4 where **L** is between 0.06 and 0.07 *M*. Further precision can be obtained as before with analogous calculations in acid–base cases. [**L** = 0.0646, $\bar{\mathbf{n}}$ = 3.541. See the next paragraph reference to Appendix B.] This examination is enough to conclude that if we want the six-complex to predominate, *excess*, free [NH_3] must be greater than 1 *M*. For the 0.01 *M* Ni(II) case, this requires total ammonia to be above 1.06 *M*. Temperature and medium (***I***) must be adjusted to those stated for the constants used in the diagrams.

In older general texts, one used to find such problems as determining the degree of dissociation when 0.1 mol $Ni(NH_3)_6Br_2$ (an easily made solid) is dissolved per liter. Often the text pretended that *x* molar Ni^{2+} and 6*x* molar NH_3 formed! By the methods here we get $\bar{\mathbf{n}}' = 6 - 10\mathbf{L}$, which leads to $\bar{\mathbf{n}} = 4.319$ and **L** = 0.168 *M*. Here α_0 is about 5.4×10^{-6}, which means [Ni^{2+}] is $\alpha_0 C_M = 5.4 \times 10^{-7}$ *M*. So much for the 1:6 ratio. See Fig. V-1*a*. (A complete spreadsheet method for this calculation is shown in Appendix B, Section B-2.)

Reverse: Solution Design for Desired Species

Suppose one wished to see if $Ni(NH_3)_4X$ can be precipitated by adding X^{2-}. How can one make a solution having 0.0500 *M* of the tetrammine ion that would be insensitive to loss of that species?

(a) The α diagram shows that $\bar{\mathbf{n}} = 4.00$ also is close to the maximum of α_4. This is generally true if the K's are evenly or widely spaced. We need $\alpha_4 C_M = 0.0500\ M$. Thus, in $\bar{\mathbf{n}} = \bar{\mathbf{n}}'$ (Eq. V-1) we have

$$4.00 = (C_L - \mathbf{L})/(0.0500/\alpha_4) \quad \text{where } C_M = 0.0500/\alpha_4$$

From the $\bar{\mathbf{n}}$ formula and system K's we get 4.00 at $\mathbf{L} = 0.112\ M$, which gives $\alpha_4 = 0.4197$, $C_M = 0.1191\ M$, and $C_L = 0.5885\ M$ as the amounts needed.

(b) Do the species shift if 0.0400 M solid $Ni(NH_3)_4X$ is removed? The remaining $C_M = 0.079$ and $C_L = 0.429$. The new balance is

$$\bar{\mathbf{n}} = (0.429 - \mathbf{L})/0.079 = 5.43 - 12.66\mathbf{L}$$

In the spreadsheet calculation, this gives $\bar{\mathbf{n}} = 4.004$ and $\mathbf{L} = 0.1127$. Thus, little change in the distribution of major species occurs in this situation.

Note that in both solutions, much more than a 4:1 ratio of C_L/C_M is required to produce $\bar{\mathbf{n}} = 4.00$. Knowledge of how to design mixtures with maxima of desired species is helpful in determining spectra, solubility, and other properties of the individual species. Here, mixing a 4:1 ratio gives more α_3 than α_4. This is a problem if the constants are unknown. Thus, K determination is vital to investigations in physical inorganic studies.

Close-Spacing Effects

The effect of close spacings of K values is especially common in ML systems. Zinc(II) and Ag(I) systems are frequent examples.[1] Their ammonia complexes have log K's only a few tenths unit apart. One of the closest is that of the Zn(II)–imidazole system shown in **Fig. V-2**.

The reader may check the $\bar{\mathbf{n}}$'s that would result if one tried to make the $Zn(Im)_4$ complex by mixing ML to make specific concentrations: (i) C's of 0.020 M Zn(II) and 0.100 M imidazole gives $\bar{\mathbf{n}}$ of about 3.6 with 0.028 M free imidazole; (ii) 0.02 M Zn(II) and 0.080 M Im gives $\bar{\mathbf{n}}$ of about 3.3; and (iii) 0.020 M Zn(II) and 0.040 M Im gives $\bar{\mathbf{n}}$ of about 1.8. In all cases there are three, and in some four Zn(II) species above 1% abundance.

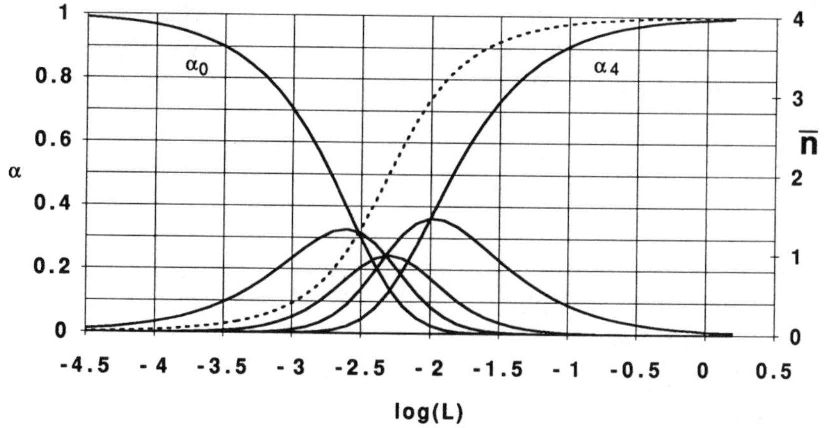

Fig. V-2. Plots of α, $\bar{n}$ of the Zn(II)–imidazole system: log $K = 2.52, 2.32, 2.32, 2.0$; $I = 0.16\ M$ at 25°C.

If the K values were all the same, the interesting curve in **Fig. V-3** results. Note how the $\bar{\mathbf{n}}$ is steep and with a single inflection for all these cases. Calculation of even approximate species distributions by single K methods is impractical.

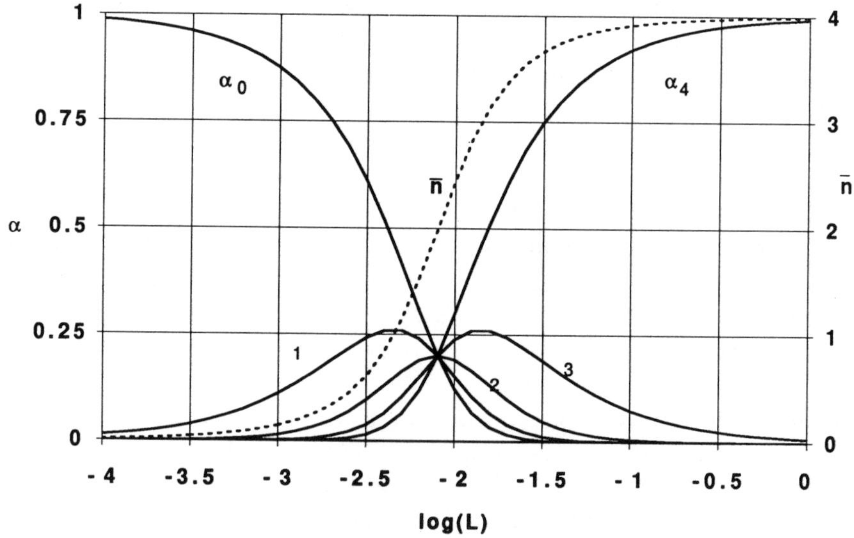

Fig. V-3. Hypothetical ML_4 system having all K values equal, 125 (log K, 2.097). All α lines cross at log[L] value of -2.097 with each $\alpha = \frac{1}{5}$.

V-3. INTERACTION OF PROTON AND ML EQUILIBRIA

The **pH** effect of a metal–ion upon solutions of basic ligands permits a major method for experimental determination of equilibrium constants, to be described in what follows. First let us examine why this is so with a simple system of known constants and without appreciable mixed complexing (L and LH both acting as ligands, considered later). This is an example of competitive reactions. The Lewis acid M^{2+} competes with the Brønsted H^+ for the basic site on the ligand:

$$M^{2+}_{(aq)} + H_nL \rightleftharpoons ML + nH^+_{(aq)}$$

If the acid constants for H_nL are known, the unknown ML constants can be found from **pH** measurements on mixtures of known C values. First let us see the reverse: with all constants known, remove protons with a strong base and deduce the predicted titration curve.

Example. We take 100 ml of a solution having 0.0300 M Ni(II), nitrate, or perchlorate and 0.1000 M HGly, a slight excess for formation of the ML_3 complex. (For clarity we hold volume constant: the adjustment of each C for $V = v_0 + v_b$ can easily be made. The equilibrium ideas are unaffected.)

The effect of **pH** on complexing by a ligand that is a weak base can be easily expressed for the given total $C_M = 0.0300\ M$ and $C_L = 0.1000\ M$. We simply put the acid–base α_{0L} expression into the preceding $\bar{\mathbf{n}}_{ML}$ expression for the complex system. To do this, the total ligand may be divided into that on the metal ion ("masked" toward protons), which is $\bar{\mathbf{n}}_{ML} C_M$, and the remainder, which is open to proton attack, $C'_L = C_L - \bar{\mathbf{n}}_{ML} C_M$. Thus, the free, unprotonated ligand is given by

$$[L] = (C_L - \bar{\mathbf{n}}_{ML} C_M)\alpha_{0L}$$

Let

$$x = C'_L = C_L - \bar{\mathbf{n}}_{ML} C_M = (C_{AB} - \mathbf{D})/\bar{\mathbf{n}}_{HL}$$

The preceding equation gives the proton balance (from the master $\bar{\mathbf{n}}$ relation) for any weak acid–base system if the H_nL species are the sole proton holders besides water. The computation strategy is an iterative one for $\bar{\mathbf{n}}$ (and α's). The **pH** values are chosen, α_{0L} is calculated, and the [L] term is iterated with the $\bar{\mathbf{n}}_{ML}$ expression to satisfactory precision (equations that follow). The constants at $\boldsymbol{I} = 0.1\ M$ and 25°C are as follows:

For glycine, $pK_{a1} = 2.36$, $pK_{a2} = 9.57$.

For ML(0–3)	$\log \beta_1$	$\log \beta_2$	$\log \beta_3$
Ni(II)–Gly$^-$	5.78	10.58	14.00

Now, $\bar{\mathbf{n}}_{HL}$ is the diprotic acid function of **H**, but $\bar{\mathbf{n}}_{ML}$ is a function of **L** (Eqs. V-1 to V-3), which in turn is $\alpha_{0HL} C'_L$. Putting these together and using x to clarify the variable C'_L and a for α_{0HL} gives us (b is the molarity of added base, after mixing, with constant volume)

$$x = C_L - \bar{\mathbf{n}}_{ML} C_M = (C_{AB} - \mathbf{D})/\bar{\mathbf{n}}_{HL}$$

$$= 0.1000 - 0.0300a \; \frac{\beta_1 x + 2\beta_2 a x^2 + 3\beta_3 a^2 x^3}{1 + \beta_1 a x + \beta_2 (ax)^2 + \beta_3 (ax)^3} = \frac{0.1 - b - \mathbf{D}}{\bar{\mathbf{n}}_{HL}} \tag{V-4}$$

where

$$\bar{\mathbf{n}}_{HL} = \frac{\mathbf{H}/K_2 + 2\mathbf{H}^2/K_1K_2}{1 + \mathbf{H}/K_2 + \mathbf{H}^2/K_1K_2} \quad \text{and} \quad \alpha_{0HL} = \left(1 + \frac{\mathbf{H}}{K_2} + \frac{\mathbf{H}^2}{K_1K_2}\right)^{-1} = a$$

Thus, we have at last an equation in **H** to a high power. By choosing values of **H**, we can solve for x, and b, iteratively since x varies from 0.1 (no ML formed) to 0.01 (0.03 ML_3 formed). Then we plot **pH** vs. b, a titration curve. The variable C_{AB} varies as base is added: $C_{AB} = C_H^0 - b = 0.100 - b$. Thus, $b = 0.1 - \mathbf{D} - \bar{\mathbf{n}}_{HL} x$. **Figure V-4** shows a curve for 100.0 ml of a solution having 0.0300 M Ni(II) and 0.1000 M HGly titrated with concentrated NaOH and assuming constant volume. The base is in millimoles, $b(100)$, since 10 mmols HGly was taken. The upper curve is for glycine alone.

The reverse case problem of finding the **pH** of a given mixture of metal ion, glycine, and any added strong acid or base is a bit more difficult, but the same approach will handle it. Given C_L, C_M, and C_H and the K values, the ligand in the metal system ($ML + 2ML_2 + 3ML_3 + \cdots$) is $\bar{\mathbf{n}}_{ML} C_M$. That in the HL system ($L + HL + H_2L + \cdots$) is C'_L (Eq. V-4): (**b** is not base, b, here):

$$\overset{\text{term } (\mathbf{a})}{C'_L = C_L - \bar{\mathbf{n}}_{ML} C_M} = \overset{\text{term } (\mathbf{b})}{(C_{AB} - \mathbf{D})/\bar{\mathbf{n}}_{HL}}$$

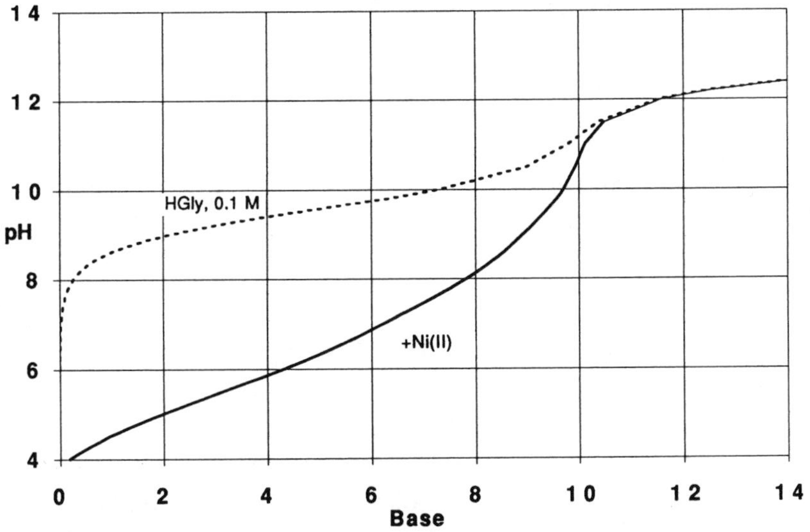

Fig. V-4. Curve for 100.0 ml of a solution having 0.0300 M Ni(II) and 0.1000 M HGly titrated with concentrated NaOH and assuming constant volume.

One chooses **pH** trials and calculates successively from it and the C's, $\bar{\mathbf{n}}_{\mathrm{HL}}$, $\alpha_{0\mathrm{L}}, C'_{\mathrm{L}}$ [from term (**b**)], $[\mathrm{L}] = \alpha_{0\mathrm{L}} C'_{\mathrm{L}}$, then $\bar{\mathbf{n}}_{\mathrm{ML}}$, and finally term (**a**) until the **pH** trials intersect [(**a**) = (**b**)].

A solution having 0.200 M HGly and 0.0100 M Ni^{2+} makes a **pH** of 4.1. Pure glycine alone had **pH** of about 6.1. The Lewis acid Ni(II) coordinated glycinate ions and freed some protons. A more detailed example follows.

Example. Find the **pH** of a solution made 0.0500 M in Ni^{2+}, 0.1000 M in glycine, (HGly), and 0.0300 M in NaOH. Here $C_{\mathrm{M}} = 0.0500\ M$, $C_{\mathrm{L}} = 0.1000\ M$, and $C_{\mathrm{AB}} = 0.0700\ M$. In Eq. V-4, this makes the Gly not bound to Ni(II):

$$C'_{\mathrm{L}} = 0.1000 - \bar{\mathbf{n}}_{\mathrm{ML}}(0.0500) = (0.0700 - 10^{-\mathrm{pH}} + 10^{\mathrm{pH}-13.80})/\bar{\mathbf{n}}_{\mathrm{HL}}$$

Programming these expressions as described previously showed (**a**) near (**b**) at **pH** 5. A few trials near that gave the following results:

pH	$\bar{\mathbf{n}}_{\mathrm{HL}}$	$\alpha_{0\mathrm{L}}$	(**b**) C'_{L}	L	$\bar{\mathbf{n}}_{\mathrm{ML}}$	(**a**) C''_{L}
5.00	1.0023	2.69×10^{-5}	0.06983	1.88×10^{-6}	0.6180	0.06910
4.98	1.0024	2.56×10^{-5}	0.06982	1.79×10^{-6}	0.6016	0.06992

A rational examination of this system is helpful:

(i) The 0.07 *M* HGly and 0.03 *M* Gly^- with the K_{a2} would suggest a buffer of **pH** 9.20.

(ii) If the Ni(II) coordinates all of the Gly^-, we have 0.03 *M* $NiGly^+$ and 0.02 *M* Ni^{2+}. We ask whether the complex will give up enough Gly^- to buffer the 0.07 *M* HGly or the Ni(II) will remove Gly^- to release protons to form more H^+ and H_2Gly^+. Roughly,

$$Ni^{2+} + 2HGly \rightleftharpoons NiGly^+ + H_2Gly^+$$

$$K_{eq} = K_{f1}K_{a2}/K_{a1} = 10^{5.78-9.57+2.36} = 10^{-1.43}$$

The K_{eq} and approximate concentrations point to the latter.

(iii) Here $\bar{\mathbf{n}}_{HL}$ is above 1, indicating there is more H_2Gly^+ than Gly^- and thus more acidic than pure HGly (**pH** 6.0).

(iv) In agreement, $\bar{\mathbf{n}}_{ML} > 0.6$, and the Ni^{2+} has taken slightly more than the 0.03 Gly^-.

EDTA–Metal Ion Solutions

Polydentate ligands usually have greater K_f values than monodentate. A few examples will show this. We compare the log K_{f1} values for one ligand, not the total positions occupied, for our purpose here. Ethylenediamine(en) and glycinate are bidentate. Nitrilotriacetic acid (NTA) is tetradentate, especially appropriate for Cu(II). Ethylenediaminetetraacetic acid (EDTA) is potentially hexadentate with two nitrogen positions and four carboxylate positions in almost perfect geometry for the six octahedral positions of many metal ions. Its α, $\bar{\mathbf{n}}$, **pH** diagram is given Fig. III-18.

The log K_{f1} values, approximate, at $I = 0.1\ M$ to $I = 1\ M$ and 20–25°C are as follows:

Metal ion	NH_3	en	gly^-	NTA	$EDTA^{4-}$
Fe^{3+}			10		25.1
Hg^{2+}	8.8	14	10.3		21.8
Cu^{2+}	4.3	10.6	8.2	13	18.8
Ni^{2+}	2.8	7.5	5.8	11.4	18.6
Zn^{2+}	2.4	5.6	5.5	10.4	16.5
Ca^{2+}	−0.2		1	7.6	10.7
Ag^+	3.37	5	3.5		7.3

Depending on the metal ion and the **pH**, EDTA may occupy fewer than six positions on the metal ions (d^2sp^3) for the octahedral d-metal ions. The usual assumption in K calculations for EDTA complexes that its α_0, α_{0y}, is the sole relevant species is thus a simplification, one that is often justified for approximate purposes with the very favorable, high K_f values involved. Thus, with Cu(II) at $K_f = 10^{18.8}$, complexing will be safely beyond the quantitative limit when **pH** is above about 5 (α_{0y} about 10^{-6}). To explain, assume an equivalent amount of total Cu(II) and of EDTA (all forms) of about 0.01 M. Let x be the concentration of uncomplexed Cu(II) and assume there is no other ligand operating on Cu(II), i.e., $\alpha_{0M} = 1$. If other ligands are present such as the NH_3 mentioned in what follows, $\alpha_{0M}x$ is used in the denominator for the corrected free Cu(II) aqua ion. The other α_0 here is α_{0Y}, the unprotonated EDTA ion fraction:

$$K_f = 10^{18.8} = \frac{[CuY^{2-}]}{[Cu^{2+}][Y^{4-}]} = \frac{0.01 - x}{x(x\alpha_0)}$$

$$x \cong \sqrt{(0.01 \times 10^{-18.8})/\alpha_0}$$

If α_0 is 10^{-6}, 10^{-4}, 10^{-2}, or 1 (**pH** about 5.2, 6.4, 8.2, or >11, see Table V-1), we can see that $\log x$ is about −7.4, −8.4, −9.4, or

TABLE V-1. α_0 Values of EDTA at pH 1–12

pH	α_0	pH	α_0	pH	α_0
1	2.39×10^{-18}	4.6	7.97×10^{-08}	8	0.0066
1.2	2.10×10^{-17}	4.8	1.98×10^{-07}	8.2	0.0105
1.4	1.70×10^{-16}	5	4.85×10^{-07}	8.4	0.0166
1.6	1.23×10^{-15}	5.2	1.17×10^{-06}	8.6	0.0261
1.8	7.81×10^{-15}	5.4	2.77×10^{-06}	8.8	0.0408
2	4.34×10^{-14}	5.6	6.35×10^{-06}	9	0.0632
2.2	2.09×10^{-13}	5.8	1.40×10^{-05}	9.2	0.0967
2.4	8.80×10^{-13}	6	2.95×10^{-05}	9.4	0.1451
2.6	3.29×10^{-12}	6.2	5.91×10^{-05}	9.6	0.2120
2.8	1.11×10^{-11}	6.4	1.12×10^{-04}	9.8	0.2990
2	3.43×10^{-11}	6.6	2.03×10^{-04}	10	0.4033
3.2	9.96×10^{-11}	6.8	3.54×10^{-04}	10.2	0.5172
3.4	2.76×10^{-10}	7	5.98×10^{-04}	10.4	0.6293
3.6	7.38×10^{-10}	7.2	9.90×10^{-04}	10.6	0.7291
3.8	1.93×10^{-09}	7.4	1.61×10^{-03}	10.8	0.8101
4	4.97×10^{-09}	7.6	2.60×10^{-03}	11	0.8711
4.2	1.26×10^{-08}	7.8	4.16×10^{-03}	12	0.9854
4.4	3.19×10^{-08}				

−10.4, increasingly smaller fractions of the total 0.01 *M* Cu(II) at these equivalence point conditions. Conversely, we can look up the **pH** at which 50% of the complex is broken up (0.005 *M* Cu^{2+} free). This requires α_0 of about 10^{-17}, which occurs near **pH** 1.2. This demonstrates the large freedom the analyst has in **pH** needed to achieve quantitative complexing of Cu(II) by EDTA. Conditions can be designed with a large safety factor by making **pH** far above 1.2 or even above 5.2. Excellent results are easily obtained even if the assumptions about complex species formed are oversimplified. [A **pH** titration of Cu(II) in the presence of EDTA, similar to the preceding Ni–glycine case, is shown in the exercises and Fig. V-15.]

This much needed method (for difficult ions like Ca^{2+} and Zn^{2+}) was developed in the 1940s. Visual indicator dyes having similar chelating groups were found to make easy titration possible. Many procedures involve added NH_3 to prevent precipitation in basic solution. Higher **pH** is then needed to overcome the competing ligand effect. More recent investigations of complexing patterns have shown many mixed OH^- and other ligand complexing occurring in the EDTA mixtures, and these are recorded in the references on stability constants and in Ringbom's book. This field illustrates most of the interacting equilibria we touch upon in this book, especially in this chapter. Many other chelating ligands have been synthesized for analytical purposes and are treated in the references.

V-4. EXPERIMENTAL DETERMINATION OF $\bar{\mathbf{n}}$ AND THE FORMATION CONSTANTS

The practical problem is usually the reverse of the previous calculations on Ni(II). One determines the **pH** of a mixture of a metal ion and a ligand (usually buffered with protonated ligand). If the ligand K_a's are known, one can calculate the equilibrium α_{0L} and then the $\bar{\mathbf{n}}_{ML}$ of the mixture from the preceding equations (C_M and C_L known). From the $\bar{\mathbf{n}}$ curve, the *K*'s, or β's, of the ML system are calculated as done for polyprotic acid–bases in previous chapters. A useful form analogous to the acid–base but using formation constant product β's is obtained by cross multiplying Eq. V-2, putting it equal to $\bar{\mathbf{n}}$ and collecting terms in powers of **L**: *N* is the maximum *n* value (6) in the Ni–NH_3 system:

$$\bar{\mathbf{n}} - (1-\bar{\mathbf{n}})\beta_1\mathbf{L} - (2-\bar{\mathbf{n}})\beta_2\mathbf{L}^2 - \cdots - (N-\bar{\mathbf{n}})\beta_N\mathbf{L}^N = 0$$

or

$$\bar{\mathbf{n}} = \sum_{n=1}^{n=N} (n - \bar{\mathbf{n}})\beta_n \mathbf{L}^n \qquad \text{(V-5)}$$

This can be solved for any β_n or any K_n. For example, for the preceding ML_3 system,

$$K_1 = \frac{\bar{\mathbf{n}}}{\mathbf{L}} [1 - \bar{\mathbf{n}} + (2 - \bar{\mathbf{n}})K_2\mathbf{L} + (3 - \bar{\mathbf{n}})K_2K_3\mathbf{L}^2]^{-1} \qquad \text{(V-6)}$$

Thus, K_1 can be found from $\bar{\mathbf{n}}, \mathbf{L}$ data if one has estimates of the other K values to use in the small overlap terms, as will be explained. (See also the citric acid calculations in Chapter IV.) For widely spaced K's and low $\mathbf{L}$ values, the last two terms may be negligible, and

$$K_1 \cong \frac{\bar{\mathbf{n}}}{\mathbf{L}(1 - \bar{\mathbf{n}})}$$

We shall test simple values to see the logic: $\bar{\mathbf{n}} = 0.5$ gives $K_1 = 1/\mathbf{L}_{0.5}$, the usual half $\bar{\mathbf{n}}$ value, the crossing of α_0 and α_1 (Fig. V-1). At $\bar{\mathbf{n}} = 0.2$, $K_1 = 0.2/0.8(\mathbf{L}_{0.2})$. These are restricted to the one stated $\bar{\mathbf{n}}$ and $\mathbf{L}$ pair of values. Equations V-5 and V-6 are forms of Bjerrum's iteration formula. [Compare the acid–base forms (Eq. IV-8).] It gives improved K_1 values for each improvement in estimates of the other K's, as will be illustrated in what follows. Similar solutions are then made for each remaining K.

General Solution

Alternatively, Eq. V-5 can be used with $\bar{\mathbf{n}}, \mathbf{L}$ data to give a set of linear simultaneous equations. They are first power in the β values. The set can be solved by common methods: determinants, or iteration computer methods. See Appendix B, where the same Cu(II)–NH_3 data used in what follows are recalculated by Gaussian elimination in the determinant. Connors[2] (pp. 70–75), also uses this data to illustrate several other methods of calculation (Connors uses $\bar{i}$ for $\bar{\mathbf{n}}$, his Eq. 2.94 is identical to our V-2.)

V-5. BJERRUM'S METHOD

The simplest experimental and calculational method was invented by Jannik Bjerrum, whose thesis[3] gives clear details of methods and actual calculations for systems he was first to describe in the step equilibrium fashion. If known C mixtures of metal ion and ammonia are prepared containing also a known high concentration of NH_4NO_3 for buffering and ionic strength control, the **pH** will be strictly set by the $[NH_3]$ left uncomplexed. Examine the K_a expressions for NH_4^+ to see that with $[NH_4^+]$ the same in a standard and in an unknown, x, we get

$$\mathbf{pH}_{(0)} - \mathbf{pH}_{(x)} = \mathrm{pNH}_{3(x)} - \mathrm{pNH}_{3(0)}$$

The $[NH_3]_x$ in the mixture can be calculated from the known buffer (0) without the metal ion, and the measured **pH** of the known and unknown buffers. Subtracting this from the total added NH_3 yields the amount bound to the metal ion and removed from the acid–base buffer system. The **pH** is lowered to the same degree as the $\log[NH_3]$. Note that this requires only correct *differences* between two measurements, not correct absolute values of **pH**. (In effect, the molarity **H** used in this system demands constant medium and determination of conditional pK_a for ammonium, or **H** in the knowns, as discussed in Section IV-6. As seen in what follows for the Cu(II) case, the actual **H** values are not used in the calculations; their Δ is used.) Thus, one can calculate the $\bar{\mathbf{n}}_{ML}$, the observed [bound $NH_3]/C_M$. A series of measurements yields an $\bar{\mathbf{n}}$ curve. If the K's are well spaced, then $\log K_1 = -\log[NH_3]_{eq}$ at $\bar{\mathbf{n}} = 0.5$, as previously explained. Check this in Fig. V-1.

Copper(II)–Ammonia Example

This example is taken from Bjerrum's work on Cu(II)–NH_3 at 30°C and in 2 M NH_4NO_3.

In a solution with $C_{Cu} = 0.02057\ M$ and $C_{NH_3} = 0.0500\ M$, the **pH** read 2.690 higher than a standard that had $p[NH_3]$ 2.002. This means that the $p[NH_3]_x$ was 4.692 and free $[NH_3]_x$ is $2.03 \times 10^{-5}\ M$. Thus, the bound $[NH_3]$ is $0.0500 - 0.0000203 = 0.00498\ M$ and $\bar{\mathbf{n}} = 0.00498/0.02057 = 0.242$. If at this low $\bar{\mathbf{n}}$, α_0 and α_1 are the only appreciable species (see the Ni case, Fig. V-1), we use the approximation from Eq. V-6 to get $K_1 \cong 0.242/0.758(0.0000203) = 15{,}727$, $\log K_1 = 4.197$. This is close to the final iterated values given in what follows.

Computing the K's

From the general $\bar{\mathbf{n}}$ equation (Eqs. V-2 to V-5) we can get an expression for each K similar to the acid–base cases in Chapters III and IV. Notice that at $\bar{\mathbf{n}} = 0.5$, the first term becomes $\mathbf{L}^{-1} = K_1$, the approximate formula for K_1. If the K's are well spaced, the next two terms are small. Also, as $\mathbf{L} \to 0$, the first term must predominate, making K_1 estimates improve as we use $\bar{\mathbf{n}} = 0.5, 0.4, 0.3, 0.2, \ldots$. See Fig. V-1. The other terms are thus small corrections for the overlap of α_2 and α_3, and rough approximations of K_2 and K_3 allow calculations of K_1. Similar rearrangements for K_2 and K_3 serve for refining their values near $\bar{\mathbf{n}} = 1.5, 2.5$ using the good K_1 and an approximate K_3, and so on. Numerous computer approaches are possible. However, a reasoned method will avoid the large error that intrudes if all experimental points are blindly entered and the results averaged. Look at what happens to the terms in Eq. V-3 if integral $\bar{\mathbf{n}}$ values are inserted. An experimentally obtained $\bar{\mathbf{n}}$ of 0.98, which has, for example, 1% uncertainty, gives us $1 - \bar{\mathbf{n}} = 0.02 \pm 0.01$, now 50% uncertain. Chemically intelligent choice of data is as vital here as in acid–base cases where one would obviously choose well-buffered regions ($\bar{\mathbf{n}}_H$ near half-integral values) for stable **pH** readings. Calculation of results is shown after the next section.

V-6. SLOPES OF $\bar{\mathbf{n}}$ CURVES

The effects of various spacings of log K values on the slopes of the resulting $\bar{\mathbf{n}}$ plots will help interpret experimental data of $\bar{\mathbf{n}} - \log \mathbf{L}$, the major means of evaluating equilibrium constants. Four sets for $N = 4$ were inserted in Eq. V-2 to make **Fig. V-5**. Obviously, the slopes give immediate information about the separations and magnitudes of K's. Note that the increasing slopes in Fig. V-5, especially for the last two cases, are rather rare and difficult to measure accurately, as will be demonstrated with the Zn(II)–NH_3 system. Note that the half-$\bar{\mathbf{n}}$ points (0.5, 1.5, 2.5, 3.5) are close to $1/\mathbf{L} = K_n$ (but not exactly because of overlap) for the first two spacings but are far off for the last two.

Cu–NH_3 Case

For systems in which successive K values decrease (normal order), the approximation formulas from Eq. V-5 converge rapidly to unique

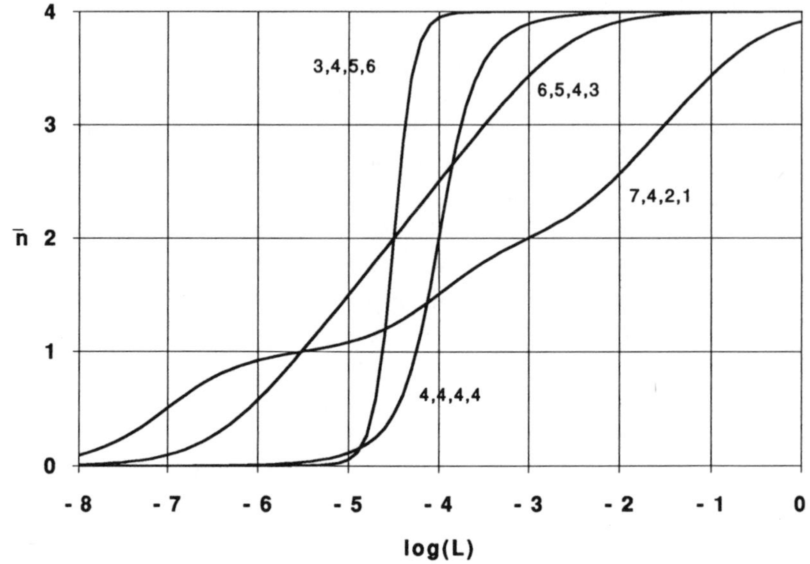

Fig. V-5. The $\bar{\mathbf{n}}$ curves for log K_f sets in order of the rise of $\bar{\mathbf{n}}$ at the left. The four ML_4 systems have log K_f values: (i) 7, 4, 2, 1; (ii) 6, 5, 4, 3; (iii) 4, 4, 4, 4; (iv) 3, 4, 5, 6.

results even if the K values are closely spaced (and if the data are consistent and if the complexes assumed are the important ones). We take Bjerrum's copper(II)–ammonia data as an example (Table V-2). The $\bar{\mathbf{n}}$ $[NH_3]$ data of the first two columns yield first-trial K estimates at $[NH_3]^{-1}$ of the half-integral $\bar{\mathbf{n}}$ values: log $K = 4.4$, 3.5, 2.9, 2.0. These are used (trial I) in the approximation equations for each K. [That is, for K_1, the rough $K_2, \ldots, K_4$ are inserted into the K_1 formula (V-6), etc.] Close and better values are found (even in trial I) only for the $\bar{\mathbf{n}}$ points at which the particular K represents major species, i.e., the first three points for K_1, the third and fourth for K_2, the fifth for K_3, and the sixth and seventh for the last K.

[The last datum, presumably with a small error giving a mathematically impossible $\bar{\mathbf{n}} = 4.002\bar{\mathbf{n}}$, produces negative K_4 values, so we omit it from the iterations. Copper(II) actually does form weak L_5 and L_6 complexes with ammonia, so this is real, but Bjerrum's data did not extend to high enough $[NH_3]$ to establish this clearly.]

The averages of these preferred points are used for each subsequent iteration (trial II). The K values become stable to a few percentage points in about 3 trials, and after 5–10 trials the computer values are stable to far more figures than demanded by the data. We end by calculating the $\bar{\mathbf{n}}$ values at the experimental points with our

TABLE V-2. Calculations of Bjerrum's Data for Cu–NH_3

Trial I. Rough trial with $K = [NH_3]^{-1}$ at the half n̄ values explained in text. The first two columns are derived from the experimental data. Approximation formulas for the K's and n̄ are entered in columns 3–7 on a spreadsheet.

First trial K's: 20,000, 3000, 800, 100

$[NH_3]$	Experimental n̄	K_1	K_2	K_3	K_4	Calculated n̄
2.030E − 05	0.244	13,885	−4,240	−76,672	−3,509,924	0.319
4.620E − 05	0.486	14,277	161	−12,213	−253,407	0.581
1.256E − 04	0.959	14,914	2,171	−523	−8,904	1.031
5.350E − 04	1.877	20,653	3,016	805	107	1.875
2.290E − 03	2.784	−1,284	6,611	930	129	2.741
8.630E − 03	3.437	−13	−259	3,632	136	3.351
2.265E − 02	3.743	−1	−26	−489	145	3.665
2.447E − 01	4.002	0	0	−8	−2,068	3.960

Trial II. In place of the trial I K's, formulas for the averages of the K's are placed in the designated lines, as described in the text. Results are after 20 iterations using *successive* averages of chosen K's.

Final K's: 13,671 (lines 1–3), 3364 (lines 3, 4), 731 (line 5), 143 (lines 6, 7)

[A]	Experimental n̄	K_1	K_2	K_3	K_4	Calculated n̄
2.030E − 05	0.244	13,678	3,377	855	6,297	0.244
4.620E − 05	0.486	13,794	3,457	1,106	8,162	0.484
1.256E − 04	0.959	13,582	3,341	698	−104	0.961
5.350E − 04	1.877	14,105	3,388	737	153	1.875
2.290E − 03	2.784	13,671	3,364	731	143	2.784
8.630E − 03	3.437	137	1,550	686	140	3.444
2.265E − 02	3.743	−7	−440	887	146	3.739
2.447E − 01	4.002	0	0	−8	−207	3.972

Final log K_n: 4.14, 3.53, 2.86, 2.16.
Bjerrum's log K_n: 4.15, 3.50, 2.89, 2.13

Weaker complexing to ML_5 and ML_6 has been observed giving K_5 and K_6 about 0.28 and 0.01.

[a]In ref. 3, pp. 126–128; 30°C, in 2 M NH_4NO_3. Bjerrum's calculations done by hand and perhaps used averages different from those chosen here. We ran 20 computer iterations after the first rough approximation. (The automatic iteration feature in the Excel spreadsheet calculation format.) Further iterations gave small change. Each successive iteration used the averages of the cells chosen as those most affected by the K being evaluated, that is, at n̄ near 0.5, 1.5, 2.5, and 3.5. Notice how the values run in these cells (roughly descending diagonal) even in the first rough trial. Starting from other rounded sets of first approximations leads to final constants within about 1% of these. Programming the $K_1, \ldots, K_4$ formulas (Section V-4) and experimenting with starting values is illuminating as to the convergence power of these favorable functions. These data are recalculated by another method in Appendix B.

best estimates of the K's in Eq. V-2. This is important in assuring that the species model chosen is a reasonable one for the data at hand. A modern investigation would use a larger number of points. Bjerrum developed this method with the then newly available glass electrode **pH** meter (1930s), allowing rapid work needed with volatile ammine solutions. He checked his data by determining $[NH_3]$ by vapor pressure measurements and by a spectral method. He also determined the deviation from ideality of solutions of $[NH_3]$ above 1 M, which were required for cases having much weaker complexing like Mg^{2+}–NH_3. He determined $\bar{\mathbf{n}}$ in constant NH_4NO_3 solutions from 0.5–5 M. This is required to check that the different **pH** (for the same $[NH_3]$ values) gave a similar complexing pattern: i.e. that OH^- complexing was not appreciable. We return to this question near the end of this chapter. The reader wishing to gain an understanding of this field is urged to examine Bjerrum's lucid book,[3] in addition to more recent literature which often leaves many details to the imagination of those well versed in the field.

Zn–NH_3 Case

When K values are not in normal order, more difficulty is experienced in finding consistent K values (Table V-3). If the noise (random uncertainty) in the $\bar{\mathbf{n}}$, [L] data is appreciable, no unique results may be obtainable. The Zn(II)–NH_3 data of Bjerrum illustrates the problems well. See explanations of starting K choices that follow.

Examine Figs. V-2 to V-5 to see that the data in Table V-3 indicate a slope high enough to signal nearly equal or inverse order of constants. Thus, we began the data analysis by setting all four constants the same, about 230 (this gives $\beta_4 = 10^{9.46}$ as used by Bjerrum since the midpoint of the $\bar{\mathbf{n}}$ data curve is about $\log[NH_3] - 2.37$, i.e., 9.46/4). (See Fig. V-6 in the next section.) This gave unequal K values increasing 1, 2, 3 and then down for the last. Averaging successive approximations in the diagonal selected cells gave oscillating results and slow divergence. By taking the K_4 as $10^{9.46}/\beta_3$, Table V-3 was obtained using two different choices of diagonal averages for the first three K's. They agree fairly well with the data, but many writers, including Bjerrum, point out that slight variations in the data will greatly affect the K's found. That is, the correction terms for overlap in the K formulas in the preceding cause oscillation when the spacing is close and the data are not perfect. Hand calculations or experimentation with computer programs are instructive as to how this happens.

TABLE V-3. Calculation of Bjerrum's Data for Zn(II)–NH_3

Experimental $[NH_3]$	Experimental $\bar{n}$	K_1	K_2	K_3	K_4
0.00121	0.434	236.0	276.1	302.2	90.7
0.002	0.889	246.0	291.2	334.8	153.4
0.00281	1.342	229.6	273.0	305.7	116.0
0.00361	1.789	234.7	278.9	315.3	127.8
0.00466	2.244	233.5	278.7	315.8	129.1
0.00604	2.669	244.0	285.9	332.8	133.7
0.00818	3.057	275.6	300.1	331.7	136.8
0.01156	3.363	480.9	341.3	346.8	139.3
0.0245	3.733	−20.4	−200.8	678.5	155.2
0.0573	3.853	1.5	18.4	135.9	114.5
0.0562	3.875	−4.1	−64.3	664.8	140.9
0.151	3.987	0	−0.6	−18.8	524.5
0.731	3.843	0	0	0.2	7.4
K's (for first trial)		240	283	320	132
log K (obtained above)		2.38	2.45	2.50	2.12
Bjerrum's log K		2.37	2.44	2.50	2.15
Second choice trial (see text below)		2.41	2.40	2.51	2.15

Calculated $\bar{n}$ check:	Calculated $\bar{n}$	Calculated β_4, only	Four equal K's (all 232)
	0.439	0.025	0.382
	0.878	0.176	0.756
	1.363	0.610	1.206
	1.798	1.315	1.650
	2.252	2.305	2.156
	2.666	3.173	2.644
	3.045	3.712	3.098
	3.346	3.924	3.442
	3.699	3.996	3.787
	3.870	4.000	3.919
	3.868	4.000	3.917
	3.950	4.000	3.971
	3.990	4.000	3.994

Further light is gained by the calculation in the preceding of $\bar{\mathbf{n}}$ with the sets of K's for the data points $[NH_3]$, "Calculated $\bar{\mathbf{n}}$ check". We include those for the first set, a β_4 set, and four equal K's. The β_4 set is calculated assuming the only equilibrium to be

$$Zn^{2+} + 4NH_3 \rightleftharpoons Zn(NH_3)_4^{2+} \quad \text{with } K = \beta_4 = 10^{9.46}$$

$$\bar{n} = 4\alpha_4 = \frac{4[Zn(NH_3)_4^{2+}]}{[Zn^{2+}] + [Zn(NH_3)_4^{2+}]} = 4/(1 + (\beta_4[NH_3]^4)^{-1})$$

This single overall equilibrium was commonly written before Bjerrum's work. He showed that stepwise mechanisms seemed truer, and in fact no examples of overall jumps could be found. They persisted for many years in elementary text treatments of complexing problems!

V-7. $\mathbf{F}_0$ OR α_0 METHODS

In the $\bar{\mathbf{n}}$ methods described in the preceding, the fraction of ligand bound is large at the start and the free metal ion concentration becomes low during the later points as excess ligand is present. If bound **L** and $\bar{\mathbf{n}}$ are not easily measurable but instead unbound $[M^{n+}]$ is experimentally available, one gets in effect (with C_M known) α_0 as a function of **L**. Unbound **L** must often be found by successive approximation, as will be shown in the following example. In these cases, large excess of ligand is often required and low C_M is used. Thus, the bound ligand is a small part of the total and $\bar{\mathbf{n}}$ almost indeterminate. Typically, potentiometric and solubility methods can determine $[M^{n+}]$. Reliable electrodes exist for Ag, Cu, Cd, Hg, and a few others. Amalgam and synthetic specific ion electrodes are also used. Solubility, with added X, works because $[M^{n+}]$ is dependent upon K_{s0}/X for a compound MX. Solubility relations and K determinations will be treated in detail in the next chapter.

Because of the form of α_0, it is convenient to use

$$1/\alpha_0 = \mathbf{F}_0 = C_M/[M^{n+}]$$

$$\boldsymbol{F}_0 = 1 + K_1\mathbf{L} + \beta_2\mathbf{L}^2 + \beta_3\mathbf{L}^3 + \cdots + \beta_N\mathbf{L}^n \qquad \text{(V-7)}$$

This is a simpler polynomial than the $\bar{\mathbf{n}}$ function (Eq. V-2). If sufficient and accurate enough data are obtained, it is mathematically straightforward to evaluate the constants. But numerical problems in estimating free **L** make this less reliable than for much $\bar{\mathbf{n}}$ data. See **Fig. V-6**, plotted from a table of the logarithm of the right side of Eq. V-7 and log **L** for two four-K sets from Fig. V-5.

The slope of log plots of α_0 or of $\mathbf{F}_0$ are easily derived from their relations to **L** as

$$\frac{\delta \log \alpha_0}{\delta \log \mathbf{L}} = -\bar{\mathbf{n}} \qquad \text{(V-8)}$$

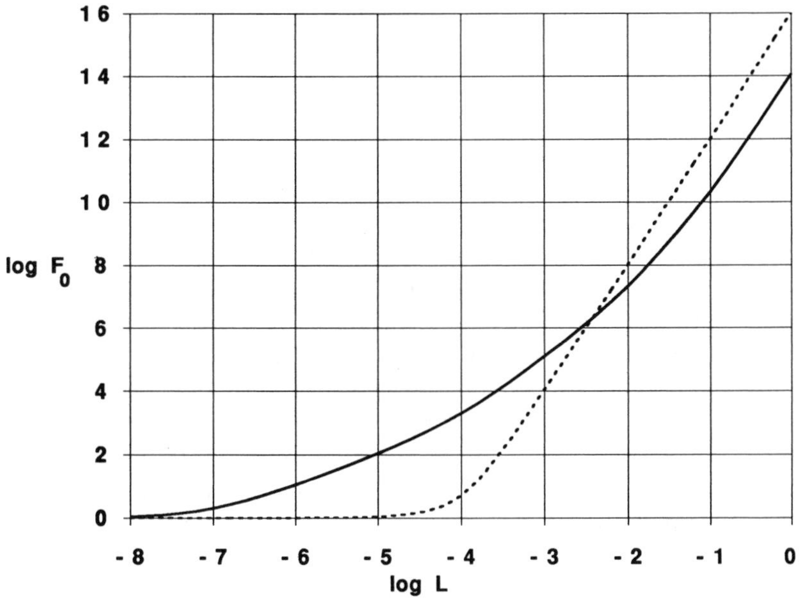

Fig. V-6. Plot of $\log \mathbf{F}_0$ for the ML_4 systems having log (constants) 7, 4, 2, 1 (solid line) and 4, 4, 4, 4 (dotted line).

This can be seen on $\log \alpha$ diagrams (Figs. V-1*b* and II-12).

In Fig. V-6 note the changes of slopes at the $\log \mathbf{L} = -\log K$ values. That is, in regions having a single species predominant (approximately halfway between $\log K$ values), the slopes must be 1, 2, 3, and 4, as seen in the preceding $\mathbf{F}_0$ (α_0) relations (Eq. V-8 and Fig. V-5). (Note that the ordinate scale is about twice the abscissa scale.) With closer spacings the curve is rounded, but it still goes through these slopes so that prediction of the number of complexing steps and the approximate $\log K$ values are possible. [Another $\log \mathbf{F}_0$ can be seen, inverted, as $\log \alpha_0$ in Fig. V-1*b*, Ni(II)–ammonia.]

Example: Silver–Allyl Alcohol. We use Eq. V-7 and its rearrangement to give

$$(\mathbf{F}_0 - 1)/\mathbf{L} = K_1 + \beta_2\mathbf{L} + \beta_3\mathbf{L}^2 + \cdots + \beta_N\mathbf{L}^{N-1} = \mathbf{F}_1 \qquad \text{(V-9)}$$

[Various arrangements of these functions are referred to as the Leden method and the Fronaeus method. A solution of the matrix formed from the equations for successive points in Eq. V-9 can be used. We shall show a computer method (illustrated graphically) similar to that in the text Hartley et al.[4] but expressed in the $\alpha\,\bar{\mathbf{n}}$ terms of the present work.]

Potentiometric data for free silver ion as allyl alcohol was added are used[4] to deduce complexing constants. The critical point to note in this method is that with such weak complexing, C_L must be much higher than C_M to obtain measurable binding. Here $\bar{\mathbf{n}}$ is low and uncertain because of the few significant figures left after subtraction to estimate a combined ligand value. Since free $[Ag^+]$ is measured (by the potential difference at a solid silver electrode between Ag(I) solutions with and without added ligand) and C_M is known, $\mathbf{F}_0$ is found, $C_M/[Ag^+]$. The first plot of the data (**Fig. V-7**) gives little indication that any but the first complex forms. Compare Fig. V-6.

This reveals that very little formation of a second complex occurred, since slope 2 would be achieved when it becomes predominant. Slope 1 is just being approached in the region available to this method. Thus, we shift to the $\mathbf{F}_1$ function of Eq. V-9 (**Fig. V-8**) since it would remain constant if only the first complex forms.

The $\mathbf{F}_1$ plot shows clearly the noise in the method. The intercept is K_1 while the slope is β_2. Thus, $K_2 = \beta_2/K_1 = 11.83/17.816 = 0.664$. Note that the highest $\bar{\mathbf{n}}$ obtained was 0.56 so that little α_2 is present, making the second constant value uncertain. If there were no second species ($K_2 = 0$), the slope in Fig. V-8 would be zero. The least squares best line can be obtained on the computer rather than by graph. In Excel on Macintosh, the LINEST(,) command for the set of results plotted prints the slope and intercept. Such a straight line was then added to the plot (Fig. V-8) for illustration.

To obtain the free ligand concentration is a problem in this

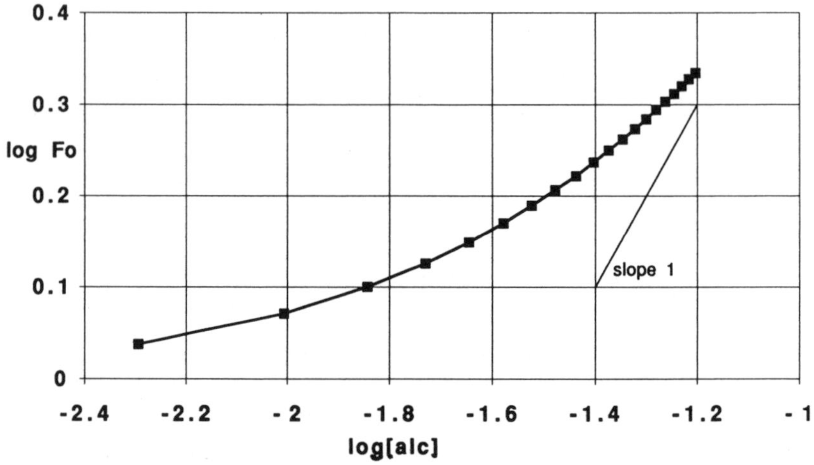

Fig. V-7. The log $\mathbf{F}_0$–log [alc] plot of data for Ag(I)–allyl alcohol.

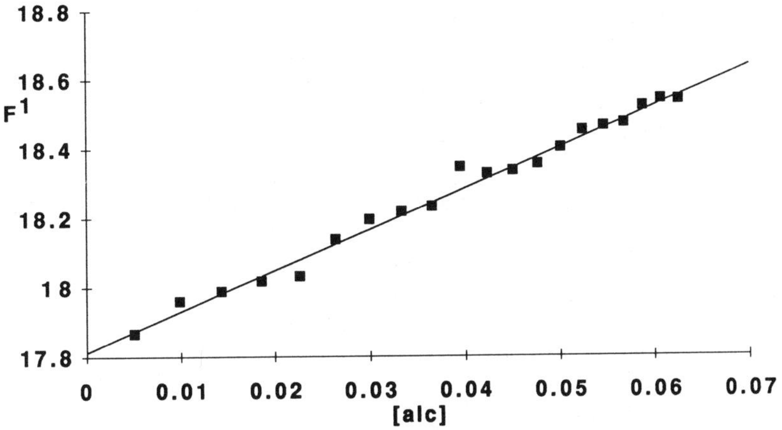

Fig. V-8. The $\mathbf{F}_1$ plot for Ag(I)–allyl alcohol: intercept, 17.816; slope, 11.83.

method. For a first approximation, it is near the total ligand added. Next, one may subtract one ligand per combined silver (i.e., $\alpha_1 = 1 - \alpha_0$), which is $C_M(1 - 1/\mathbf{F}_0)$. Then a plot like that in Fig. V-8 yields approximations to the constants, which can be used in the $\bar{\mathbf{n}}$ equation to calculate a better approximation to the bound ligand, $\bar{\mathbf{n}}C_M$, and thus the free ligand, $C_L - \bar{\mathbf{n}}C_M = [\text{alc}]$. Successive estimates of constants and [alc] are continued as shown in Table V-4 until no change results. The data and final iteration are also shown. More figures than justified are carried to prevent rounding error. The total Ag(I) was kept constant at 0.00982 *M*. The total silver was divided by the $[Ag^+]$ measured in each step to obtain the first column, $\mathbf{F}_0$.

From the constants and the α equations, one may find that at $\mathbf{L} = 0.1\ M$, the values are

$$\alpha_0 = [1 + 17.8(0.1) + 11.83(0.01)]^{-1} = 0.344$$

and then

$$\alpha_1 = 0.616 \qquad \alpha_2 = 0.041$$

One may wonder whether the K_2 has much meaning since all the data are below this $\mathbf{L}$ value. Hartley et al.[4] should be consulted for a detailed treatment of error evaluation since the linear least squares used here is not strictly valid for this data. The authors conclude that K_1 is good to about ±0.05% but K_2 is accurate only to ±10%.

TABLE V-4. Silver–Alcohol Data and Results[a]

		Iterated Results		
F_0	alc total M	$\bar{n}$	[alc]	$F_1 = (F_0 - 1)/[\text{alc}]$
1.0906	0.00589	0.08338	0.005071	17.866
1.177	0.01134	0.15115	0.009856	17.959
1.2584	0.0164	0.20723	0.014365	17.988
1.3355	0.02112	0.25445	0.018621	18.017
1.408	0.02552	0.29462	0.022627	18.032
1.4788	0.02963	0.32918	0.026397	18.138
1.5452	0.03349	0.35932	0.029961	18.197
1.6071	0.03711	0.38576	0.033322	18.219
1.6657	0.04053	0.40925	0.036511	18.233
1.725	0.04374	0.43010	0.039516	18.347
1.7766	0.04678	0.44887	0.042372	18.328
1.8265	0.04965	0.46577	0.045076	18.336
1.8747	0.05238	0.48116	0.047655	18.355
1.9219	0.05496	0.49514	0.050098	18.402
1.9673	0.05741	0.50792	0.052422	18.452
2.009	0.05975	0.51972	0.054646	18.464
2.0485	0.06197	0.53055	0.056760	18.473
2.0887	0.06409	0.54059	0.058781	18.521
2.1256	0.06611	0.54989	0.060710	18.541
2.1597	0.06804	0.55854	0.062555	18.539

[a]Total Ag(I) is 0.00982 M.

Further Examples: Generation of L from Protonated Ligand

Clearly, a chemically equivalent situation to the direct ammonia addition of Bjerrum's method might be achieved by titrating a metal ion–NH_4NO_3 solution with NaOH to make the NH_3 in situ, although this makes changes in the ionic materials. Bjerrum and others have used the direct method just described with a number of amines. But many other ligands are treated by the strong base method. The principles are the same, but calculations are different enough to require explanation. Addition of NaOH to a mixture of glycine (HL) and Ni^{2+} will cause the mixture to progress through the species as shown in Fig. V-4 from the low **pH** side. We next show how $\bar{n}$ can be obtained from the NaOH–**pH** data.

The $\bar{n}$ method deals neatly with a system like ethylenediamine (en) and Ni^{2+} as a function of **pH**, the experimental method for finding the three formation constants for the ML system. One starts with the conjugate acid enH_2^{2+} and adds the base, measuring **pH**. Total analytical concentration of ligand (en) present is known, i.e., C_L.

Take the K_a's of en as known so that **pH** will fix the proton speciation allowing calculation of $\bar{\mathbf{n}}$ and α's as needed:

$$\bar{\mathbf{n}}_{HL} = \frac{[HL] + 2[H_2L]}{[L] + [HL] + [H_2L]} = (C_H - \mathbf{D})/C'_L$$

As before, we divide the total ligand, C_L, into ML and HL systems,

$$C_L = [L] + [HL] + [H_2L] + [ML] + 2[ML_2] + 3[ML_3]$$
$$= C'_L + \bar{\mathbf{n}}_{ML} C_M = (C_H - \mathbf{D})/\bar{\mathbf{n}}_{HL} + \bar{\mathbf{n}}_{ML} C_M$$

We can find $\bar{\mathbf{n}}_{ML}$ values from the **pH** measurements and the total C's of the ligand and metal in the mixtures if the acid constants of the ligand, and thus $\bar{\mathbf{n}}_{HL}$, are known. (With a new ligand of unknown K_a's, these are measured in a run without metal ion present; Fig. V-4). Then,

$$\bar{\mathbf{n}}_{ML} = [C_L - (C_H - \mathbf{D})/\bar{\mathbf{n}}_{HL}]/C_M \qquad \text{(V-10)}$$

From the full $\bar{\mathbf{n}}_{ML}$ expression (Eqs. V-1 and V-2), we can solve for the K's of the ML system if we have at least three $\bar{\mathbf{n}}$, **pL** data in suitable regions of the system. To obtain the free ligand at each point, we use the $\bar{\mathbf{n}}_{HL}$ and α_{0L} equations with the acid constants:

$$\mathbf{L} = \alpha_{0L}(C_H - \mathbf{D})/\bar{\mathbf{n}}_{HL}$$

The $\bar{\mathbf{n}}, \log \mathbf{L}$ data are treated to get constants as shown in the preceding for ammonia. If this seems difficult, reexamine the Ni(II)–Gly reverse case (Fig. V-4).

V-8. PROTONATED LIGAND, POLYNUCLEAR, AND MIXED COMPLEXING

Frequently complexes other than ML_n form. When these complications are present, the K's calculated by the method just given are not constant, and the calculated $\bar{\mathbf{n}}$ may rise and then fall as ligand or NaOH is added. This is due to the proton acidity being present in some of the complexes as well as in the HL system when $M(HL)_n$ complexing occurs. This makes the bound **L** calculation invalid. An iteration similar to the preceding but far more tedious and prone to uncertainty is used. These problems can occur with polydentate ligands, where protonation of some sites does not prevent coordination of other sites to metal ions. Lysine, arginine, EDTA, citrate, and similar ligands are prime candidates. Examination of the constant compilations of such ligands (see Sillen and Martell and Martell and Smith in the Annotated Bibliography), especially with *d*-metal ions, will illustrate the varied complications reported in recent years. Accurate and high precision data are usually required to establish the species. The material balances now must include M, L, and **H** for a system that may involve H_2L, ML_n, $M(HL)_m$, $M(OH)L_n$, and polynuclear M_yL_n complexes:

$$C_M = [M] + [ML] + [ML_2] + \cdots + [MHL] + [M(HL)_2] + \cdots$$
$$+ 2[M_2L] + \cdots$$
$$C_L = \mathbf{L} + [HL] + [H_2L] + \cdots + [ML] + 2[ML_2] + \cdots$$
$$+ [MHL] + 2[M(HL)_2] + \cdots$$
$$C_H = \mathbf{D} + [HL] + 2[H_2L] + \cdots + [MHL] + 2[M(HL)_2] + \cdots$$
$$+ 2[MH_2L] + \cdots$$

Inserting the **H**, **L**, and [M] dependencies from the *K* expressions for each species gives a set of equations depending upon the three variables in rather complex relations. Correct choice of species and often independent methods for estimating species other than **H** alone are required to deduce *K*'s. One might start with a model system using a few guessed species and see if constant *K* values can be found and the data fitted. Then additional species are postulated and tested until a "best" fit is obtained. The computer programs for doing this are long but essentially like those described in this chapter. The reader is referred to references 5–10.

Usually measurements other than p**H** are required to establish the nature of species present. Spectral, metal ion potentiometric or specific ion electrodes, and solvent extraction can give evidence of the species. Nickel(II) and Cu(II), for example, change from pale blue or green aqua species to deeply colored complexes that sometimes have different spectra for successive numbers of ligands. The p**H** alone cannot give enough information if several L- and HL-bonded species are present at once. Numerous papers in the *Journal of the American Chemical Society* and elsewhere since about 1950 illustrate the methods and calculations, notably for citrates of Cu(II) and Fe(II) and Fe(III), which were studied by a combination of p**H**, polarographic, and solubility measurements. Other polyamino acids and polycarboxylic acids have been studied. Papers are listed in the complications of stability constants in the Annotated Bibliography as in the previous paragraph.

Example: Ni(II)–Lysine Complexes

Lysine is L-2,6-diaminohexanoic acid, an HL that can protonate on both NH_2 groups to form H_3L^{2+} and is thus a triprotic acid system. Besides the HLys, both HCl salts are available as stable solids. If the colored *d*-metal ions Co(II), Ni(II), or Cu(II) are added to the HLys solution, the deep-colored complexes are formed. The question is whether the complexes are like those of glycinate ion (HLys has the carboxyl and α-amino groups deprotonated and free to give bidentate five-membered ring complexes as does Gly^-) or if the metal ion also binds the second amino group, releasing a proton and forming tridentate or bridged complex. The latter might have an eight-membered ring if it is mononuclear or a bridge to another metal ion might form. Some **pH** tests might illuminate the question but might also be ambiguous. If one adds Ni^{2+} to a pure HLys solution, proton release should lower the **pH** if appreciable $Ni(Lys)_n^{2-n}$ forms. If $Ni(HLys)_n^{2+}$ forms instead, dilution of the HLys (an ampholyte) might cause very little **pH** change. If both form, intermediate **pH** change would depend on the equilibrium constants of complexing. The **pH**, α, $\bar{\mathbf{n}}$ diagram of lysine is shown in **Fig. V-9**, where the pK_a values at 25°C and $\boldsymbol{I} = 0.1$ *M* are 2.19, 9.15, and 10.68.

Pure lysine (HLys) is the α_1 ampholyte, and its dilute solutions would have **pH** of about 9.8. Here significant amounts of Lys^- are available for complexing, and the **OH** might cause precipitation of some ions or formation of M–OH–Lys complexes. This is a difficult system to unravel. The **pH** data alone will not serve if many of the possibilities occur together. Starting from lower **pH**, using H_2Lys^+ solutions, adding Ni^{2+}, and generating ligands by addition of NaOH has shown that simple bidentate complexing by HLys serves to correlate the **pH** data below **pH** 9. The H_2Lys^+ has **pH** of about 5.8. When appreciable Ni^{2+} is added, the **pH** becomes about 4 [as with Ni(II)–Gly in the preceding], indicating possible complexing to release protons:

$$Ni^2 + H_2Lys^+ \rightleftharpoons Ni(HLys)^{2+} + H^+$$

The protons enter the lysine system and put it back toward the

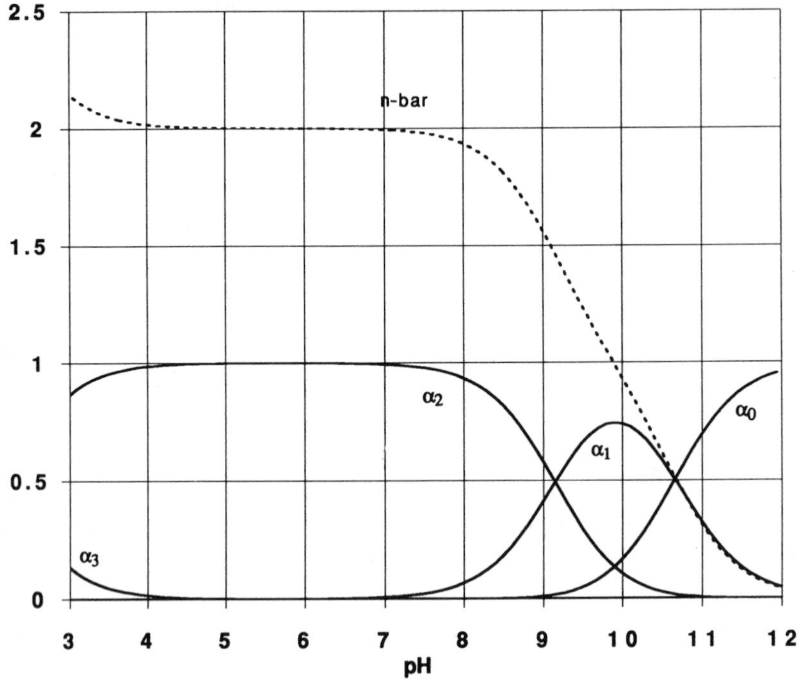

Fig. V-9. The **pH**, α, $\bar{\mathbf{n}}$ diagram of the lysine system.

H_3Lys^{2+} buffer. Then, as NaOH is added, more HLys can form more Ni(II) complexes. Calculation of $\bar{\mathbf{n}}$ by the Bjerrum methods shown above gives consistent values for three constants for complexes up to $Ni(HLys)_3^{2+}$. At **pH** above 10, these deprotonate, but OH^- mixed species and precipitation have prevented reliable constant determinations.[6]

Details of a more complicated case are given by Daniele and co-workers[7] for mixtures of Cu(II), Ni(II), and 5-hydroxy-2,6-diamino hexanoic acid (hydroxylysine). A number of polynuclear and mixed-metal complexes were deduced. A full explanation of the computer programming is given in a previous article.[8]

V-9. HYDROXIDE COMPLEXES AND POLYNUCLEAR CASES

Metal ion–hydroxide species are abundant in water solutions except for the alkali metal ions. A few are simple step examples like Zn(II), which forms up to $Zn(OH)_4^{2-}$. A **pH** (or **pOH**) scale serves to show the regular exponentially varied speciation $\alpha_0, \ldots, \alpha_4$ just as was done before, except that the complication of precipitation occurs in most cases. However, many metal ions form a mixture of several mono- and polynuclear species. The α fractions of these vary both with log **L** and $\log[M^{n+}]$, where **L** is the **OH** (or 1/**H**) variable. In these systems the form $K_{x,y}$ uses x for the M ion number and y for the number of OH^-. For mixed OH and L complexing, the full formulas and K expressions are used to assure clarity.

Example: Experimental Approach to the Cu(II)–OH System

Even though CuO and/or $Cu(OH)_2$ may precipitate between **pH** 6 and 8, (pK_{s0} about 19), **pH** data can be taken by adding dilute NaOH up to $\bar{\mathbf{n}}$ of about 0.2 to show a clear difference in the curves if polynuclear complexing is appreciable. Then K_{xy} values for various postulated models of the system can be calculated to fit the data. A simple laboratory case is the following for Cu(II).

Note: In this section $\bar{\mathbf{n}}$ is $\bar{\mathbf{n}}_{OH}$, bound OH^- per metal ion.

We made 100 ml of 0.025, 0.0125, and 0.0051 *M* Cu(II) nitrate with KNO_3 to make ***I*** about 0.1 for all. These were titrated with small portions of 0.0200 *M* standard NaOH, and the pa_H read. The data are in the tables that follow.

For the reaction

$$Cu^{2+}(H_2O) \rightleftharpoons CuOH^+ + H^+_{aq}$$

we see that the net bound OH^- must be any free net acid **D** (**H–OH**) plus the added strong base. For example, before any base is added, the Cu(II) solution is acidic, say, **pH** 4. This means that 10^{-4} *M* net H^+ has been formed by proton loss from the Cu(II) aqua ion. If total Cu(II) is 0.02 *M*, then 10^{-4} *M* $CuOH^+_{aq}$ out of 0.02 *M* gives us $\bar{\mathbf{n}} = 0.005$. If the **pH** is neutral after adding NaOH, the OH^- must have attached to Cu(II), etc. Thus, $\bar{\mathbf{n}} = (\mathbf{D} + C_{sb})/C_M$, the average number of OH^- bound per metal ion. Calculate $\bar{\mathbf{n}}$ for each point and plot them. The data and calculations of $\bar{\mathbf{n}}$, $K_{1,1}$, and $K_{2,2}$ are also shown in the table that follows. We test a model using postulated species Cu^{2+}_{aq}, $CuOH^+$, and $Cu_2(OH)_2^{2+}$; water molecules, as usual, are not shown.

Calculations

Take $\bar{\mathbf{n}} = (\mathbf{D} + C_{sb})/C_M$. (Use millimoles and new total volume at each step.)

Hypothesis I: If $K_{1,1}$ alone is acting, $[CuOH^+] = \bar{\mathbf{n}}C_M$, $[Cu^{2+}] = (1 - \bar{\mathbf{n}})C_M$, and

$$Cu^{2+}_{aq} \rightleftharpoons CuOH^+ + \mathbf{H}$$

$$K_{1,1} = \mathbf{H}[CuOH^+]/[Cu^{2+}] = \mathbf{H}\bar{\mathbf{n}}/(1 - \bar{\mathbf{n}})$$

Hypothesis II; For $K_{2,2}$ alone, $[Cu_2(OH)_2^{2+}] = \bar{\mathbf{n}}C_M/2$, $[Cu^{2+}] = (1 - \bar{\mathbf{n}})C_M$, and

$$2Cu^{2+}_{aq} \rightleftharpoons Cu_2(OH)_2^{+} + 2\mathbf{H}$$

$$K_{2,2} = \mathbf{H}^2[Cu_2(OH)_2^{2+}]/[Cu^{2+}]^2 = \mathbf{H}^2\bar{\mathbf{n}}/2(1 - \bar{\mathbf{n}})^2C_M$$

(The pa_H was corrected to **pH** by subtracting 0.081 during the $\bar{\mathbf{n}}$ calculation on the spreadsheet that follows.)

(**A**) 2.545 mmol Cu(II), in 100 ml, adding V_b milliliters of 0.0194 *M* NaOH:

V_b	pa_H	$\bar{\mathbf{n}}$	K_{11}	K_{22}	
0	4.681	0.00099	2.482E − 08	1.226E − 11	
1.04	5.086	0.00832	8.294E − 08	1.641E − 11	
2	5.228	0.01553	1.125E − 07	1.632E − 11	
3	5.317	0.02310	1.374E − 07	1.652E − 11	
4.02	5.387	0.03084	1.573E − 07	1.640E − 11	
5	5.44	0.03830	1.742E − 07	1.635E − 11	
6	5.483	0.04590	1.906E − 07	1.649E − 11	
8	5.553	0.0611	2.196E − 07	1.674E − 11	
10	5.61	0.0763	2.445E − 07	1.692E − 11	
12	5.658	0.0916	2.670E − 07	1.713E − 11	
14.03	5.694	0.1071	2.923E − 07	1.788E − 11	
15	5.705	0.1144	3.072E − 07	1.863E − 11	
16	5.713	0.1220	3.244E − 07	1.965E − 11	
18	5.721	0.1373	3.646E − 07	2.245E − 11	
20	5.728	0.1526	4.058E − 07	2.545E − 11	Precipitation starts

(**B**) 0.5094 mmol Cu(II) in 100 ml vs. 0.0194 *M* NaOH.

V_b	pa_H	$\bar{\mathbf{n}}$	K_{11}	K_{22}
0	5.046	0.00213	2.31E − 08	2.46E − 11
0.5	5.61	0.01963	5.92E − 08	1.76E − 11
1	5.76	0.0385	8.38E − 08	1.81E − 11
2	5.921	0.0765	1.20E − 07	1.87E − 11
3	6.025	0.1145	1.47E − 07	1.91E − 11
4	6.095	0.1525	1.74E − 07	2.03E − 11
5	6.12	0.1906	2.15E − 07	2.51E − 11
6	6.137	0.2287	2.61E − 07	3.09E − 11
7	6.148	0.2668	3.12E − 07	3.83E − 11

Note the larger drift of $K_{1,1}$ rising over a factor of 10, while K_{22} is more stable over the middle buffered range and well before precipitation (and possibly polymerization just before), which invalidates the calculation. Here, $K_{2,2}$ averages about 1.7×10^{-11}. The published value at this I is 2.5×10^{-11} and has at least 20% uncertainty. These are shown in **Figs. V-10** and **V-11**.

Deductions from the Data. Species and K deductions from experimental data are treated in detail in the references by Baes and Mesmer[9] and by the Rossottis.[10] They are simple in concept but may be difficult to use because of the uncertainties in the data imposed by the low C and small **pH** changes attained before precipitation occurs. We shall list observations that can be deduced from the preceding $\bar{\mathbf{n}}$ expressions in this Section. Relative values of the K's determine how pronounced these effects will be. Fortunately, most systems have fewer than four species in the **pH** range up to 7. Often, only a few of these can be determined with precision better than a power of 10:

1. If only mononuclear species are present, the $\bar{\mathbf{n}}$ determined upon solutions of different starting C_M will coincide: $\bar{\mathbf{n}}$ is a function of **H** alone.
2. If only polynuclear species are present in appreciable amounts, the separate $\bar{\mathbf{n}}$ plots will not meet above their zero value. (See Figs. V-10 and V-13.) In the preceding $\bar{\mathbf{n}}$ equation the 2, 2 term shifts $\bar{\mathbf{n}}$ to smaller values (for a given **H**) as C is decreased.

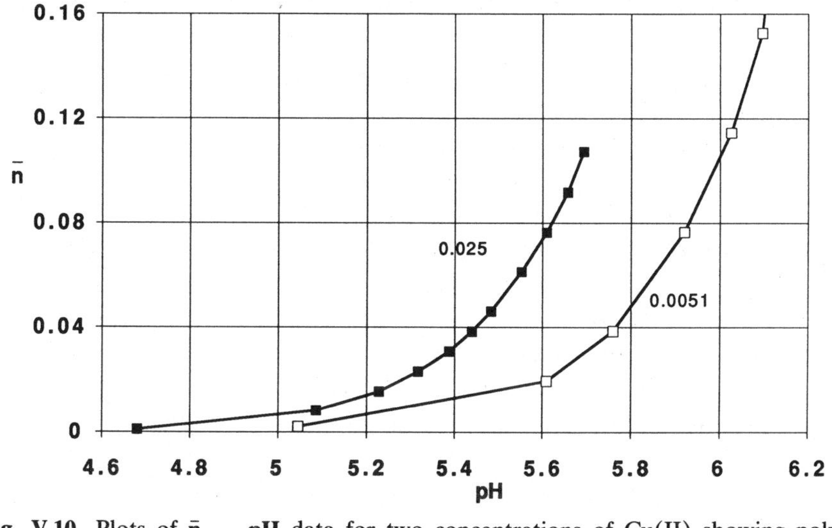

Fig. V-10. Plots of $\bar{\mathbf{n}}_{OH}$, **pH** data for two concentrations of Cu(II) showing polynuclear effects.

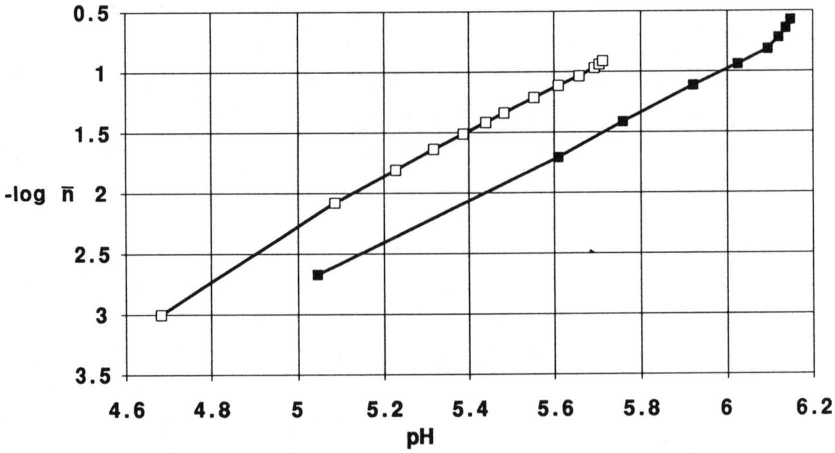

Fig. V-11. Plots of log **n̄** vs. **pH** of data, Cu^{2+}–OH^-, in the text show slopes near 2 with no drift toward 1. See discussion.

3. If both types are important near the start of neutralization, **n̄** will be a common line at first, splitting into separate lines as the polynuclear species become predominant. See the equation: The **n̄** line depends on the 1, 1 term at the start if the 2, 2 term is negligible because of the $1/\mathbf{H}^2$ dependence. Then as **H** decreases, the 2, 2 term overtakes the 1, 1 term.
4. As $C \to 0$, the polynuclear terms in **n̄** approach zero and the mononuclear terms form a single line (called the mononuclear wall).
5. A plot of log **n̄** vs. **pH** may help deduce species. Where **n̄** is small (<0.2) and the α_0 stays near 1, if the 2, 2 predominates, the slope will be 2. If the 1, 1 term predominates, the slope will be 1. Refer to the **n̄** equation that follows for the lower **pH** range where higher OH complexes are negligible. In the preceding Cu(II) example (Fig. V-11) the slopes are about 2, implying that only the 2, 2 species is important under the conditions used: $K_{1,1}$ is too small to affect the **n̄** as determined. This is further clarified by the full complexing picture that follows. The 1, 1 complex may exist and become important at high dilution of the C_M (see Fig. V-12*b*). At higher **n̄**, lower α_0, the full equations must be used. The complete methods are described by the Rossottis[10] (Ch. 17) and Baes and Mesmer[9] (Ch. 3).

Plots from the Published Constants

Using the K expressions in the C_M balance

$$C_M = [Cu^{2+}] + [CuOH^+] + 2[Cu_2(OH)_2^{2+}] + [Cu(OH)_2] + [Cu(OH)_3^-] + [Cu(OH)_4^{2-}]$$
$$= [[Cu^{2+}] + K_{1,1}[Cu^{2+}]/\mathbf{H} + 2K_{2,2}[Cu^{2+}]^2/\mathbf{H}^2 + K_{1,2}[Cu^{2+}]/\mathbf{H}^2 + K_{1,3}[Cu^{2+}]/\mathbf{H}^3 + K_{1,4}[Cu^{2+}]/\mathbf{H}^4$$

the equilibrium condition α's and $\bar{\mathbf{n}}$ are

$$\alpha_0 = [Cu^{2+}]/C_M = (C_M/[Cu^{2+}])^{-1}$$
$$= (1 + K_{1,1}/\mathbf{H} + 2K_{2,2}[Cu^{2+}]/\mathbf{H}^2 + K_{1,2}/\mathbf{H}^2 + K_{1,3}/\mathbf{H}^3 + K_{1,4}/\mathbf{H}^4)^{-1} \quad \text{(V-11)}$$

But $[Cu^{2+}] = \alpha_0 C_M$. Thus, we have C_M dependence and a quadratic in α_0 to solve first when plotting the α curves in (**Figs. V-12*a*, *b***) for given K's, C, and $\mathbf{H}$. From Eq. V-11, cross multiplying and inserting $[Cu^{2+}] = \alpha_0 C_M$, we get the α quadratic:

$$(2C_M K_{2,2}/\mathbf{H}^2)\alpha_0^2 + (1 + K_{1,1}/\mathbf{H} + K_{1,2}/\mathbf{H}^2 + K_{1,3}/\mathbf{H}^3 + K_{1,4}/\mathbf{H}^4)\alpha_0 - 1 = 0$$

The value α_0 is then found in the spreadsheet for each value of $\mathbf{H}$ and C_M by use of the quadratic formula. Then the other α's and $\bar{\mathbf{n}}$ are evaluated:

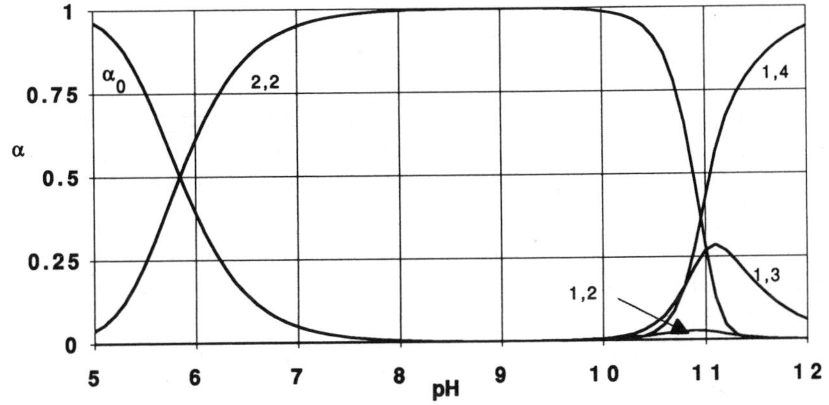

Fig. V-12*a*. The Cu(II)–OH^- species in 0.1 M total Cu(II) at $I = 1\ M$, at 25°C.

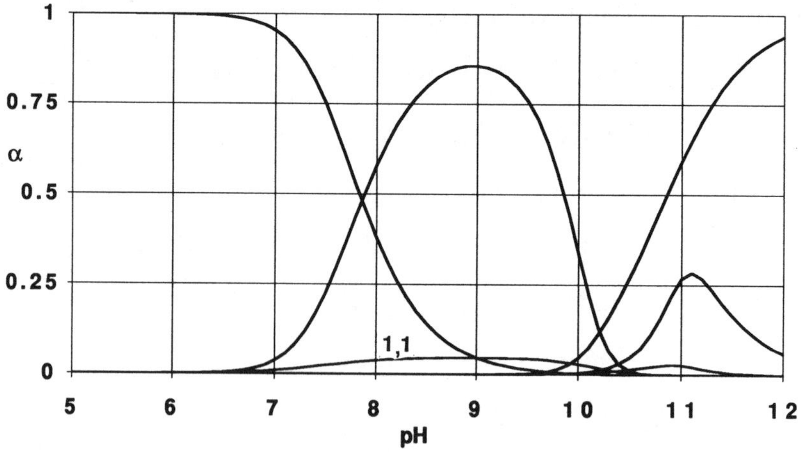

Fig. V-12*b*. The Cu(II)–OH^- diagram at $C_M = 10^{-5}$ *M*. The α's are in the same order as in Fig. V-12*a* with addition of the 1, 1 line.

$$\alpha_{1,1} = \alpha_0 K_{1,1}/\mathbf{H} \qquad \alpha_{2,2} = 2\alpha_0^2 C_M K_{2,2}/\mathbf{H}^2 \quad \alpha_{1,2} = \alpha_0 K_{1,2}/\mathbf{H}^2$$

$$\alpha_{1,3} = \alpha_0 K_{1,3}/\mathbf{H}^3 \qquad \alpha_{1,4} = \alpha_0 K_{1,4}/\mathbf{H}^4$$

$$\bar{\mathbf{n}} = \alpha_{1,1} + 2\alpha_{2,2} + 2\alpha_{1,2} + 3\alpha_{1,3} + 4\alpha_{1,4} \tag{V-12}$$

Surprisingly, the $Cu_2(OH)_2^{2+}$, rather than $CuOH^+$, is a major species. Note in Figs. V-12*a*, *b* the unsymmetrical α plots the metal ion dependence causes. We calculated the α's from Eq. V-11, 12 using these *K* values for 25°C and **I**1 *M*. (We drop the comma in the *K* subscripts in the following.)

$$\log K_{11} = -9 \qquad \log K_{22} = -10.7 \qquad \log K_{12} = -18$$
$$\log K_{13} = -28 \qquad \log K_{14} = -38.8$$

In contrast with mononuclear diagrams, these change as the total metal ion *C* is changed. High *C* favors the polynuclear species, $\alpha_{2,2}$. Compare **Fig. V-12*a*** with the 10^{-5} *M* Cu diagram in **Fig. V-12*b***.

Note that dilution favors mononuclear species so that the 1, 1 becomes visible in the last case and has the same range as the higher 2, 2 since 1, 1 is a function of $[Cu^{2+}]/\mathbf{H}$ and the 2, 2 is a function of $[Cu^{2+}]^2/\mathbf{H}^2$; they rise and fall together.

Both Cu(II) diagrams are metastable since such solutions are supersaturated with basic Cu salts over much of the **pH** range. The solubility diagrams are shown in Chapter VI.

The variation of n̄ with total C of Cu(II) is calculated from Eq. V-12 and the same published K values and is shown in **Fig. V-13**. Remember that n̄ vs. **pH** is the same for any total Cu(II) for mononuclear systems. Thus, experimental determination of n̄ for several solutions of different C_M will distinguish the mono- and polynuclear cases. The long region of $\alpha_{2,2}$ dominace shrinks with C_{Cu}. Note that $\bar{n} = 1$ for the 2, 2 species and $\bar{n} = 4$ for $Cu(OH)_4^{2-}$. The predicted n̄ curves in Fig. V-13 reflect the dinuclear OH species predominance at low **pH** where $\alpha_0 \cong 1$. Compare $\alpha_{2,2}$ in Figs. V-12*a* and V-12*b*.

The data obtained in the preceding seem to agree with this formulation in the separate n̄ lines at the start of **pH** increase in the data in Fig. V-10. However, the 1, 1 species seems not truly absent but appears so at higher total Cu(II) concentrations.

Pure Water pH of Cu(II) Solutions

Note: In this section n̄ is $\bar{n}_{OH}$, bound OH^- per metal ion.

One might ask what species and **pH** are found in pure water solutions of "neutral" Cu(II) compounds, say, $Cu(NO_3)_2$. (See the first datum in each solution, A and B, of the experimental results presented in the table earlier in this section.) Proton balance requires that each OH^- taken from water must supply new hydronium to the solution so that **D** is a measure of this, and $\bar{n}_{OH} = \mathbf{D}/C_M$, since there is no other source of protons. Referring to the preceding equations for K and n̄, the bound OH^- is

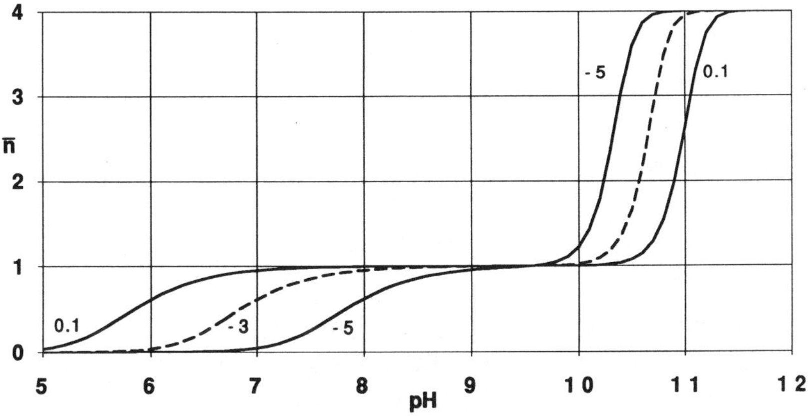

Fig. V-13. Predicted n̄ curves for three total Cu(II) concentrations: 10^{-1}, 10^{-3}, and 10^{-5} *M*.

$$\mathbf{D} = [\mathrm{CuOH^+}] + 2[\mathrm{Cu_2(OH)_2^{2+}}] + 2[\mathrm{Cu(OH)_2}] + 3[\mathrm{Cu(OH)_3^-}] + 4[\mathrm{Cu(OH)_4^{2-}}]$$
$$= \alpha_0 C_\mathrm{M}\{K_{1,1}/\mathbf{H} + 2K_{2,2}\alpha_0 C_\mathrm{M}/\mathbf{H}^2 + 2K_{1,2}/\mathbf{H}^2 + 3K_{1,3}\mathbf{H}^3 + K_{1,4}/H^4\}$$

For any chosen C_M, we need only calculate the { } on the spreadsheet showing the α_0 already used for the preceding plots α (Figs. V-12*a*, *b* and find the **pH** at which the right side bound $\mathrm{OH^-}$ equals **D**. Doing that for several C values including the A and B experimental ones, we get the following:

C_M	pH	α_0	$\alpha_{2,2}$	$\mathbf{pH}_\mathrm{approx}$
0.100	4.13	0.99926	0.000725	4.13
0.025	4.53	0.9988	0.00114	4.53
0.010	4.80	0.9984	0.0016	4.80
0.0051	4.99	0.9980	0.00194	5.00
0.0001	6.10	0.9876	0.0107	6.13

Note the small amount of reaction, most $\mathrm{Cu^{2+}}$ remains aqua ions, $\alpha_0 \cong 1$. Also note that we add $0.08(-\log f_+)$ to these **pH** values to get $\mathrm{pa_H}$, the pH meter readings in the experimental A and B solutions for cases 2 and 4 which become $\mathrm{pa_H}$ 4.61 and 5.07, in rather good agreement with the 4.68 and 5.05 for poorly buffered solutions.

Looking back to the plots and equations of this Section, we note that for these cases, at rather acidic **pH**, the major Cu(II) product is the 2, 2 complex. Trying this as an approximation produces the relation

$$\mathbf{D} = 2[\mathrm{Cu_2(OH)_2^{2+}}] = 2K_{2,2}\alpha_0^2 C_\mathrm{M}^2/\mathbf{H}^2$$

Taking **D** as **H** and α_0 as 1.00 yields

$$\mathbf{H}^3 = 2K_{2,2}C_\mathrm{M}^2 \qquad \mathbf{pH}_\mathrm{approx} = (10.4 - 2\log C_\mathrm{M})/3$$

The values in the last column were found from this. It was no trouble to use the full function since the α terms vs. **H** were already on a spreadsheet. However, with different values of constants and C's, examination would be required to determine which species are truly negligible before approximations could be justified.

Comments

Determination of these species and constants is made difficult by precipitation of basic solids, $Cu(OH)_2$, CuO, and $Cu_2(OH)_2CO_3$ (natural malachite), for example. The Zn(II) and Cu(II) precipitates become more soluble in very basic solution forming the anionic species. To examine the intermediate **pH** region, very dilute solutions must be used. The deciphering of many complex cases has been possible in recent years with reliable **pH** instrumentation. Both **pH** and solubility must be determined. The methods and calculation of results are summarized in the book by Baes and Mesmer[9], which shows α and log S plots for all the elements that had been unraveled up to about 1970.

Some old friends that were treated as simple ions until recently in chemistry turn out to be complex: Al(III) is a large complex, possibly $Al_{13}(OH)_{32}^{7+}$, from **pH** 4 to 9 in 0.1 M total Al, while at total 10^{-5} M Al the 1, 1, 1, 2, 1, 3, 3, 4, and 1, 4 species predominate. Other species that proved to be important in aqueous solutions are $Bi_6(OH)_{12}^{6+}$, $Fe_2(OH)_2^{4+}$, and $Fe_3(OH)_4^{5+}$. Coordinated water molecules are omitted in these species formulas. That is, the Al(13, 32) is likely $Al_{13}O_4(OH)_{24}(H_2O)_{12}^{7+}$, a symmetrical species having 12 AlO_6 octahedra (oxygen shared) that is found in precipitated basic solids such as a sulfate.

Some idea of the educated guesses needed and the disagreement of interpretations can be gleaned from recent reexaminations of the Al problem[11] and comparison with the chapter on Al(III) in the book by Baes and Mesmer.[9] Some work by these authors favors the 14, 34 form, ratio 2.43.[12] The recent paper[11] illustrates the data fit with the 13, 32 species but points out that several other large species with OH–Al ratio about 2.46 (13, 32 is 2.46) will work as well ranging in Al numbers from 6 to 14. Note that $\bar{\mathbf{n}} = 2.46$ for this species so that **pH** ($\bar{\mathbf{n}}$) data alone will not identify a single polynuclear species, only the average ratio.

One might ask how such species are identified. The term for the 13, 32 species in the $\bar{\mathbf{n}}$ expression is, for the reaction written,

$$13Al^{3+} + 32H_2O \rightleftharpoons Al_{13}(OH)_{32}^{7+} + 32H^+_{aq}$$

$$\alpha_{13,32} = 13K_{13,32}(\alpha_0 C_M)^{13} / C_M \mathbf{H}^{32}$$

One sees that **H** dependence is extremely sensitive compared with the 1–4 power dependence seen in the previous Cu(II)–OH^- example.

The rate of rise of a ploynuclear species will provide an estimate of this power dependence and of its possible formulas. For Al(III), **pH** ($\bar{\mathbf{n}}$) data were obtained in acid solutions identifying species Al^{3+}, $AlOH^{2+}$, $Al_2(OH)_2^{4+}$, and $Al_3(OH)_4^{5+}$. Then, by solubility of alumina in neutral and basic solutions and by ultracentrifugation, evidence was obtained for $Al(OH)_3$, $Al(OH)_4^-$, and one large polynuclear species having 10–15 Al ions. The 13, 32 and 14, 34 species have been used to fit the data. The slow approach to equilibrium by the large species and by precipitation of alumina have led workers to study the system at high temperatures up to 124°C, where materials are more soluble, reactions faster, and equilibrium constants smaller. See Baes and Mesmer[9] for further details and references on these methods. One further point should be stressed. Reactions involving the large polynuclear complexes are quite slow so that study of Al(III) equilibria with ligands other than OH^- may be severely distorted in the **pH** region where large complexes and supersaturated solutions of hydroxides–oxides form.

V-10. THE MIXED SYSTEM: Cu(II)–NH_3–OH

Bjerrum's deductions on the Cu(II)–NH_3 system detailed in Section V-5 succeeded because he used high *C* (0.5–3 *M*) NH_4NO_3 media that kept **pH** low enough to minimize the interference of OH^- complexing. Here let us combine both ammonia and OH^- complexing. The mixed constants needed were reported by Fisher and Hall[13]. In keeping with the preceding discussion, they had to use measurement of two species in the solutions in order to solve the simultaneous equations for constants. To prevent precipitation, they used very dilute Cu(II) nitrate, $C = 0.0005\ M$, in 0.5 *M* KNO_3, adding ammonia and NaOH to vary ligand and **pH** values. They measured the **pH** (glass electrode) and free $[Cu^{2+}]$ (polarographically, by use of a dropping mercury electrode). This allows use of $\mathbf{F}_0$ methods (Section V-7). The nine species in the model that we shall use are Cu^{2+}, $Cu_2(OH)_2^{2+}$, $Cu(NH_3)^{2+}$, $Cu(NH_3)_2^{2+}$, $Cu(NH_3)_3^{2+}$, $Cu(NH_3)_4^{2+}$, $Cu(NH_3)_3(OH)^+$, $Cu(NH_3)_2(OH)_2$, and $Cu(NH_3)(OH)_3^-$.

[Examination of *K* values at the **pH** range here is required to prove that the higher OH^- complexes of Cu(II), 1, 3 and 1, 4 (Section V-9) are negligible in this particular case.] We express the formation of the 2, 2 complex as

$$2Cu^{2+} + 2H_2O \rightleftharpoons Cu_2OH_2^{2+} + 2H^+$$

Here C_M is the sum of the concentrations of the nine species with the second one doubled. The constants estimated for ionic strength 0.5 are

$$K_{2,2} = \frac{[Cu_2(OH)_2^{2+}]\mathbf{H}^2}{[Cu^{2+}]^2} = 10^{-10.6}$$

The log β_1 to log β_4 values for the four Cu(II) amines are 4.1, 7.6, 10.4, and 12. For example,

$$\frac{[Cu(NH_3)_4^{2+}]}{[Cu^{2+}][NH_3]^4} = 10^{12.4} = \beta_4$$

then

$$\beta_{1,3,1} = \frac{[Cu(NH_3)_3OH^+]}{[Cu^{2+}][NH_3]^3[OH^-]} = 10^{14.9}$$

$$\beta_{1,2,2} = \frac{[Cu(NH_3)_2(OH)_2^+]}{[Cu^{2+}][NH_3]^2[OH^-]^2} = 10^{15.7}$$

$$\beta_{1,1,3} = \frac{[Cu(NH_3)(OH)_3^+]}{[Cu^{2+}][NH_3][OH^-]^3} = 10^{16.3}$$

Solve each for the complex species and enter into the nine-term C_M sum divided by $[Cu^{2+}]$ to get the $\mathbf{F}_0$ equation, $\mathbf{F}_0 = 1/\alpha_0 = C_M/[Cu^{2+}]$ ($[NH_3]$ is designated as **A**):

$$\mathbf{F}_0 = 1 + 2K_{22}[Cu^{2+}]/\mathbf{H}^2 + 10^{4.1}\mathbf{A} + 10^{7.6}\mathbf{A}^2 + 10^{10.4}\mathbf{A}^3 + 10^{12.4}\mathbf{A}^4 + 10^{14.9}\mathbf{A}^3\mathbf{OH} + 10^{15.7}\mathbf{A}^2\mathbf{OH}^2 + 10^{16.3}\mathbf{AOH}^3$$

Thus, their experimental data for $C_M/[Cu^{2+}]$ and **pH** would give access to the constants. The free **A** must be estimated from the starting amount and preliminary constants, and **A** and **OH** must be varied independently to evaluate the last three constants because of the direct relation of **A** and **OH** via K_a (see next example). Here, we want to examine the prediction of species in various mixtures.

For example, let us take the given initial 0.0500 *M* $[NH_4^+]$ and examine three cases of total Cu(II): 0.001, 0.005, and 0.01 *M*. We add NH_3 and make the equilibrium $[NH_3]$ a chosen variable for calculation. The total ammonia that must have been added can then be calculated from the equilibrium species found. By use of the K_a for the ammonia system ($10^{-9.2}$) and pK_w 13.8, we can express **H** or

OH in terms of free ammonia **A** with the constant ammonium ion concentration (0.05 M) included in the exponents (as -1.30) for the first-approximation calculation:

$$\mathbf{H} \cong 10^{-10.5}/\mathbf{A} \qquad \mathbf{OH} \cong 10^{-3.3}\mathbf{A}$$

We include a corrected $[NH_4^+]$ calculation column in the spreadsheet to check variations caused by the OH^- complexes: each OH^- complexed can produce a proton that must then appear as new NH_4^+ or in **D**:

$$\mathbf{F}_0 = 1 + 2(10^{10.4}\mathbf{A}^2[Cu^{2+}]) + 10^{4.1}\mathbf{A} + 10^{7.6}\mathbf{A}^2 + 10^{10.4}\mathbf{A}^3 + 10^{12.4}\mathbf{A}^4$$
$$+ 10^{11.6}\mathbf{A}^4 + 10^{9.1}\mathbf{A}^4 + 10^{6.4}\mathbf{A}^4$$

Then dividing each term successively by $\mathbf{F}_0$ gives the nine α values wanted for the plot in **Fig. V-14*a–c***. We finesse the copper term by iterating it as $\alpha_0(C_M)$. The spreadsheet calculation will handle this here (but not for $\mathbf{H}^{32}$ of the Al problem) if the line of α values is first calculated with 0.01 for $[Cu^{2+}]$. Then change the preceding formula to $\alpha_0(0.01)$ for $[Cu^{2+}]$ refering to α_0 later in the same spreadsheet line. The ITERATE command (a loop device) will then recalculate $\mathbf{F}_0$ and the rest until a constant set is found. Depending on powers and the K values, oscillation may occur and other subterfuge must be

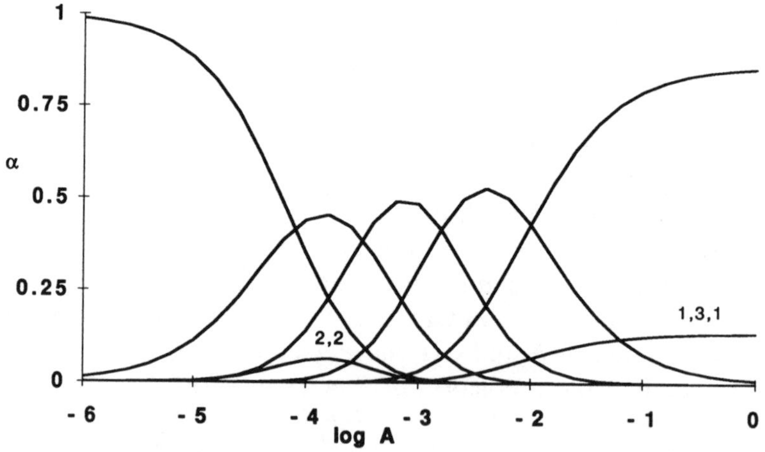

Fig. V-14*a*. Plot of α's vs. the logarithm of free ammonia for 0.001 M total Cu(II), 0.05 M NH_4^+. The upper curves are the five simpler copper species, Cu^{2+} through the four ammonia complexes; $Cu_2(OH)_2^{2+}$ is marked 2, 2 and $Cu(NH_3)_3(OH)^+$ is 1, 3, 1.

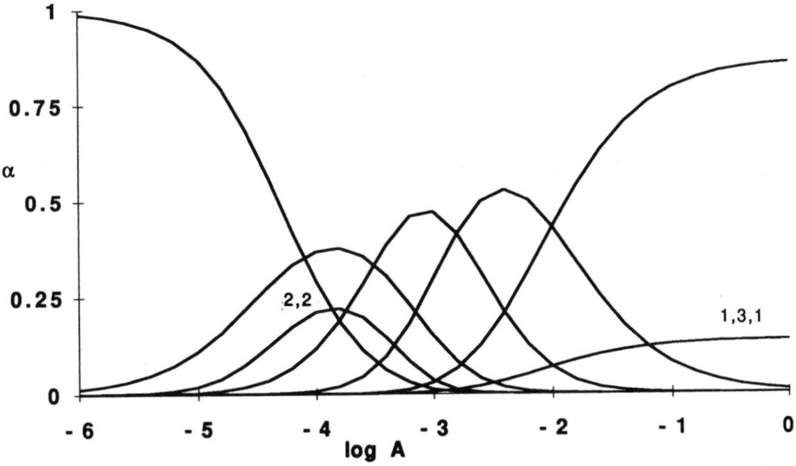

Fig. V-14*b*. Total copper(II) is 0.005 *M*. Initial NH_4^+ is 0.05 *M*. Note higher rise of dinuclear metal species.

found. (This refers to Microsoft Excel on Macintosh Plus.) Figures V-14*a*–*c* show the trends.

[This approximate calculation shows qualitatively the shifts of species as ammonia is added. Ammonia added is equal to the total bound ammonia plus **A**. The ammonia added toward the right side of the plot is nearly the same as unbound **A** since added ammonia is so much larger than the Cu(II) available to bind. Here the Cu^{2+}

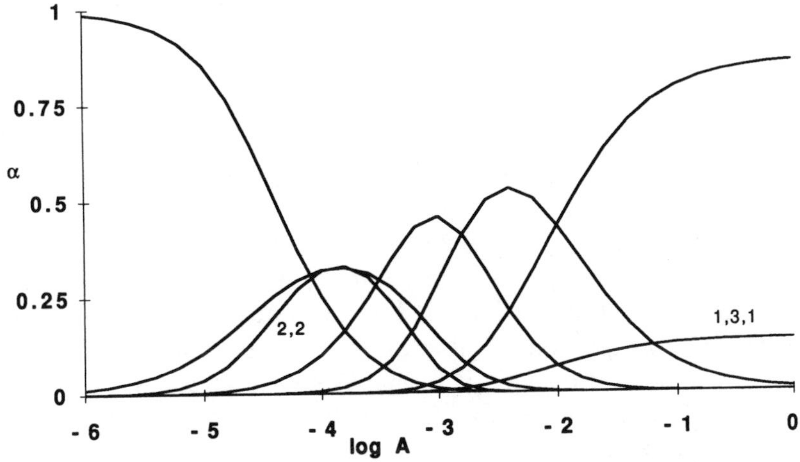

Fig. V-14*c*. Copper(II) is 0.01 *M*; initial NH_4^+ is 0.05 *M*.

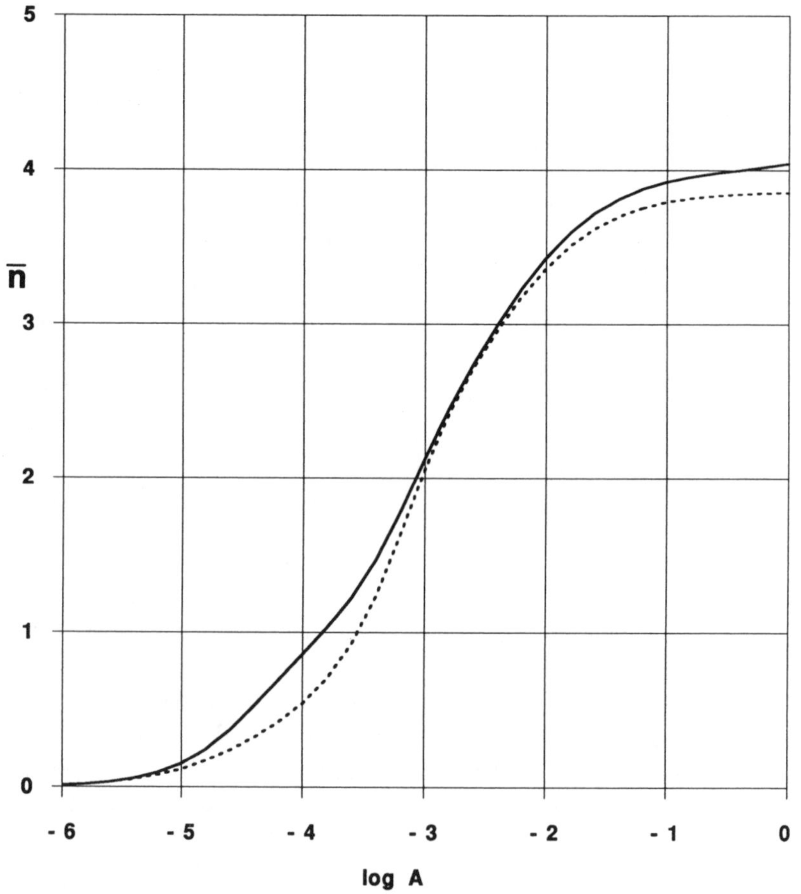

Fig. V-14*d*. Apparent $\bar{n}$ calculated as described in text: thin line, ammonia bound to species; bold line, bound ammonia as would be found by **pH** which counts new NH_4^+ as ammonia lost by binding.

remains 100% in the two highest bound ammonia complexes. The lower mixed complexes of NH_3 and OH^- never achieve 1% of the total and are not shown. The approximation for **A** and **OH** is not exact while the OH^- complexing is high since the protons released by this make more NH_4^+. See the exercises at the end of the chapter. However, our choice of $[NH_4^+]$ significantly greater than $[Cu^{2+}]$ means that production of new ammonium is always much smaller than the 0.05 *M* present. This makes the approximate **H** and **OH** adequate for graph calculations. A better approximation for ammonium ion could use

$$2[Cu^{2+}] + 2NH_3 + 2H_2O \rightleftharpoons Cu_2(OH)_2^{2+} + 2NH_4^+$$

for which $K_{eq} = K_{22}/(K_{a,N})^2 = 10^{7.8}$ using K_a for the ammonia system, $10^{-9.2}$. Then, assuming x and $2x$ for the two products will allow iteration to find total ammonium, $0.05 + 2x - \mathbf{D}$.]

The three cases graphed in Figs. V-14*a–c* show the rise in the dinuclear copper species as total Cu(II) increases.

As ammonia is added, a sequence of log(free **A**) choices on the abscissa are used to calculate species and the resulting total ammonia and **pH**. Note that the 2, 2 species forms at the expense mainly of CuA^{2+}. As **A** rises, the ammonia complexing reduces free $[Cu^{2+}]$, which affects the dimeric species as a square. In these spreadsheets, **pH** ran from about 4 to 10.5 and the truer ammonium from 0.0497 to 0.053 *M* in the region of high 2, 2 formation. The first occurs because Cu^{2+} can bind some of the small $[NH_3]$ in equilibrium with the initial ammonium present, lowering ammonium below the 0.05 *M* put in.

The apparent $\bar{\mathbf{n}}$ for bound ammonia per Cu(II) was calculated in two ways and plotted in **Fig. V-14*d***. First is the total NH_3 put in as given by the sum of α's of ammonia containing species minus free ammonia over the denominator $0.01C_M$ for the third mixture (Fig. V-14*c*). Second is the corrected ammonium ion (in the preceding paragraphs) minus 0.05 added to the numerator to give the $\bar{\mathbf{n}}$ that would be found by **pH** measurements. This goes above 4.00, showing how a Bjerrum-type experiment might overestimate $\bar{\mathbf{n}}$. Note that the $\bar{\mathbf{n}}$ is overestimated by the **pH** method in the regions of high OH^- complexing. (Compare Fig. V-14*c*.)

Now we can address the question of whether OH^- complexes should be present in Bjerrum's early work on the Cu(II)–NH_3 complexes (data in Section V-5) and produce false results for the purely ammonia complexing model assumed. In the first and last data points in the table of his results we calculate α_0 and $[Cu^{2+}]$ to use in the 2, 2 and 1, 3, 1 K expressions (those containing OH^-) to make rough estimates of the equilibrium concentrations of the complexes. For **pH**, the K_a expression for ammonium was used as before but with Bjerrum's 2 *M* NH_4NO_3 and the experimental $[NH_3]_{eq}$ he found, $\mathbf{A}_{expt}$:

Cu_{tot}	$\mathbf{A}_{expt}$	α_0	$[Cu^{2+}]$	pH	pOH	$Cu_2(OH)_2^{2+}$	CuA_3OH^+
0.02057	2×10^{-5}	0.6	0.01	4.2	9.6	$10^{-6.2}$	10^{-11}
0.1254	0.2447	10^{-10}	10^{-11}	8.2	5.6	10^{-16}	$10^{-3.7}$

Thus, we find negligible interference from the $\mathbf{OH}^-$ complexing over the range of Bjerrum's experiments. However, the effect of the 1, 3, 1 complex on the free **A** level has a fourth-power dependence as shown in the preceding $\mathbf{F}_0$ equations. A calculated $\bar{\mathbf{n}}$ of 4 would result from **pH** measurements in a Bjerrum experiment if the mixture contained only CuA_4^{2+} and CuA_3OH^+. Although the effect of the 1, 3, 1 complex is not so small at high $[NH_3]_{eq}$, the slightly high $\bar{\mathbf{n}}$ of 4.002 for the last point reported by Bjerrum seems possible as explained before by the $\bar{\mathbf{n}}$ plots.

SUMMARY

The analogy and differences between protonic acid–base step equilibria and Lewis acid–base, metal ion–ligand complexing are shown. The more common situation has metal ions and protons both competing for the ligand. A straightforward combination of $\bar{\mathbf{n}}$'s of both the HL and ML systems simultaneously treats such mixtures exactly. Determination of $\bar{\mathbf{n}}_{ML}$ and the formation constants is detailed. Polynuclear and mixed complexes are described mathematically and graphically.

EXERCISES

1. A **pH** titration of metal ions via their effect on a solution of pure Na_2H_2Y(EDTA) has been proposed. Calculate the titration curves for 50 ml of 0.05 *M* EDTA by 0.2 *M* NaOH after adding 10.00 ml of 0.15 *M* pure nitrates of (i) Cu(II), (ii) Mn(II), and (iii) Ca(II).

Hint: Here is another example where choice of the independent variable as **pH** followed by calculation (Eq. V-10) of $\bar{\mathbf{n}}_{HL}$ and $\bar{\mathbf{n}}_{ML}$ (here α_1 for CuY^{2-}) and thus the C_H left (initial − NaOH added) is simpler than seemingly direct determination of the **pH** for each addition of NaOH. Iteration is required. The plot for the **pH** titration of Cu(II) case is shown in **Fig. V-15**. We assume that only CuY^{2-} is important, that protonated species are not appreciable though they are known to exist (see the figure).

Take $\log K_f = 18.78$ for $CuEDTA^{2-}$. The pK_a's for EDTA used were 2.00, 2.68, 6.11, and 10.17. (The other two very acid *K*'s are not effective in the range needed here.) See the α diagram of EDTA (Fig. III-18). Note the buffering regions in this case. What EDTA species account for the upper and lower flat regions of the titration curve? Note that, even at 20 ml NaOH, buffering still occurs, showing that NaOH is not yet in excess of some weak acid species (5 ml excess NaOH should produce 1 mmol OH^-/70 ml, 0.0143 *M*, for **pH** about 12 instead of the 10 seen.)

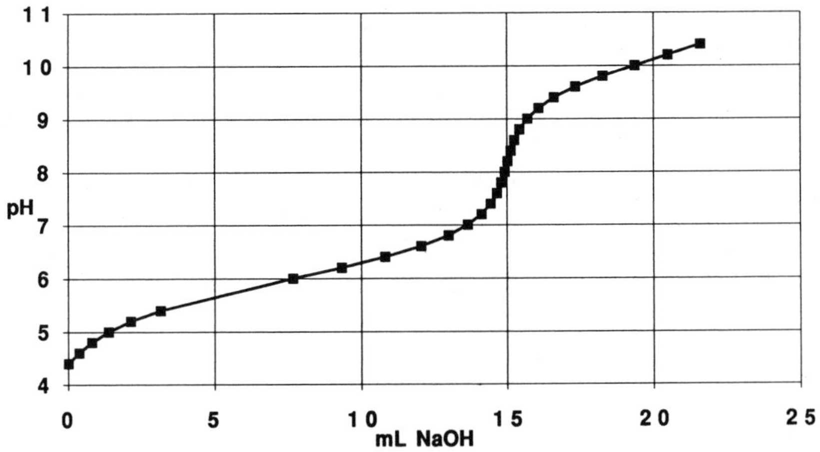

Fig. V-15. Copper titration for Exercise 1: p**H** titration of Cu(II) by NaOH in the presence of excess EDTA.

2. Refer to the $\mathbf{F}_0$ equations in this chapter for the Cu–NH_3–OH diagram and add a full **OH** term using a corrected $[NH_4^+]$ that is the total of the original 0.05 *M* plus 1 for each bound $\mathbf{OH}^-$. Reprogram the calculation and plot your answers. What would be the consequences of having no initial ammonium salt?

REFERENCES

1. Guenther, W. B., *Chemical Equilibrium*, Plenum, New York, 1975. Chapter 9 shows $\alpha, \bar{\mathbf{n}}$ diagrams for 18 systems of *d*-metal ions with halides, CN-, SCN-, and ammonia.
2. Connors, K. A., *Binding Constants*, Wiley, New York, 1987.
3. Bjerrum, J., *Metal Ammine Formation In Aqueous Solution*, P. Haase, Copenhagen, 1941, 1957. Experimental and calculation details of many metal ion–ammonia and other ammine complex systems.
4. Hartley, F. R. et al., *J. Chem. Soc.* (*Dalton*), 469 (1977). Also treated in the text: Hartley, F., Burgess, C., and Alcock, R., *Solution Equilibria*, Wiley, New York, 1980, Ch. 11.
5. Martel, A. E. and Motekaitis, R. J., *Determination and Use of Stability Constants*, VCH Publishers, New York, 1988.
6. Brubaker, G. R. and Busch, D. H., *Inorgan. Chem.*, **5**, 2110 (1966). Discusses the evidence for the species.
7. Daniele, P. G. et al., *J. Chem. Soc., Dalton Trans.*, 1745 (1989).
8. Gans, P. et al., *J. Chem. Soc., Dalton Trans.*, 1195 (1985).

9. Baes, C. F., and Mesmer, R. E., *The Hydrolysis of Cations*, Wiley, New York, 1976.
10. Rossotti, F. J. C. and Rossotti, H., *The Determination of Stability Constants and Other Equilibrium Constants in Solution*, McGraw-Hill, New York, 1961.
11. Brown, P., Sylva, R., Batley, G., and Ellis, J., *J. Chem. Soc. Dalton*, 1967 (1985).
12. Baes, C. F. and Mesmer, R. E., *Inorgan. Chem.*, **10**, 2290 (1971).
13. Fisher, J. F. and Hall, J. L., *Anal. Chem.*, **39**, 1550 (1967).

VI

SOLUBILITY: SATURATED SOLUTIONS OF SLIGHTLY SOLUBLE COMPOUNDS AND DEDUCTIONS FROM SOLUBILITY DATA AND OTHER DISTRIBUTIONS

VI-1. DEFINITIONS AND SIMPLE CASES

Homogeneous solutions, treated in the previous chapters, contain all species at equilibrium in one phase. If such a solution is placed in contact with another phase (solid, liquid, or gas) that can dissolve to supply one or more of the species in the first solution, the equilibrium is displaced and the new phase may become in effect a probe for the transferable species. Most common, a solid like AgCl placed in contact with an ammonia solution will dissolve to saturation with AgCl and will simultaneously form equilibrium activities of all the possible Ag(I) species. The constant activity of the pure substance $AgCl_{(s)}$ forces a balancing constant activity of silver species in the solution. This can be expressed by several equivalent means described in what follows.

The new phase may also be a liquid immiscible with the first solution but that dissolves one or more of its species. For example, the highly associated $HgCl_2$ in aqueous HCl solution forms Hg(II)–Cl^- complexes up to $HgCl_4^{2-}$. Benzene added to the mixture extracts only the uncharged species, mainly $HgCl_2$, until a constant distribution ratio is achieved. This previously determined ratio allows one to determine this species concentration in aqueous mixtures by analysis of the benzene and remaining aqueous solution for Hg(II). The distribution ratio determined using benzene and pure $HgCl_2$ solutions is about 10:1 for the water-to-benzene concentrations.

Thus, by finding C of Hg(II) in benzene, one can say that the actual molecular species $[HgCl_2]$ in the aqueous solution is $10C$. This offers added information for K determination. In addition, some way to estimate the free ligand must be found. Methods like those that follow in solubility studies are used.

The probe phase might be a gas. Bjerrum measured NH_3 pressure over metal–amine solutions (Chapter V). The CO_2 pressure determines equilibrium carbonate species in saturated solutions. Geochemical study of carbonates in seas and caves depends upon this fact. Gas chromatography has been employed for determination of volatile ligands. The probe phase may be reactive solids like ion exchange resins and other chromatographic media. While each of these has valuable applications, solubility is often of higher precision and reliability so that we shall focus on it. The mathematical methods learned will be applicable to the other distribution methods. The techniques are given in detail in the equilibrium texts of the previous chapters and general references listed in the Annotated Bibliography. Our interest here is to outline the broad applicability of α, $\bar{\mathbf{n}}$ mathematical treatment of these heterogeneous equilibrium cases.

Solubility

The familiar K_{s0} (IUPAC), the solubility product K_{sp} in former notation, is a product of uncomplexed ions (0) (Eq. VI-1) and is satisfied in all saturated solutions. However, it is rarely the only condition determining solubility. The AgCl dissolves to AgCl molecues (K_{s1}, or better K_{s11}, the first complex) as well as to ions (K_{s0}). With added NaCl, $AgCl_2^-$ ($K_{s12} = K_{s0}\beta_{12}$) and higher complexes form. In $AgNO_3$ solutions, Ag_2Cl^+ forms (K_{s21}). The solubility of AgCl therefore is the sum of all these forms. In NH_3 solution, AgCl produces some mixed forms like $Ag(NH_3)_nCl_m^{-(m-1)}$ in addition to the others. Only use of all appreciable "side reaction" K terms will produce correct solubility values. These are neatly indicated with α terms, for example, for AgCl,

$$K_{s0} = [Ag^+][Cl^-] = C_M\alpha_{0M}C_{Cl}\alpha_{0,Cl} \qquad \text{(VI-1)}$$

where the C terms are the total analytical concentrations of metal and Cl ions (the solubility S plus any added common ion), and the α_0's are the uncomplexed fraction of each ion. The **pH** may effect ions, and its role may be appreciable. We shall look at successive increasingly complex situations. Precise quantitative descriptions of solubility in terms of equilibrium constants are restricted to compounds of low solubility for which dilute aqueous solution assumptions apply. Saturated NaCl, 6 *M*, is a definite system, but with addition of $HCl_{(gas)}$ the saturation condition varies in complex fashion owing to large change of the ionic medium.

Intrinsic Solubility. This is a constant K_s. An example where this is the major term, K_{s12}, is with HgX_2 [$X = Cl^-$, Br^-, I^-, SCN^-, or CN^-]. Under 1% of $HgCl_2$, for example, gives ions in saturated solution of the compound alone in water.

Simple Ionic Solubility. This is rather rare, requiring ions having low Brønsted and Lewis acid–base strengths. The perchlorates of K^+, Rb^+, and Cs^+ seem near ideal cases.[1] The bromates, iodates, and sulfates of Ag^+, Ca^{2+}, Sr^{2+}, and Ba^{2+} are often treated as simple, but these involve low to moderate ion pairing.[2,3] (The terms *association*, *ion pairing*, and *complexing* can be used without distinction in equilibrium mathematics. Any difference of mechanism has no effect here.) Even the lowest solubility perchlorate has a complication: $RbClO_4$ has $K^0_{s0} = 2.87 \times 10^{-3}$. Thus, its saturated solutions have appreciable ionic strengths. This requires iteration with the log f expressions to achieve consistent solubilities. An example follows. [Throughout this chapter, activity coefficient (f) tables and equations refer to Appendix A, Section A-2.]

Example 1. Find the solubility S of $RbClO_4$ in 0.0400 *M* $HClO_4$ assuming that only the simple ions need be considered:

$$K_{s0} = [Rb^+][ClO_4^-] = S(S + 0.0400) \quad \text{and} \quad \boldsymbol{I} = S + 0.0400$$

$$K_{s0} = K^0_{s0}/f^2_\pm \qquad \log f_\pm = -0.51\sqrt{\boldsymbol{I}}/(1 + \sqrt{\boldsymbol{I}})$$

$$S = -0.0200 + (0.0004 + K_{s0})^{1/2}$$

The last equation is the positive root from the quadratic formula.

One may start with the pK^0_{s0} value, 2.541, insert the first S in the $\boldsymbol{I}$ and log $f_\pm$, and iterate to a final S. The spreadsheet columns can be labeled log $f_\pm$, $f^2_\pm$, K_{s0}, S, and $\boldsymbol{I}$.

The formulas under each will thus refer to as yet uncalculated terms. Choose the MANUAL calculation and ITERATE modes from the Calculation menu. The first $S = 0.037$ *M* (at $\boldsymbol{I} = 0$). Four iterations reach $S = 0.0510$ *M* at $\boldsymbol{I} = 0.091$ *M*. Other expressions for log $f_\pm$ like the Davies equation may yield results different by a few percentage points. (See Appendix A for activity tables and equations.) The log $f_{\pm}$ Debye-Hückel (DH) expression is also close to the Kielland table since it takes the size factor close to 1 for these ions. A less favorable case having final I rather high follows.

Example 2. Find the resulting situation after mixing equal volumes of 0.2000 M RbCl and 0.2000 M $HClO_4$. Note this mixing halves each starting M since volume is doubled. Let x molar precipitate; $S = 0.1000 - x$. This is identical with finding the S in 0.1000 M HCl:

$$K_{s0} = (0.1000 - x)^2$$

Iterating between the ionic strength and K values as before leads to $S = 0.0711$ M using the Davies equation and to $S = 0.0758$ with the (DH) (Guntelberg) equation for log $f_{\pm}$. Now I is about 0.17 M.

VI-2. EXPERIMENTAL DETERMINATION OF K_{s0}

The data[1] for solubility in water and in solutions of various salts, with and without common ions, are shown in Table VI-1. Note the small effects on S of inert ions (NaCl on ionic strength) and larger effects of common ions. Note agreement among mixtures of lower I ($RbClO_4$) and poorer agreement for $KClO_4$ where both too high I (for simple DH treatment) and specific ion effects seem present. Here mixtures with the same I ($HClO_4$ vs. NaCl) give K's that differ well outside experimental uncertainty.

Plots of the $CsClO_4$ data and of the $RbClO_4$ points up to $I = 0.12$ M are shown in **Fig. VI-1**. The apparent empirical pK_{s0} is plotted vs. $-\log f^2$ (or $2pf$) calculated from the Kielland equation. The zero point is added calculated from the first point (pure water solubility) and the Kielland equation. This shows how futile would be an attempt to extrapolate to zero I in these high I solutions. However, the extension seems a plausible slope. The lesson is that the data remain factual, but the K^0_{so} is an interpretation based on a choice of extrapolation. The Kielland (Guntelberg) function does give good results for mean activity coefficients of similar small, singly charged ions in solutions of a single electrolyte at I below 0.1 M. The pure water cases for both of these compounds fall below this limit. However, $KClO_4$ does not, and it shows greater deviations with specific added electrolytes.

TABLE VI-1. Solubility (*M*) of Perchlorates at 25°C

Medium	S	I	$-\log K_{s0}$	Calculated $-\log K_{s0}^0$
		$RbClO_4$		
H_2O	0.0683	0.0683	2.3307	2.5416
0.0200 $LiClO_4$	0.0602	0.0802	2.3164	2.5409
0.0492 RbCl	0.0505	0.0997	2.2981	2.5422
0.0200 NaCl	0.0702	0.0902	2.3073	2.5425
0.0300 NaCl	0.0708	0.1008	2.3000	2.5453
0.0400 NaCl	0.0717	0.1117	2.2890	2.5445
0.05036 NaCl	0.0726	0.1229	2.2788	2.5431
0.0200 $HClO_4$	0.0598	0.0798	2.321	2.546
0.0500 $HClO_4$	0.0500	0.1000	2.301	2.546
0.1000 $HClO_4$	0.0386	0.1386	2.272	2.548
		$CsClO_4$		
H_2O	0.0840	0.0840	2.1514	2.3802
0.0334 CsCl	0.0715	0.1049	2.1249	2.3739
0.0617 CsCl	0.0626	0.1243	2.1088	2.3742
0.0984 CsCl	0.0525	0.1509	2.1003	2.3851
		$KClO_4$		
H_2O	0.1476	0.1476	1.6618	1.9444
0.0506 KCl	0.1277	0.1783	1.6425	1.9447
0.0925 $HClO_4$	0.1111	0.2036	1.6455	1.9620
0.0504 NaCl	0.1523	0.2026	1.6345	1.9505
0.0990 NaCl	0.1540	0.2530	1.6249	1.9656
0.2937 NaCl	0.1626	0.4563	1.5777	1.9881

From the basic equilibrium expression,

$$K_{s0}^0 = [M^+][ClO_4^-]f_{\pm}^2$$

Taking negative logs gives

$$pK_{s0}^0 = pK_{s0} + 1.018\frac{\sqrt{I}}{1+\sqrt{I}}$$

using the Kielland expression (size factor 1) for f^2, as in preceding examples. For $RbClO_4$, a log activity plot using two slightly different f equations is shown in **Fig. VI-2**. From the K_{s0}^0 relation in terms of

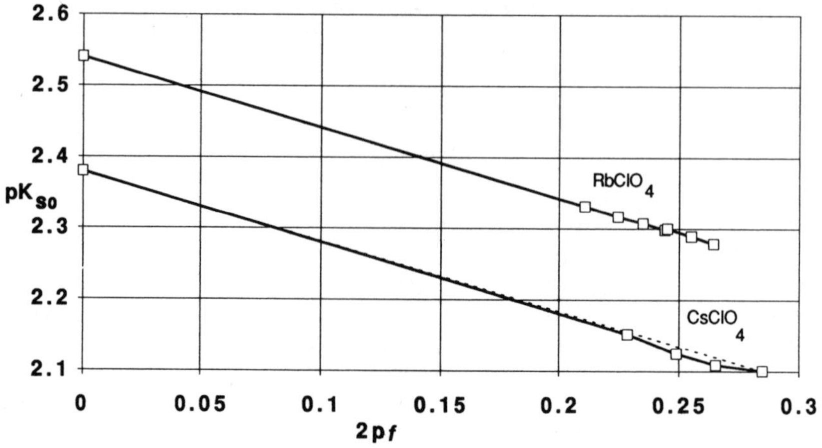

Fig. VI-1. Plots of solubility functions for rubidium and cesium perchlorates.

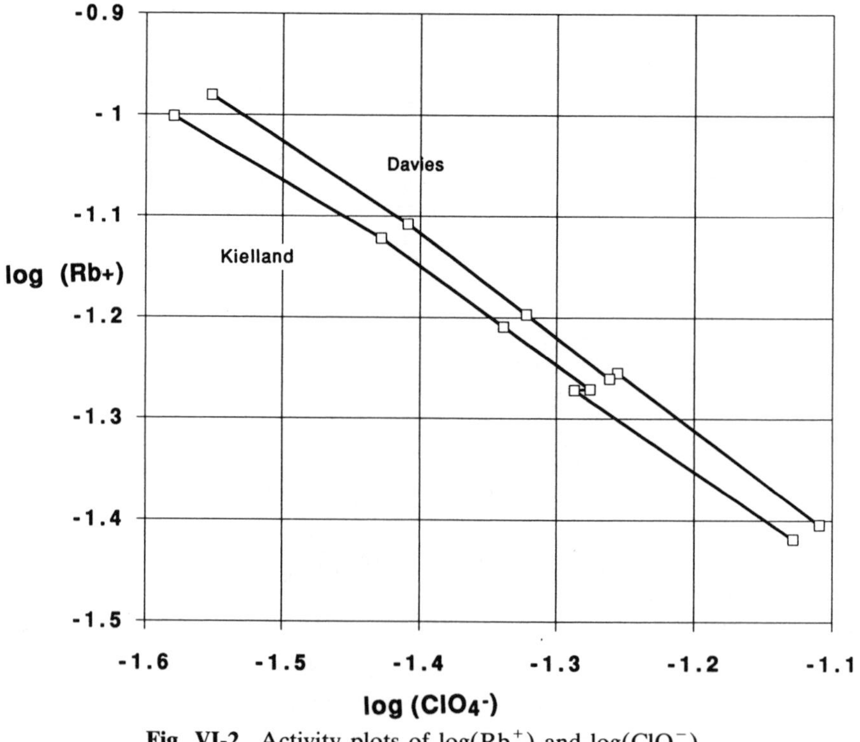

Fig. VI-2. Activity plots of $\log(Rb^+)$ and $\log(ClO_4^-)$.

activities, taking logs yields

$$\log K_{s0}^0 = \log(\text{Rb}^+) + \log(\text{ClO}_4^-)$$

The two variables on the right should plot with slope -1 and should have a constant sum, $\log K_{s0}^0$. The slopes are close to -1, and the sums are 2.55 (Kielland) and 2.52 (Davies), omitting the highest ionic strength point ($\boldsymbol{I} = 0.14$). The lowest ionic strength (0.0683, pure water S) gives a $\text{p}K_{s0}^0$ of 2.542 when extrapolated to zero with the Kielland equation (Fig. VI-1). The experimental K's were found by analysis of Rb content of saturated solutions by a tetraphenylborate gravimetric method precise to about ± 3 parts per thousand. The assignment of K_{s0}^0 is based on the author's estimation of the likely best f equation. In this case (singly charged small ion electrolytes at $\boldsymbol{I}$ below 0.1 M), the Kielland method gives good agreement with experimental mean activity coefficients of pure electrolytes (see Appendix A).

The material $KClO_4$ is more soluble (0.1476 M in water, 25°C) than $RbClO_4$ (0.0683 M). One may conclude that *if* a chosen f equation is valid, *then* the $\log K_{s0}^0$ is about 1.9. The Kielland function plot gives 1.944 and the Davies plot gives 1.899, calculated from the lowest $\boldsymbol{I}$ point (water) with the slope assumed. However, if the

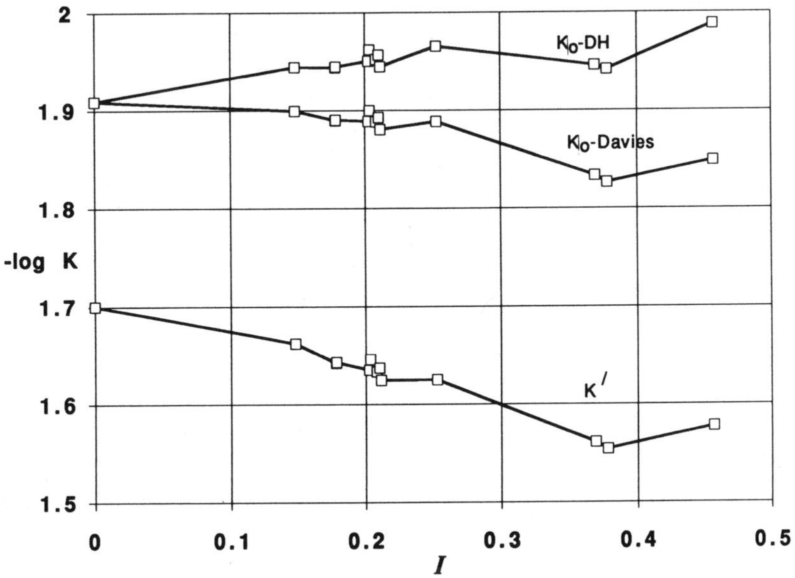

Fig. VI-3. Solubility of $KClO_4$: $\log K_{s0}$ vs. $\boldsymbol{I}$.

highest 3 of the 12 points of data are omitted, the least squares best lines through the 9 lower points extrapolate to −1.91, as shown in **Fig. VI-3**. The data molar K_{s0} (lower curve) and the (activity) K^0_{s0} are calculated from Davies and simple Debye–Hückel equations. The corrected K^0_{s0} would be constant at all $\boldsymbol{I}$ values if the f's were consistent for the different media. The Davies equation shows a more constant vaue. The *slope* of the data log K_{s0} vs. $\boldsymbol{I}$ is not meaningful.

VI-3. IONIC AND MOLECULAR SOLUBILITY

In the presence of varied initial concentrations C of $CaCl_2$, $CaSO_4$ solubility data do not follow simple K_{s0} behavior (see Table VI-2; from ref. 2, Seidell, Vol. I, p. 699), but can be rationalized by assuming that only one complex forms.[4] The intrinsic solubility $[CaSO_4]$ (ion pairs or molecules) is K^0_{s1}, i.e., it cannot vary in saturated solutions of $CaSO_4$ assuming its f remains 1. This goes to equilibrium with ions. Thus the solubility is the sum of forms of sulfate. The needed relations are

$$S = [CaSO_4] + [SO_4^{2-}] \qquad K_{s0} = K^0_{s0}/f^2_{\pm} = [Ca^{2+}][SO_4^{2-}]$$

$$S = K^0_{s1} + (K^0_{s0}/f^2_{\pm})/[Ca^{2+}] \qquad [Ca^{2+}] = S + C - K^0_{s1}$$

$$\boldsymbol{I} = 2[Ca^{2+}] + 2[SO_4^{2-}] + [Cl^-]/2 \qquad [SO_4^{2-}] = S - K_{s1}$$

$$\boldsymbol{I} = 2(S + C - K^0_{s1} + S - K^0_{s1}) + C \qquad [Cl^-] = 2C$$

$$\boldsymbol{I} = 4(S - K^0_{s1}) + 3C$$

$$\log f_{\pm} = -0.51(2^2)[\sqrt{\boldsymbol{I}}/(1 + \sqrt{\boldsymbol{I}}) - 0.3\boldsymbol{I}] \quad \text{(Davies)}$$

[Note that $K^0_{s1} = K^0_{s0}K^0_{f1} = (CaSO_4) \cong [CaSO_4] = K_{s1}$. Thus, the superscript is not significant on K^0_{s1} owing to the lack of charge.] Here K_{f1} is not an independent K:

$$K_{f1} = [CaSO_4]/[Ca^{2+}][SO_4^{2-}] = 1/K_{dis} = K_{s1}/K_{s0}$$

TABLE VI-2. Solubility of $CaSO_4$ in C Molar $CaCl_2$ Solutions 25°C

M $CaCl_2$	S found	M $CaCl_2$	S found	M $CaCl_2$	S found
0	0.0153	0.0150	0.0118	0.0300	0.0104
0.0050	0.0137	0.0200	0.0112	0.0400	0.0097
0.0100	0.0126	0.0250	0.0108	0.0500	0.0093

We ask what values of the constants satisfy the S, C experimental data in Table VI-2. One might iterate by letting $K^0_{s1} = xS$, where x (the fraction of ion pairs) runs from 0 to 1, observing the variations in K^0_{s0} (**Fig. VI-4**). Each change of K^0_{s1} produces a different ionic strength to insert to get K^0_{s0} for each datum in the table. A tedious hand calculation becomes simple on a spreadsheet. Various estimates of K^0_{s1} range from 0.007 to 0.003 M. The value 0.005 for this set of data puts 33% of the dissolved material as molecular in the pure water datum. This rises to 54% in the last case. This greatly modifies the assumed ionic strengths. An earlier calculation (in ref. 4) using the shorter (Kielland) form of the DH log f expression gave poorer consistency with the higher ionic strength points (up to 0.2 M). The plot in Fig. VI-4 was made using the Davies form with an added $\boldsymbol{I}$ term that puts in an upward trend in f above 0.1 M. This produced the crossing in the plot. Checking the result, the plot of the K^0_{s0} expression in the S equation gives excellent agreement, as shown in **Fig. VI-5*a***. Note that in the figure the pure water point is at far right.

Examination of the S as a function of the added C of $CaCl_2$ will reveal generalities that we seek to codify for a wide variety of solubility functions. From the K_{s0} expression with the preceding $[Ca^{2+}]$ and $[SO_4^{2-}]$ relations, we can derive the solubility function for

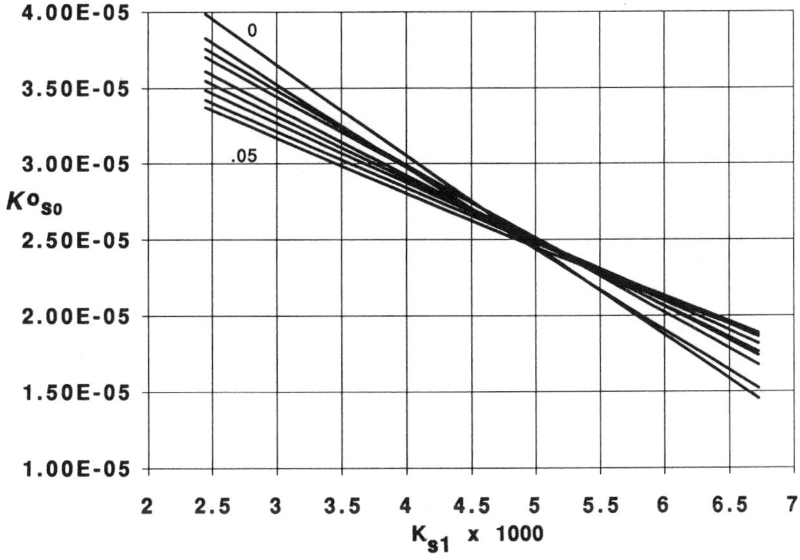

Fig. VI-4. Plot of K^0_{s0} values for $CaSO_4$ calculated using K^0_{s1} values from 0.0025 to 0.007 for each S datum in Table VI-2. The self-consistent area is near 0.0050. Nine experimental S values in the presence of $CaCl_2$ from 0 to 0.05 M are shown.

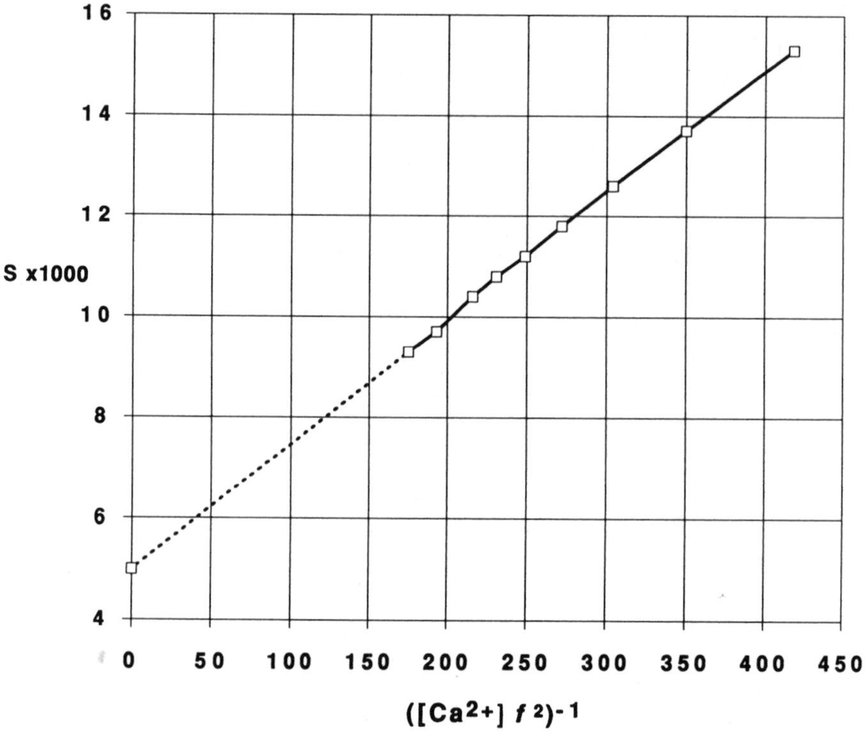

Fig. VI-5*a*. Linearity of K^0_{s0} plot. Slope is 2.48×10^{-5} (K^0_{s0}) and intercept is 0.0050 M, the K_{s1}.

these data:

$$K_{s0} = (S + C - K_{s1})(S - K_{s1})$$
$$S = \sqrt{C^2/4 + K_{s0}} - C/2 + K_{s1}$$

We solved for S taking the positive root of the quadratic formula. Look at the limits: As $C \rightarrow \infty$, $S \rightarrow K_{s1}$ (i.e., ionized forms are repressed). As $C \rightarrow 0$, $S \rightarrow \sqrt{K_{s0}} + K_{s1}$. Testing the last with the pure water case in the preceding (where $f^2_\pm = 0.23$ and $K_{s0} = 1.078 \times 10^{-4}$) gives $S = 0.0104 + 0.0050 = 0.0154$. This is the general case for compounds forming ions and only the first complex. As C increases, S falls to a constant value, K_{s1}. But if further complexing is added (MX_2, MX_3, etc.), the S rises. Examination of the S (or log S data thus allows deductions about the number of complexing steps occurring. Plots of the preceding S equations are shown in **Figs. VI-5*b*,*c***. If there were no $CaSO_4$ molecule formation, simple K_{s0} behavior would give a straight line of slope -1 in the log-log plot (Fig. VI-5*c*).

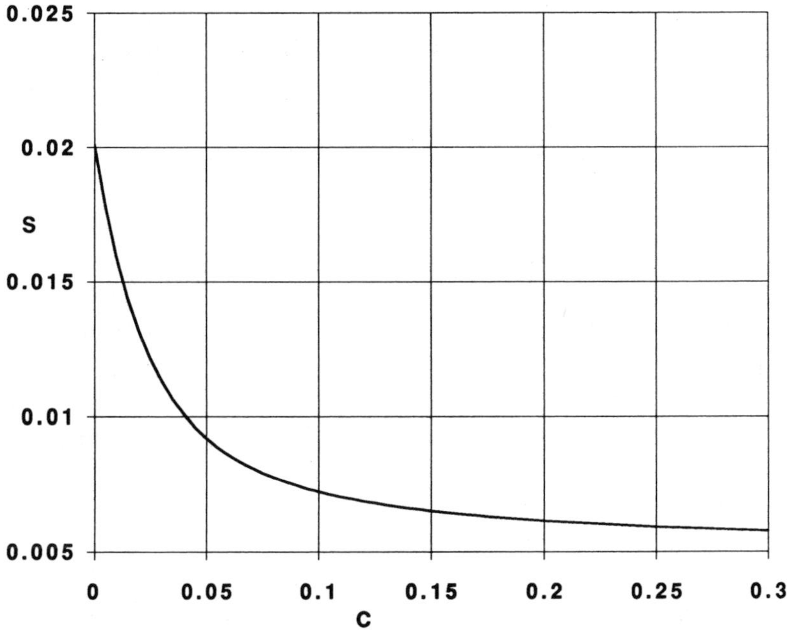

Fig. VI-5*b*. Solubility of $CaSO_4$ vs. added $CaCl_2$.

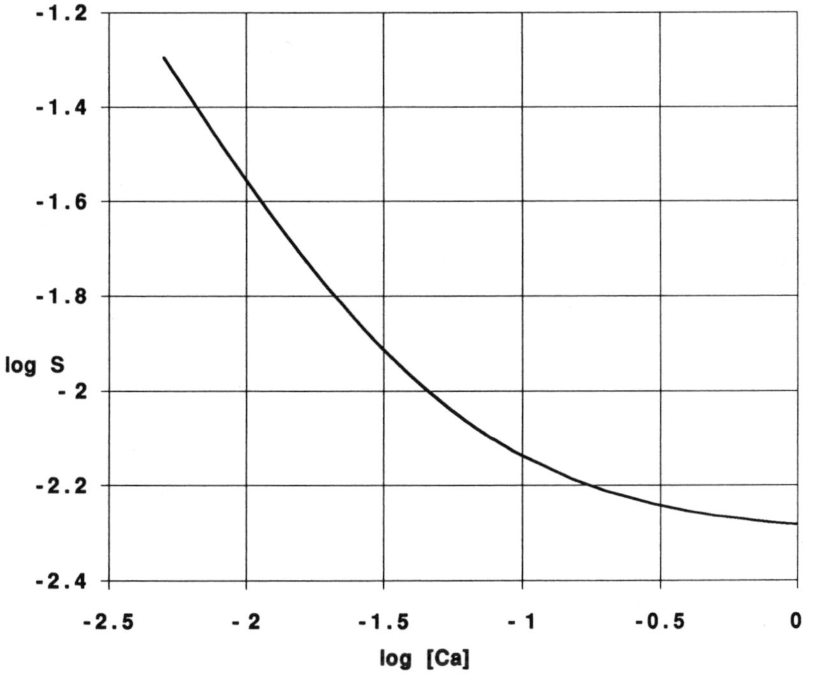

Fig. VI-5*c*. $CaSO_4$: log S vs. log$[Ca^{2+}]$ approaches log(0.005 M), the K_{s1}. The slope approaches -1 at left.

The seeming fine resolution of the early $CaSO_4$ data in the preceding may be deceiving. A study at much higher ionic strength does not neatly add to that data.[5] Yeatts and Marshall varied $NaNO_3$ and Na_2SO_4 keeping $\boldsymbol{I}$ constant: the lowest $\boldsymbol{I}$ group was 0.23 M. The solubility of the calcium sulfate found was rationalized using the following assumptions:

(i) Molecular $CaSO_4$ varied slightly with ionic strength being 0.0047 M at this ionic strength and 0.0038 M at zero $\boldsymbol{I}$.

(ii) Complexing to form $NaSO_4^-$ or $CaNO_3^+$ was assumed negligible.

(iii) Linear plots were obtained using the Kielland function rather than the Davies added term we found workable before.

Using five ionic strength groups from 0.23 to 6 M, they obtained an extrapolated $K^0_{s0} = 3.03 \times 10^{-5}$, pK^0_{s0} 4.519. (4.606 in the preceding, about 20% difference). Considering the high $\boldsymbol{I}$ and 2 − charged common ion, the linearity of their plots is admirable and would seem to justify their assumptions. They titrated total Ca with EDTA. Their paper offers valuable instruction in approaches to this difficult field.

VI-4. SOLUBILITY WITH MULTIPLE COMPLEXING: COMPLEXING WITH ADDED COMMON ION

Many slightly soluble heavy metal ion compounds are made increasingly soluble when high concentrations of the common anion are present. Examples are halides, thiocynates, cyanides, and acetates of silver Ag(I), Hg(II), lead, Cu(I), etc.

Cave and Hume[6] determined AgSCN solubility in water with added KSCN (water solubility about 10^{-6} M). The data were consistent with complexes up to $Ag(SCN)_4^{3-}$, with constants so large that S becomes about 0.1 M in 1 M KSCN.

As before, we can express the dissolving in α, $\bar{n}$ terms. Let C be the added KSCN (KX) and S the moles per liter of AgSCN (AgX) dissolved at saturation equilibrium. Total X is $C + S$. The SCN^- bound in the Ag(I) complexes is $\bar{n}S$:

$$K_{s0} = [Ag^+][X^-] = (S\alpha_{0,Ag})(C + S - \bar{n}S)$$

Assuming the four-complex model, the fraction of Ag(I) uncomplexed is by Eq. V-3

$$\alpha_{0,\mathrm{Ag}} = [1 + \beta_1 \mathrm{X} + \beta_2 \mathrm{X}^2 + \beta_3 \mathrm{X}^3 + \beta_4 \mathrm{X}^4]^{-1}$$

where X, free SCN^- ion concentration at equilibrium, is $C + S - \bar{\mathbf{n}}S$, the total put in minus that complexed to Ag(I) ($\bar{\mathbf{n}}$ times S). For much of the data table, both S and $\bar{\mathbf{n}}S$ are much smaller than C, so that we can use $\mathrm{X} \cong C$ and solve for S to get (with $K_{s1} = K_1 K_{s0}$)

$$S = K_{s0}/\mathrm{X}\alpha_{0,\mathrm{Ag}} = K_{s0}/\mathrm{X} + K_{s1} + \beta_2 K_{s0}\mathrm{X} + \beta_3 K_{s0}\mathrm{X}^2 + \beta_4 K_{s0}\mathrm{X}^3 \quad \text{(VI-2a)}$$

where X is to be corrected as $C + S - \bar{\mathbf{n}}S$ when needed. Reviewing the $\mathbf{F}_0$ methods for evaluating K's in Chapter V (Eqs. V-7 and V-9) will show that similar treatment can be used here ($\mathbf{F}_0 = \alpha_0^{-1}$):

$$S = K_{s0}/\mathrm{X}\alpha_{0,\mathrm{Ag}} = (K_{s0}/\mathrm{X})[\mathbf{F}_0] = K_{s0}[1/\mathrm{X} + K_1 + \beta_2 \mathrm{X} + \beta_3 \mathrm{X}^2 + \beta_4 \mathrm{X}^3] \quad \text{(VI-2b)}$$

The plot of S vs. X experimental data (not logs here) should produce this polynomial curve. The lowest X datum, however, is so high that we are already past the minimum of the curve (**Figs. VI-6** and **VI-7**): $1/\mathrm{X}$ is much smaller than any of the other terms. The last four terms determine the rest of the curve. It turns out to start so near zero S

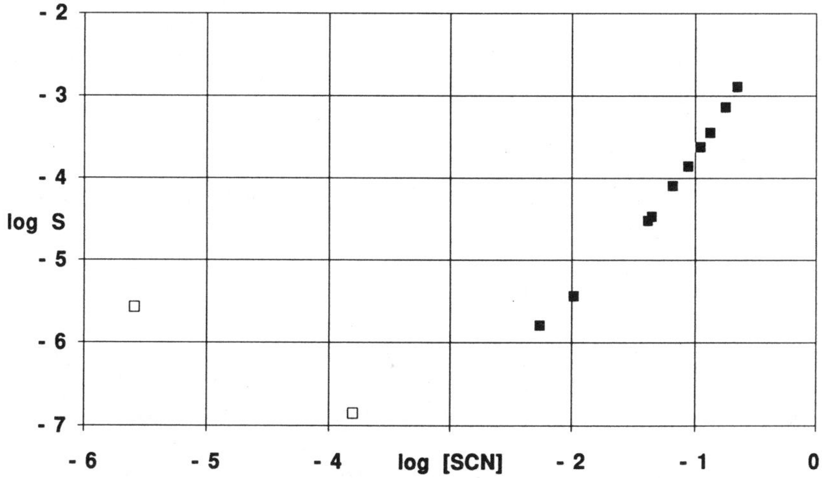

Fig. VI-6. Solubility of AgSCN in KSCN solutions at $I = 2.2\ M$ (KNO_3), 25°C.[1] The two low points (left) were calculated from other estimates of the constants since only the higher [SCN^-] values were detectable in the Cave and Hume experimental method.

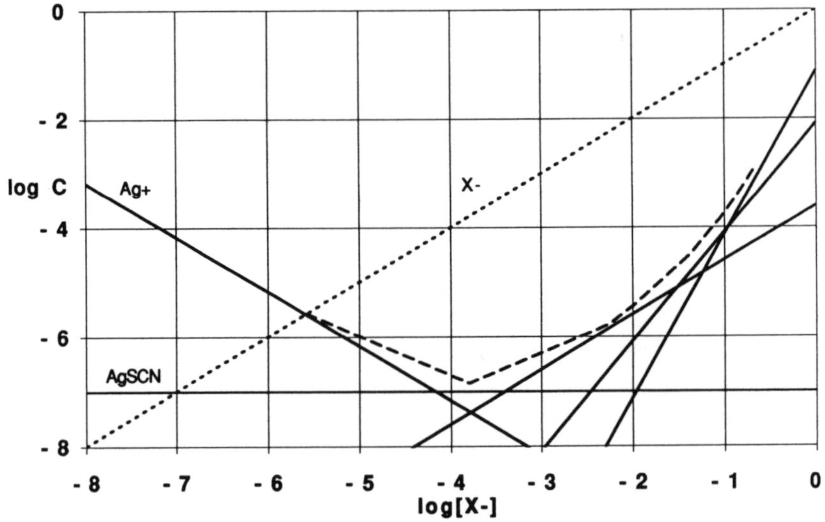

Fig. VI-7. Equilibrium species in saturated AgSCN solutions as a function of $[SCN^-]$, X. The solubility is the sum of all Ag species; as indicated by the small bars in Fig. VI-6. The separate log species lines cannot be deduced until the *K*'s have been determined. The fact that they form an envelope for the experimental solubility points supports the species choices. The solid lines of positive slope are slope 1 for AgX_2^-, 2 for AgX_3^{2-}, and 3 for AgX_4^{3-}.

that K_1 is not determinable. Its slope is approximately β_2 at first (while AgX_2 is predominant). Then as in the method of Chapter V (Section V-7), dropping the first two small terms, one rearranges to

$$S/K_{s0}X = \beta_2 + \beta_3 X + \beta_4 X^2$$

The plot of the determined values of the left term vs. X is a parabola intersecting zero at β_2. Then

$$\frac{S/K_{s0}X - \beta_2}{X} = \beta_3 + \beta_4 X$$

A plot of the left side vs. X is a straight line with intercept β_3 and slope β_4. By this method, Cave and Hume deduced the values

$$\beta_2 = 3.7 \times 10^7 \qquad \beta_3 = 1.2 \times 10^9 \qquad \beta_4 = 1.2 \times 10^{10}$$

This required knowledge of a value of K_{s0} from other work, that is, 6.75×10^{-12}.

Iterations are required since K's and β's are required for $\bar{\mathbf{n}}$ which is then needed for calculating the constants to use for the diagrams. Five constants, K_{s0} and the four formation constants of the AgX_{1-4} species, can be obtained from the solubility data slopes as described in the references and in other examples that follow (Fig. VI-6). The log-log solubility data would show only a straight line (the Ag^+ line in Fig. VI-7) if simple K_{s0} ionic behavior applied. The fact that it goes through a minimum and displays several slopes proves that several successive complexes form, as will become clear in the following treatment and plots.

Examination of the logs of the S equation shows that log S should go through slopes of -1, 0, 1, 2, and 3, corresponding to the five silver species in the solution, if their formation constants are well separated. The fairly continuous curvature suggests that the constants are not well separated. **Figure VI-8** shows a metal–ligand species diagram for the Ag^+–SCN^- system (not necessarily saturated with AgSCN). Note how the major species areas control the solubility (Figs. VI-6 and VI-7) by reducing α_0, which increases S (Eqs. VI-2a,b). The log formation constants relevant to the Cave and Hume data were 4.17, 3.40, 1.51, and 0.96. Note that the crossings on the α diagram and on the log C and log S diagram agree.

A brief exercise in differentiation and substitution with the preceding S, α, and $\bar{\mathbf{n}}$ formulas will show that

$$\frac{\delta \log S}{\delta \log X} = \bar{\mathbf{n}} - 1$$

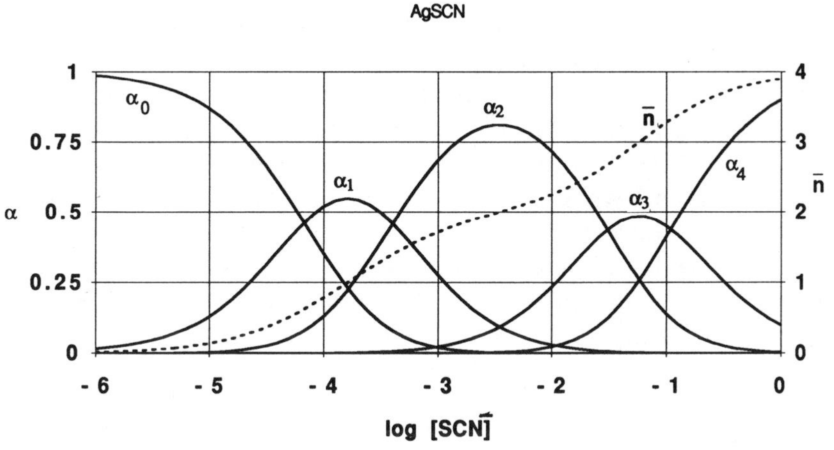

Fig. VI-8. Metal–ligand species diagram for the Ag^+–SCN^- system.

This confirms the more intuitive conclusion in the preceding that the slopes conform to the powers of the X in the S equation. This is seen to be the case on the log plots. Where AgX predominates, $\bar{\mathbf{n}}$ is near 1 and the slope is zero (Figs. VI-6 and VI-7). It then goes through slopes 1 for AgX_2^-, 2 for AgX_3^{2-}, etc. For $\bar{\mathbf{n}}$ near zero, to the left side of the log diagram, the slope goes toward -1.

VI-5. GENERAL SOLUBILITY DIAGRAMS

The number and type of complexing steps determine the shapes of solubility curves. For the single complex in the preceding $CaSO_4$ case S plotted vs. $[SO_4^{2-}]$ shows a decrease to a constant value, that of the K_{s1}, and slope zero. With Ag(I)–SCN^-, a change in curvature occurs as each step becomes predominant. Review them in the light of the following.

Silver acetate solubility in sodium acetate solutions shows the forms in **Fig. VI-9*a*** and **VI-9*b***. If one studied S up to NaAc of about 0.6 M, one might conclude that it behaved very much like $CaSO_4$, and only molecular AgAc and its simple ions are present. The higher acetate concentrations and the log plot shows that more happens. Here we take the published constants for K_{s0} (0.00310), K_{s1}(0.0107), K_{s2}(0.00717), and K_{s3}(0.00246). These are for ionic strength 0.5 M, and the last two are vague estimates. The solubility function is derived as in Eq. VI-2 and is

$$S = 0.00310/[Ac^-] + 0.0107 + 0.00717[Ac^-] + 0.00246[Ac^-]^2$$

The calculations at the high M values exceed the ionic strength of the constants, so they are given only to show what experimental conditions must be reached to detect the last two constants. When AgAc, $AgAc_2^-$, and $AgAc_3^{2-}$ are the major species, the log slopes must become 0, 1, and 2 as shown. One would have to revise the experiment and work at higher $\boldsymbol{I}$. Some results are published for $\boldsymbol{I} = 3$ M. The uncertainties with weak complexing makes the deduction of K values difficult in such cases. Most of the ligand remains free, and the unbound M must be determined with high precision. The reliability of silver electrodes helps here.

General Solubility Curves with Varied Complexing Strengths of the Common Ion

For illustration, let us take AgX, something like AgI or AgCN with pK_{s0} 16, and pure water solubility about 10^{-8} M. Let us examine the

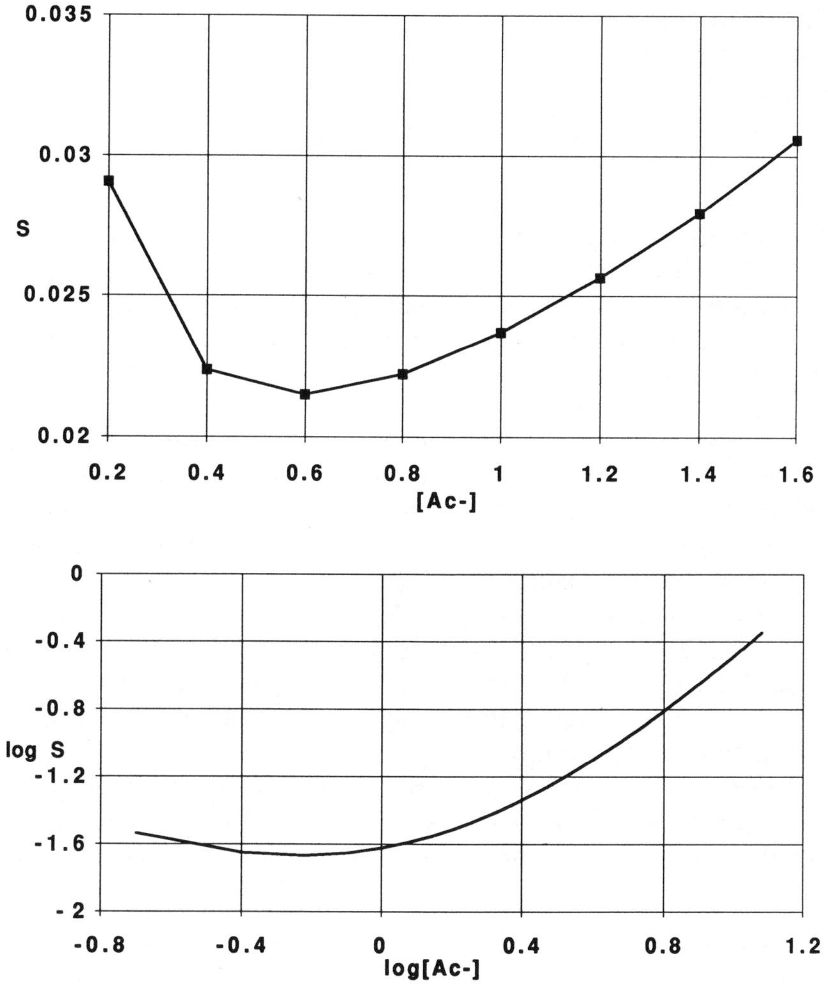

Fig. VI-9. Silver acetate solubility in sodium acetate solutions: (*a*) molarity scales; (*b*) log S vs. log$[Ac^-]$ plot calculated up to 12 M $[Ac^-]$.

S curves for X complexing constants, for example, $K_{s1} = \beta_1 K_{s0}$:

	$\log K_1$	$\log K_2$	$\log K_3$
Case A	6	3	0
Case B	4	0	−1
Case C	11	10	1

Putting these into the solubility equation (VI-2) gives the S functions

$$S = [Ag^+] + [AgX] + [AgX_2^-] + [AgX_3^{2-}]$$
$$= K_{s0}/X + K_{s1} + \beta_2 K_{s0} X + \beta_3 K_{s0} X^2$$

Let $X = [X^-]_{eq}$.

A. $S = 10^{-16}/X + 10^{-10} + 10^{-7}X + 10^{-7}X^2$.
B. $S = 10^{-16}/X + 10^{-12} + 10^{-12}X + 10^{-13}X^2$.
C. $S = 10^{-16}/X + 10^{-5} + 10^{+5}X + 10^{+6}X^2$.

These are shown as **Figs. VI-10*a–c***. Figure VI-10*a* shows the solubility of AgX (pK_{s0} 16) with three steps of X^- complexing with log constants 6, 3, and 0. The slopes relate to the major species as shown on α diagrams: slope -1 below log[X] = -6, slope 0 at -4.5, and slope 1 at -1.5 and slope 2 above 0. Compare AgSCN in the preceding and the following cases of this chapter. Figure VI-10*b* shows AgX solubility with different X^- complexing with log constants 4, 0, and -1. The slopes are -1 at left, changing to 0 after log[X] $-$ 4 and to 1 and 2 at 0, 1. Figure VI-10*c* shows AgX solubility with complexing with log constants 11, 10, and 1. Note that slopes in the regions of species maxima are $-1(\alpha_0)$, $0(\alpha_1)$, $+1(\alpha_2)$, and $+2(\alpha_3)$. The species maxima, as on all α diagrams, occur halfway between $-\log K$ values: at log[X] -10.5, -5.5 for α_1 and α_2. The end species do not have maxima but approach limits below $-11(\alpha_0)$ and above $-1(\alpha_3)$. (*The very high S* to the right are unreal and are shown for the functional relation only.)

The previous examples and AgSCN had pure water solubility close to $\sqrt{K_{s0}}$. Here (Fig. VI-10c), however, complexing constants are high enough to make appreciable AgX and AgX_2^- form. That is, AgX

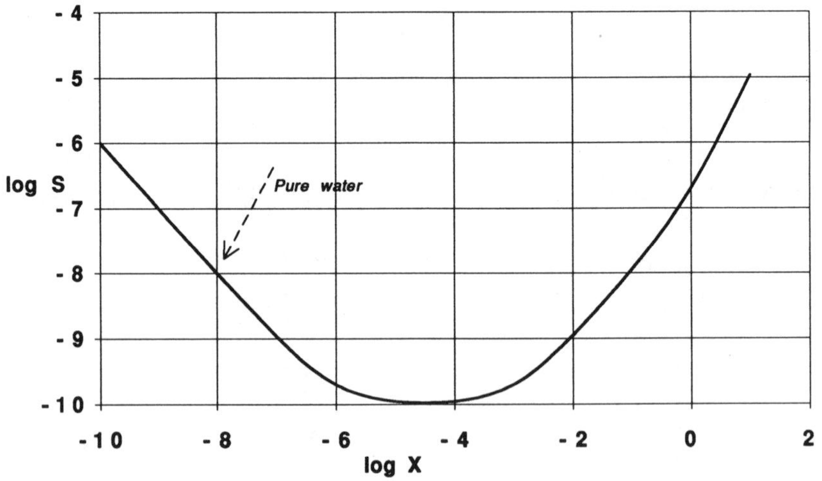

Fig. VI-10*a*. Solubility of AgX (pK_{s0} 16) with three steps of X^- complexing with log constants 6, 3, and 0.

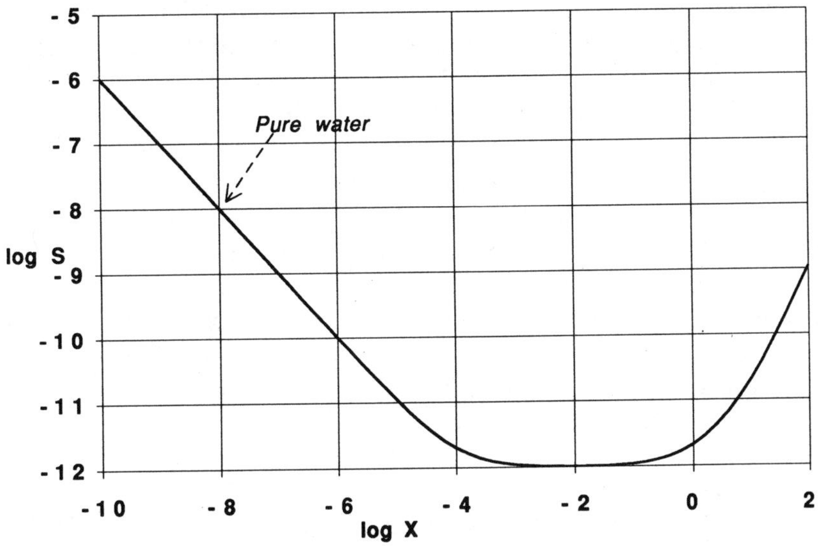

Fig. VI-10*b*. AgX solubility with different X^- complexing with log constants 4, 0, and −1.

dissolves to give $10^{-5.0}$ *M* AgX and $10^{-5.5}$ *M* of both Ag^+ and AgX_2^-. This is deduced from the pure water conditions that total [Ag] = total [X] = *S*. This occurs at $S = 1.63 \times 10^{-5}$ *M* at $[X] = 10^{-10.5}$ *M*. Calculations of *S* and all species lines by the means shown for previous cases is left to the reader for these AgX cases. Note that log Ag and log X lines on all three must cross at −8.

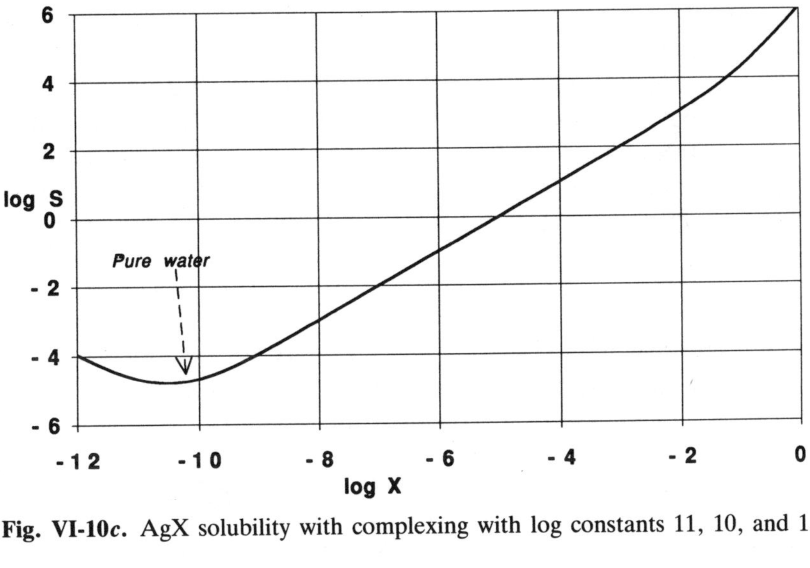

Fig. VI-10*c*. AgX solubility with complexing with log constants 11, 10, and 1.

Thus, we see the effects of high, low, and intermediate complexing by the common ion X. Extension to fewer or more steps of complexing are simple (see, e.g., $CaSO_4$ in the preceding). In the AgSCN data of Cave and Hume (Figs. VI-6 and VI-7), the Δ slopes calculated between pairs of data points (extending to higher added KSCN, to 2.2 M) vary somewhat erratically, in the order −0.7, +0.7, 1.3, 1.5, 1.6, 2.1, 1.9, 2.4, 2.2, 2.5, 2.6, 2.7, 2.97, 2.7, 3.0, and 3.19. A maximum slope of 3 is expected for their complexing model. Slopes are $n - 1$, where n is the number of X groups on the species. They do not offer a reason. In high concentration of ions and charged species like AgX_4^{3-}, some ion pairing may be expected. The difficulties here and in the previous $CaSO_4$ high ionic strength study are typical of efforts to decipher equilibria in such mixtures.

VI-6. SOLUBILITIES OF MX WITH ADDED LIGAND L THAT IS NOT THE COMMON ION X

Eq. VI-I provides a complete solubility function for MX in terms of [L]. An example case is the solubility of AgCl in ammonia solutions. The Ag(I) is successively complexed as the mono and diammine complexes that increase the solubility by lowering $[Ag^+]$. The Ag^+–Cl^- complexes are considered later. The equation is

$$\begin{aligned} K_{s0} &= [Ag^+][Cl^-] \\ &= (S - K_{s1})\alpha_{0,ML}(S - K_{s1}) \end{aligned}$$

To explain the terms, AgCl dissolves to saturation as AgCl molecules (at constant value $K_{s1} = 3.6 \times 10^{-7}$ M) and its ions. If NH_3 is added, its complexes form, but the totals of silver and of chloride are always S. The K_{s1} part is removed from the silver ammonia system leaving $S - K_{s1}$ of free Cl^- and $S - K_{s1}$ of Ag(I) in the silver–ammonia system. The latter has $\alpha_{0,ML}$ of the $S - K_{s1}$ as free Ag(I) ion. Solving for $S - K_{s1}$ gives (with $\beta_1 = K_{f1}$)

$$S - K_{s1} = \sqrt{K_{s0}/\alpha_{0,ML}} = \sqrt{K_{s0}(1 + K_{f1}[L] + \beta_2[L]^2)} \qquad \text{(VI-3)}$$

Note the contrast with Eq. VI-2 in the square root dependence. This is commonly derived from the material balance statement that S is just the sum of all forms of Ag(I) in the solution:

$$S = [Ag^+] + [AgCl] + [Ag(NH)_3^+] + [Ag(NH_3)_2^+]$$

The reader can show that using the K_{s0} and K complex formation expressions transforms this to the function of [L], or $[NH_3]$, given in Eq. VI-3.

One can examine the equation for predictions: At very low ligand concentration,

$$S - K_{s1} \rightarrow \sqrt{K_{s0}}$$

As [L] becomes larger, the $S - K_{s1}$ increases, as shown in what follows.

System Constants

At 25°C and ionic strength 1 M (NH_4NO_3 required to maintain low **pH**, 8–9, so that AgOH will not form by base action of ammonia)

$$\text{AgCl:} \quad pK_{s0}\ 9.30\,, \quad pK_{s1}\ 6.44$$

$$\text{Ag(I)–NH}_3\text{:} \quad \log \beta_1 = 3.37\,, \quad \log \beta_2 = 7.15$$

$$S - K_{s1} = 10^{-4.65}\sqrt{1 + 10^{3.37}[L] + 10^{7.15}[L]^2}$$

As $[L] \rightarrow 0$, the [AgCl] is under 1% of S, which is close to $10^{-4.65}$ Testing intermediate (10^{-3}) and higher ($>10^{-2}$ M) $[NH_3]$ will show that the S becomes proportional to [L] at high values where the last term predominates. This is shown in the plot of the $\log(S - K_{s1})$ function. As before, a differentiation will confirm these slopes, which for these non-common ion complexes become $\bar{n}/2$. **Figure VI-11** shows the solubility and species log plots in the system: saturated AgCl in ammonia solutions with 1 M NH_4NO_3 present: A is the uncomplexed ammonia.

The lines are found by taking logs of $S - K_{s1} \cong S = [Cl^-]$ calculated from the preceding equation for a series of log[A] values running from -6 to 0 in steps of 0.1 log unit. The log[species] are then found from the terms in the preceding species sum. Let A be NH_3:

$$\log[\text{AgCl}] = -6.44 \text{ constant} \qquad \log[\text{Ag}^+] = -9.30 - \log S$$

$$\log[\text{AgA}^+] = 3.37 + \log[\text{A}] + \log[\text{Ag}^+]$$

$$\log[\text{AgA}_2^+] = 7.15 + 2\log[\text{A}] + \log[\text{Ag}^+]$$

Note how the species of Fig. VI-11 follow all equilibrium conditions which link them as follows:

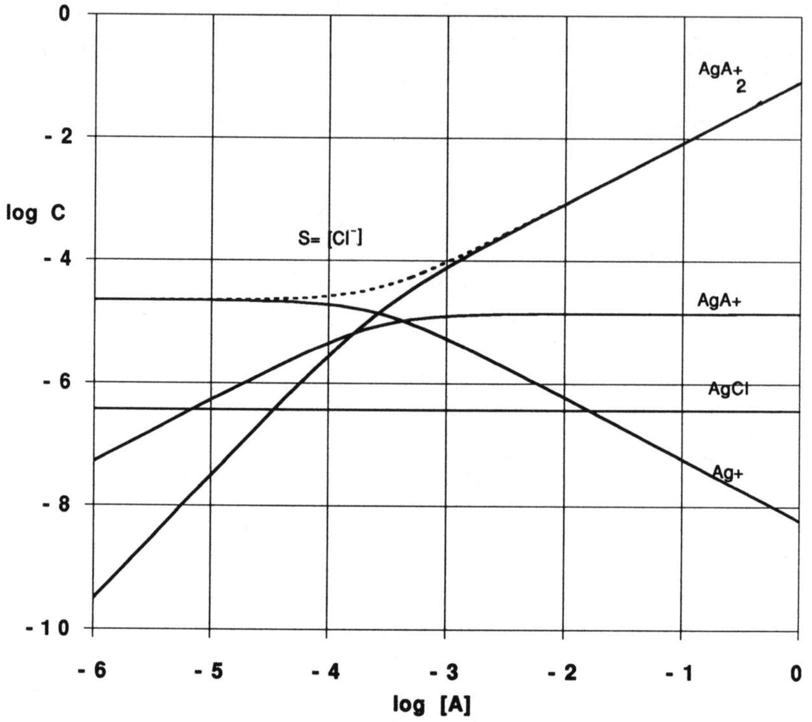

Fig. VI-11. Solubility and species log plots.

(i) The sum of the $\log[Ag^+]$ and $\log[Cl^-]$ lines is everywhere -9.30 log units, as required by the K_{s0}.

(ii) The slope of the $\log[AgA_2^+]$ line becomes 1 at high [A] where S is determined by the $\sqrt{[A]^2}$ term.

(iii) The distance between silver species lines is exactly that on a log ratio or a log α diagram of the Ag(I)–NH_3 system.

The usual caution in all equilibrium work is as follows: The diagram and conclusions from it are only as valid as the assumption that all important equilibria have been included. For example, in the literature we find the constant for

$$AgCl_{(s)} + 2NH_3 \rightleftharpoons AgA_2Cl$$

$$K_{s21} = \frac{[AgA_2Cl]}{[A]^2} = 10^{-2.24}$$

This is for constant [AgCl] (saturated solution). Remember that S

and $[Cl^-]$ rise to about 0.1 M at 1 M NH_3. Does this make enough of this new complex to increase the S? Yes, $10^{-2.24}$ M, or 0.0058 M. Thus, if $[NH_3]$ is increased above 1 M, the mixed complex becomes a few percent of the total. Constants for other species show that $AgACl_2^-$, AgACl, and $AgCl_2^-$ are successively smaller than the first one mentioned. Their presence would be indicated by the experimentally measured S line going higher than the model plot at the right extreme of Fig. VI-11.

VI-7. SOLUBILITY WITH PROTON REACTIONS

Silver Acetate

We consider the solubility of AgAc with varied **pH** (varied by adding, say, HNO_3) but no added common ions. Please see the "Note on α Designations," p. 229. Thus, as inspection of the preceding acetate complexing data shows (Figs. VI-9*a* and VI-9*b*). AgAc is barely significant (K_{s1} adds 0.0107 M at every point), but the di- and tricomplexes are negligible. As before, use $\boldsymbol{I} = 0.5$ M at 25°C. Silver is present as the ion or as AgAc (0.0107 M). Acetate is spread among Ac^- ions and AgAc and HAc. Thus,

$$K_{s0} = [Ag^+][Ac^-] = (S - K_{s1})\alpha_{0,Ag}(S - K_{s1})\alpha_{0,Ac}$$

where $\alpha_{0,Ag}$ is taken as 1 below **pH** 10 because the acidity effect on Ag^+ to form AgOH ($K_{a,Ag} = 10^{-12}$) is weak. The $\alpha_{0,Ac}$ is $(1 + \mathbf{H}/K_a)^{-1}$. Take p$K_a$ 4.48 for HAc at these conditions. So, the solubility function is

$$S - K_{s1} = \sqrt{K_{s0}(1 + \mathbf{H}/K_a)} \qquad \text{(VI-4)}$$

$$S = 0.0107 + \sqrt{0.00310(1 + 10^{4.48-p\mathbf{H}})}$$

The plot of this function in **Fig. VI-12** shows the simplest **pH** effect on a solubility. Note slope −0.5 on the left as predicted by the S function equation. The high **pH** limit on the right is $\log(K_{s1} + \sqrt{K_{s0}}) = \log(0.0664) = -1.178$. The rising line is the S_0 in pure water.

If we ask what the solubility is in water alone, it is a point on this log S line that is at the **pH** produced by that solubility, S_0. Here, $S_0 - K_{s1}$ acetate is released. The **pH** of that solution is given by the usual weak base alone relation (Chapter II), and the HA formed equals −**D**, the new $[OH^-]$ produced:

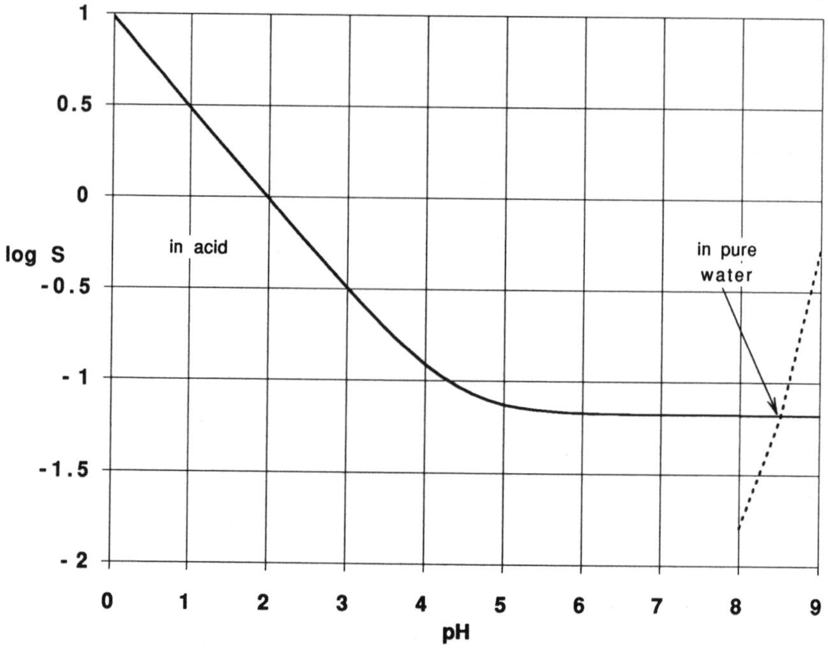

Fig. VI-12. The pH effect on solubility of silver acetate.

$$\alpha_1(S_0 - K_{s1}) = -\mathbf{D}$$

$$S_0 - K_{s1} = -\mathbf{D}(1 + K_a/\mathbf{H})$$

Plotting this S_0 on the graph shows the point that also satisfies the solubility equation. It is 0.0664 M, as expected. For more basic anions, this might appear on the curved portion, as will be seen in another case.

The reasons for the observed S are clarified by sketching in the log lines for HAc, Ac^-, Ag^+, and AgAc following the previous examples (**Fig. VI-13**):

$$\log[\mathrm{AgAc}] = -1.971 \qquad \log[\mathrm{Ac}^-] = \log(S - K_{s1})\alpha_0$$

$$\log[\mathrm{HAc}] = \log(S - K_{s1})\alpha_1 \qquad \log[\mathrm{Ag}^+] = \log(S - K_{s1})$$

At the left, the major reaction is

$$\mathrm{AgAc}_{(s)} + \mathrm{H_3O^+} \rightleftharpoons \mathrm{Ag}^+ + \mathrm{HAc}$$

At the right it is

$$\mathrm{AgAc}_{(s)} \rightleftharpoons \mathrm{Ag}^+ + \mathrm{Ac}^- + \mathrm{AgAc}_{(aq)}$$

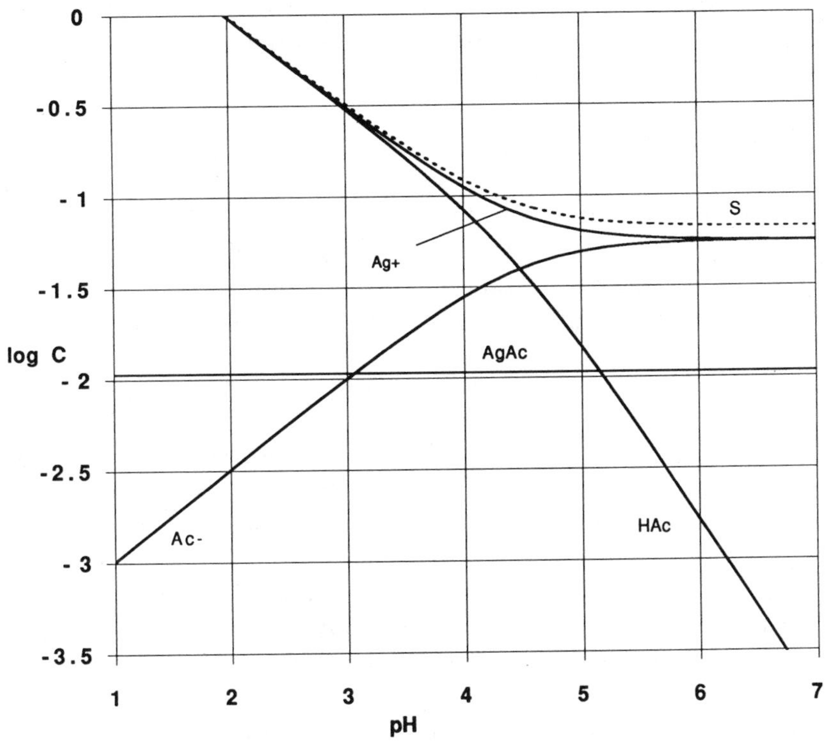

Fig. VI-13. Species in saturated silver acetate solutions with varied **pH**.

Note that the distance between the acetate species lines is exactly that on a log C or a log α diagram of acetic acid. The lines are bent owing to the increasing *total* acetates as one goes to the left.

Calcium Carbonate: The Limestone Problem

Limestone, largely $CaCO_3$, is the most abundant nonsilicate mineral. Its rather high solubility in water (about $10^{-3.6}$ M) and its very basic anion make its solubility highly **pH** dependent in natural waters. The CO_2 affects dissolving of carbonates by supplying protons, causing cave formation and deposit of limestone beds under water as CO_2 escapes. The relations are analogous to the preceding monoprotic example of silver acetate (see Eq. VI-4) and are shown in many of the references. We summarize the equations for this important case. First, in a closed system allowing neither addition nor escape of CO_2 but varying the **pH** by adding acid or buffers without common ions,

$$S - K_{s1} = \sqrt{K_{s0}/\alpha_{0,\mathbf{H}}} = \sqrt{K_{s0}(1 + \mathbf{H}/K_2 + \mathbf{H}^2/K_1K_2)} \quad \text{(VI-5)}$$

$K_{s0} = 10^{-7.53} = [Ca^{2+}][CO_3^{2-}]$, and the α_0 for carbonic acid are used for the $[CO_3^{2-}]$. Small corrections for formation of $CaOH^+$ and $CaHCO_3^+$ can be applied but are omitted here. Here $[CaCO_3] = K_{s1} = 10^{-5.2}$, so $[Ca^{2+}] = S - K_{s1}$. All constants are given for 0.1 M ionic strength. (The K_a's for H_2CO_3 and its α diagram are shown in Fig. IV-7.) This solid does not reach a low constant solubility until above **pH** 11. Therefore, its pure water solubility S_0 is affected by protonation of the carbonate ions. As before, this S_0 is given by the point satisfying the first S relation (Eq. VI-5) and that for the **pH** of dissolved ionic CO_3^{2-} $(S_0 - K_{s1})$ in water alone, with no added acids:

$$S_0 = K_{s1} - \mathbf{D}/\bar{\mathbf{n}}_H$$

The reader might solve and plot these as a spreadsheet exercise analogous to the silver acetate case. (This is shown in ref. 4, pp. 182–190.) Details for the open system, a CO_2 mediated plot, follow.

Open System. Consider equilibrium mixtures of $CaCO_{3(s)}$, $CO_{2(g)}$, and water. Here the pressure of CO_2 gas determines the $[H_2CO_3]$ by the direct proportionality constant $K_pP = [H_2CO_3]$,

$$K_p = 10^{-1.46} = 0.0347\ M/\text{atm}$$

Thus, 0.01 atm P_{CO_2} (P left at equilibrium) would force the $[H_2CO_3]$ to be 3.47×10^{-4} M ($K_pP = [H_2CO_3]$). This restraint adds to that of the K_{s0} relation to determine the total carbonate **pH**, and S, with no added acids or bases. The formation constant of $CaHCO_3^+$ at these conditions has been estimated as 4 (see Eq. VI-6e). It is minor, but we include it to show the effect it has at high P and how to work around the algebraic complexity it introduces. [Ionic strength $I = 0.1$ M is used for all ionic constants.] Take total calcium at equilibrium as

$$[Ca^{2+}] + [CaCO_3] + [CaHCO_3^+] = S$$

Using these relations, $K_pP = [H_2CO_3]$, and the acidity constants K_1 and K_2 for carbonic acid, we get

$$[CO_3^{2-}] = K_{s0}/[Ca^{2+}] \qquad \text{(VI-6a)}$$

$$K_1 = 10^{-6.16} = \mathbf{H}[HCO_3^-]/K_pP \qquad \text{(VI-6b)}$$

$$K_2 = 10^{-9.93} = \mathbf{H}[CO_3^{2-}]/[HCO_3^-] = \mathbf{H}K_{s0}/[Ca^{2+}][HCO_3^-] \qquad \text{(VI-6c)}$$

$$K_1K_2 = \mathbf{H}^2K_{s0}/[Ca^{2+}]K_pP \tag{VI-6d}$$

$$[CaHCO_3^+] = 4[Ca^{2+}][HCO_3^-] \tag{VI-6e}$$

$$[Ca^{2+}] = (S - K_{s1})/(1 + 4[HCO_3^-]) \tag{VI-6f}$$

$$\mathbf{H} = \{[Ca^{2+}]K_1K_2K_pP/K_{s0}\}^{1/2} \tag{VI-6g}$$

The last equation shows that **H** is not independent of P and S and will allow derivation of an equation for S in terms of P. This is achieved by total acidity and total carbon balances. Total C is the sum of carbon from $CaCO_3$ dissolved and from CO_2 dissolved, both at equilibrium. That is, we can describe any solution in this system as having been made from S moles per liter of solid $CaCO_3$ and G moles per liter of CO_2 gas. Set equal to the equilibrium species,

$$\text{Total C} = S + G = [CaCO_3] + [CaHCO_3^+] + [H_2CO_3] + [HCO_3^-] + [CO_3^{2-}] \tag{VI-6h}$$

For total H, dissolved $CO_{2(aq)}$ may be considered to start as H_2CO_3, which then reacts to equilibrium. (The K_a's include this convention.) Thus, in this mixture, the total acidity, C_H, is $2G$ since there is no other source of acid or any strong base added. In terms of the equilibrium acid species,

$$C_H = 2G = 2[H_2CO_3] + [HCO_3^-] + [CaHCO_3^+] + \mathbf{D} \tag{VI-6i}$$

Combining Eqs. VI-6h and VI-6i to eliminate G gives the relation

$$S = [CaCO_3] + [HCO_3^-]/2 + [CaHCO_3^+]/2 + [CO_3^{2-}] - \mathbf{D}/2 \tag{VI-6j}$$

Now we arrive at an S function of P_{CO_2} by inserting relations VI-6a to VI-6g. For algebraic convenience let $S - K_{s1} - [CaHCO_3^+]$ be $\boldsymbol{x}$, the ionic $[Ca^{2+}]$ term in (Eq. VI-6f and the Ca balance), and use **D** as $\mathbf{H} - K_w/\mathbf{H}$ [the sign of $[CaHCO_3^+]/2$ that follows results when $[CaHCO_3^+]$ is subtracted from both sides to make the first $\boldsymbol{x}$ term]:

$$\boldsymbol{x} = K_1K_pP/2\mathbf{H} - 2\boldsymbol{x}K_1K_pP/\mathbf{H} + K_{s0}/\boldsymbol{x} - (\mathbf{H} - K_w/\mathbf{H})/2 \tag{VI-7a}$$

Inserting relation VI-6g to eliminate **H** then gives a relation of $\boldsymbol{x}$ and P. It contains $\boldsymbol{x}$ to the powers 2, $\frac{3}{2}$, and $\frac{1}{2}$ and P to the powers $\frac{1}{2}$ and

$-\frac{1}{2}$. Putting in the values of the constants at 25°C and **I** = 0.1 *M* gives

$$x^2 = (10^{-2.91}\sqrt{P} + 10^{-9.09}/\sqrt{P})x^{1/2} - (10^{-2.31} + 10^{-5.31})\sqrt{P}x^{3/2} + 10^{-7.53} \quad \text{(VI-7b)}$$

Choosing values of *P* and solving iteratively for x (described in what follows) allows a plot of solubility as a function of CO_2 pressure. Then use of the relations for the individual species (Eqs. VI-6) allows showing all of them on the equilibrium diagram in **Fig. VI-14*a***, as has been explained in previous cases. In the figure the H_2CO_3 line is shown for completeness of carbon species even though it is implicit in the abscissa.

Spreadsheet and Plot Details. A representative portion of the spreadsheet follows. There is latitude for the order of entries for iteration. Set Calculation to MANUAL with 10 iterations. Here x^2 is calculated using the form of Eq. VI-7b. This requires values of *P* and x. We put in the set of *P* values desired and a dummy x of 10^{-3} in the first

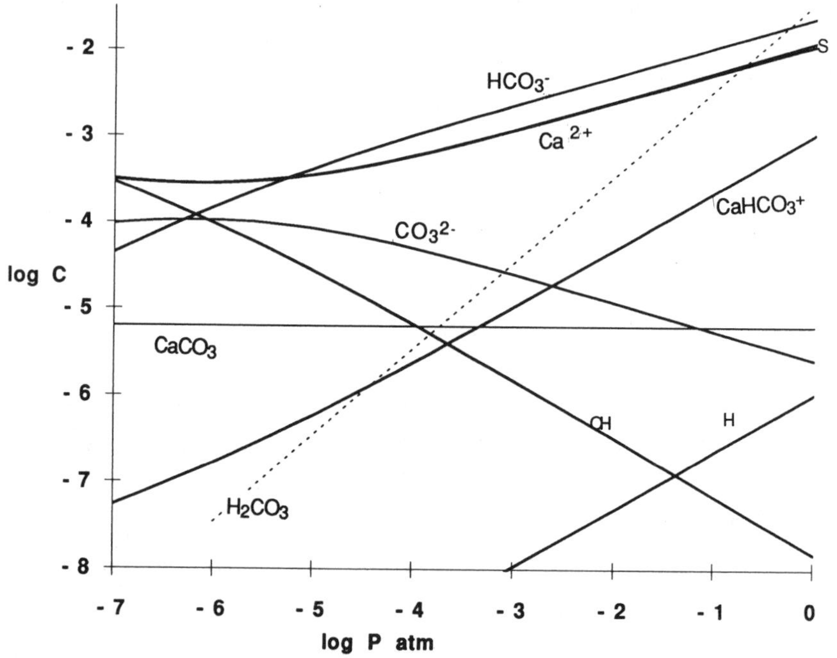

Fig. VI-14*a*. The system $CaCO_{3(s)}$, water, and CO_2: $[H_2CO_3] = K_pP$, $\log[H_2CO_3] = -1.46 + \log P$; *P* is P_{CO_2}. The *S* curve is clearer in the next plot to higher *P*.

column and then calculate the second column as Eq. VI-7b. Next we put in the formulas for any desired species (combining Eqs. VI-6) and replace the first column with $x = \sqrt{\text{column 2}}$. Fill these down and initiate the iteration calculations. (Four iterations gave constant values to better than 1%, while 10 gave five-to-six-figure constancy. The iteration may be repeated to demonstrate constancy.)

The table is in the order needed to plot all the log species vs. log P, selecting the columns from 3 on for plot. To plot line graphs in Excel, choose PASTE SPECIAL for the data selected after the new chart window opens and then choose SCATTER diagram in the linear form (since logs were already in the data). The G values are used in the discussion that follows.

x	x^2	log P	log S	log **H**	log **OH**
3.494E − 04	1.221E − 07	−5	−3.4482	−9.2383	−4.5617
1.171E − 03	1.370E − 06	−3	−2.9253	−7.9758	−5.8242
5.260E − 03	2.767E − 05	−1	−2.2603	−6.6495	−7.1505
1.114E − 02	1.241E − 04	0	−1.9142	−5.9865	−7.8135
2.319E − 02	5.377E − 04	1	−1.5540	−5.3274	−8.4726
4.649E − 02	2.161E − 03	2	−1.1696	−4.6763	−9.1237

log[Ca^{2+}]	log[CO_3^{2-}]	log[HCO_3^-]	log[$CaCO_3$]	log[$CaHCO_3^+$]	log G
−3.4567	−4.0733	−3.3817	−5.2	−6.2363	−3.7110
−2.9316	−4.5984	−2.6442	−5.2	−4.9737	−2.9305
−2.2790	−5.2510	−1.9705	−5.2	−3.6474	−2.0491
−1.9530	−5.5770	−1.6335	−5.2	−2.9845	−1.3296
−1.6347	−5.8953	−1.2926	−5.2	−2.3253	−0.4264
−1.3326	−6.1974	−0.9437	−5.2	−1.6743	0.5484

Figure VI-14*b* illustrates the high pressure CO_2 extreme of the previous solubility function showing an increase of S as $CaHCO_3^+$ becomes significant.

The species solubility diagrams (Figs. VI-14*a*,*b*) allow many predictions and explanations:

1. The species relations to S in Eq. VI-6j are followed (antilogs, not logs).

2. The K_{s0} (Eq. VI-6a) is followed as seen in the symmetry of CO_3^{2-} and Ca^{2+} about a horizontal between them. The sum of their logs is −7.53, the log K_{s0}.

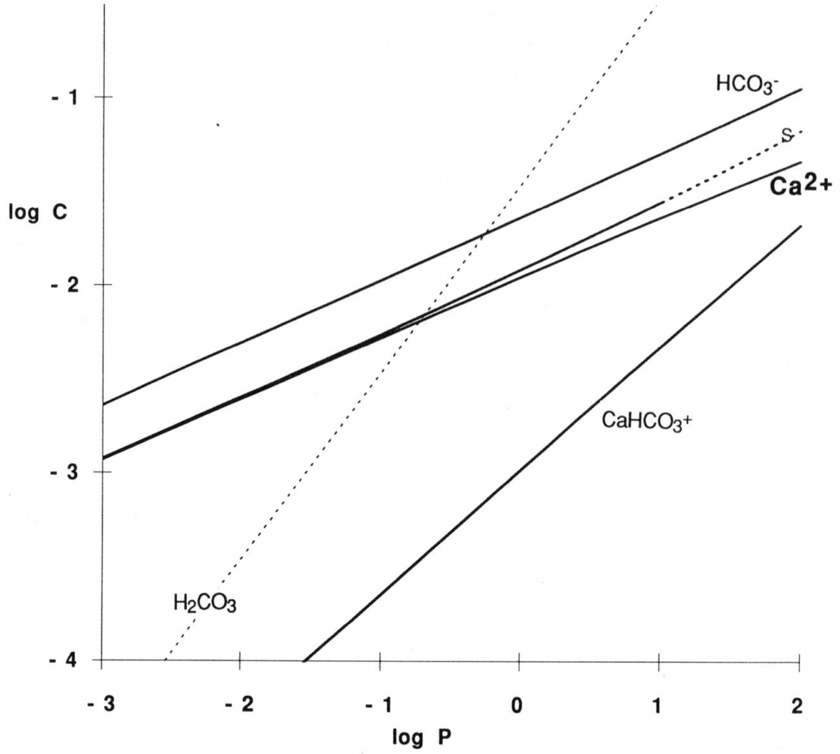

Fig. VI-14*b*. High pressure CO_2 extreme of previous solubility function showing increase of *S* as $CaHCO_3^+$ becomes significant.

3. At the left, where CO_2 and $[H_2CO_3]$ are very small, one can verify that the following equation approximately describes the dissolving:

$$2CaCO_{3(s)} + H_2O \rightarrow 2Ca^{2+} + HCO_3^- + CO_3^{2-} + OH^-$$

The last three are nearly equal at **pOH** 3.9, which is near **pH** 9.93, the pK_{a2} of H_2CO_3 at which $[CO_3^{2-}] = [HCO_3^-]$. Since H_2CO_3 and **H** are negligible, proton balance requires that HCO_3^- and OH^- ions form in equal numbers. (Note that **H** and **OH** lines are not quite straight at the left side.)

This reaction equation also implies that CO_2 has small effect here, and thus, total carbon equals total Ca^{2+}. This is near the point of pure water solubility of $CaCO_3$. But this might imply that no CO_2 is present. The diagram (and equilibrium constants) require that such a solution in pure water not have $P_{CO_2} = 0$. Indeed some dissolving

$CaCO_3$ will form H_2CO_3, and thus CO_2, at equilibrium and begin to exert a gas pressure of about 10^{-6} atm:

$$2HCO_3^- \rightleftharpoons CO_2 + CO_3^{2-} + H_2O$$

Some CO_2 will escape an open container if the opposing CO_2 pressure is lower. In the closed-system dissolution, no CO_2 is allowed to escape or dissolve (no gas space is present). There, $S = 2.45 \times 10^{-4}$ M. In the open system, the minimum S is slightly lower, about 2.3×10^{-4} M, owing to the effect of increased carbonate exceeding the K_{s0}. To the left, the net increase in S can be related to the Le Chatelier shift removing CO_2 gas and thus CO_3^{2-},

$$CO_3^{2-} + H_2O \rightleftharpoons CO_2 + OH^-$$

Note that $[OH^-]$ and $[Ca^{2+}]$ increase at lower pressures while $[CO_3^{2-}]$ decreases.

4. At the right side, dissolving is described well by

$$CaCO_3(s) + H_2CO_3 \rightarrow Ca^{2+} + 2HCO_3^-$$

The HCO_3^- line is about 0.3 units (log 2) above the Ca^{2+} (and S) lines. This is the case well toward the middle at pressures as low as 10^{-4} atm of CO_2. In normal air, about 10^{-3} atm of CO_2, this equation describes the weathering of limestone.

5. In agreement with $S = [Ca^{2+}] + [CaCO_3] + [CaHCO_3^+]$, the S line is close to the Ca^{2+} line but rises slightly above it at either end caused by $CaCO_3$ (left) and by $CaHCO_3^+$ (right).

6. Cave formation is clearly explained on the diagram. Groundwaters running through decaying vegetation with respiring organisms producing CO_2 may become charged with the gas and dissolve limestone at the right. Such solutions dripping out in fresh air with lower P_{CO_2} shift left, and the S falls below its previous value supersaturating $CaCO_3$, which may then deposit on suitable surfaces.

Example 1. If a limestone-saturated water was formed underground at $P_{CO_2} = 0.100$ atm and then was exposed to air at $P_{CO_2} = 0.001$ atm, what quantity of $CaCO_3$ (relative formula mass 100) can precipitate and what volume of CO_2 gas escapes per liter? The S change is 0.00430 M (spreadsheet, subtract antilogs), which makes 0.43 g of the solid. The change in G (Eqs. VI-6h to VI-6j) is 0.00776 M, or about 194 ml gas at 20°C. Note that this is not just the carbon lost by precipitation or the change in total carbon. Using the total carbon balance has simplified this picture. An alternative method of deriving Eq. VI-6j from charge balance confuses the picture and makes the CO_2 loss more complex to deduce (see ref. 4, p. 187).

Example 2. Laboratory confirmation of the *S* relations: Place 1.500 l of water and about 10 g of pure $CaCO_3$ powder in a 5-pint screw-capped acid bottle. Displace the air space with pure CO_2 gas, which has been bubbled through water, and cap tightly. Stir often for a week at 25°C. Note that determination of any one variable species and the p**H** will serve to locate the vertical on the graph (Fig. VI-14 and Eqs. VI-7) and thus indicate the *P* and other dissolved species. Here *P* is the only independently variable quantity if the solution is saturated with $CaCO_3$ and CO_2 at the final equilibrium *P*.

With the pH meter calibrated and ready before opening the bottle, pipet 100 ml of the settled top solution into a tall vessel and read the p**H**. Immediately titrate the same portion with EDTA for total Ca^{2+}, *S*, following standard methods in analysis texts. Calculate whether the measurements agree (fall on the same P_{CO_2} value, now less than 1 atm). Another portion might be pipetted into excess NaOH and back titrated with HCl to determine C_H and *G*. We are assuming that the pipetted solution will not lose gas fast enough to change the p**H**. A method for measuring p**H** directly within the original bottle would be better. (Actual tests of this method have given satisfactory results.) Other points can be tested by adding CO_2 or air for the opposite direction, recapping, and finding a new equilibrium point.

An interesting test might reveal whether the rise of solubility at the left is observable. The $CaCO_3$ solubility in pure water bubbled often with N_2 or air freed of CO_2 could reduce P_{CO_2} to very low values. If the preceding mathematical model is correct, the solubility should rise. However, slowness or formation of $CaOH^+$ (*K* about 20) might affect this.

Ambiguity of the Solid State. In all previous examples we have assumed that the solid state dissolving is a reproducible crystal form. If several crystal states and hydrates exist, they usually have different solubilities. This has accounted for seemingly wide disparities in solubility investigations. In general, one wants the most stable (lowest free energy) solid, which will have the lowest solubility. Higher states will slowly change in solubility as the lower, equilibrium form is produced (see Leussing[7]). For example, $CaCO_3$ exists in at least three crystal forms: calcite, aragonite, and vaterite. The last is rare.

Aragonite is about 10% more soluble than calcite. At some temperatures and in the presence of catalyzing ions, aragonite precipitates from solution rather than the ground state calcite. The other forms very slowly change to calcite in contact with saturated solutions.

Some early solubility data on $CaCO_3$ (in water with measured pressures of CO_2) in Seidell[2] agree with the predicted values derived from preceding equilibrium constants at $I = 0.1\ M$ values, considering the differences in ionic strength. Higher I causes salts in ions.

I	$\log P_{CO_2}$	−3	−1	0	1
0.1 M	$-\log S_{calc}$	2.93	2.26	1.91	1.55
Low (?)	$-\log S_{expt}$	3.11	2.41	2.05	1.65

Magnesium Ammonium Phosphate

$$MgNH_4PO_4 \rightleftharpoons [Mg^{2+}] + [NH_4^+] + [PO_4^{3-}] \qquad pK_{s0}^0 = 13.15$$

$$K_{s0} = [Mg^{2+}][NH_4^+][PO_4^{3-}] = S(\alpha_{1,Mg})S(\alpha_{1,NH_4})S(\alpha_{0,PO_4})$$

Using Davies equation activity coefficients for $I = 0.1$ gives pK_{s0} 11.65 (estimate of 11.5 for 1 M). The interest of this compound (important in analysis) is that all three ions are affected by side reactions dependent upon p**H**:

	$I = 0.1\ M$	$I = 1\ M$ for graph
$Mg^{2+} + H_2O \rightleftharpoons Mg(OH)^+ + H^+$	$pK_{aM} = 11.65$	12.0
$NH_4^+ \rightleftharpoons NH_3 + H^+$	$pK_{aN} = 9.29$	9.5
$HPO_4^{2-} \rightleftharpoons PO_4^{3-} + H^+$	$pK_{a3} = 11.67$	11.0

The preceding table is for the solubility in buffers containing no added common ion (only p**H** is varied). Assuming just the three acid steps shown limits this to the p**H** range in which $H_2PO_4^-$ is negligible, above p**H** 8, which is where the compound precipitates. A fuller treatment would use the complete α_0 expression for the phosphate system. Increased p**H** decreases the concentration of the first two species and increases the $[PO_4^{3-}]$. Thus, we expect the S function to go through a minimum. Inserting the α expressions into the K_{s0} equation and solving for S gives

$$S = \sqrt[3]{K_{s0}/\alpha_{1,Mg}\alpha_{1,NH_4}\alpha_{0,PO_4}}$$

The term α, for Mg^{2+} signifies the protonated form of this HA. The

$I = 0.1\ M$ constants make several species lines coincide, so we shall use estimated (less reliable) constants for I of about 1 M to make a clearer graph (**Fig. VI-15**) to illustrate the trends involved:

$$\log S = \tfrac{1}{3} \log[10^{-11.5}(1 + 10^{\mathbf{pH}-12})(1 + 10^{\mathbf{pH}-9.5})(1 + 10^{11-\mathbf{pH}})]$$

We show just the K_{a3} term in α_{0PO_4} since the others are negligible at high **pH**. The log S plot is made at 0.1-unit **pH** intervals and the species calculated from the α terms:

$$[Mg^{2+}] = S(\alpha_{1,Mg}) \qquad [NH_4^+] = S(\alpha_{1NH_4}) \qquad [PO_4^{3-}] = S(\alpha_{0PO_4})$$

Note that the conjugate species relations (log differences) are those of any equilibrium diagram involving them. The α's for the phosphate, magnesium, and ammonia conjugates cross and diverge as they should.

The dotted line h is the log of the sum $[OH^-] + [NH_3] + [MgOH^+]$, which must equal $[HPO_4^{2-}] + \mathbf{H}$ (the proton balance: gain equals loss) in the case of the solid dissolved in pure water. The intersection with S is thus the solubility in pure water, which does not occur at the minimum because appreciable (about 90%) PO_4^{3-} is protonated by acids NH_4^+ and Mg^{2+}. To unprotonate 90% of the phosphate requires enough NaOH to reach **pH** about 13, which also removes much of the other two salt ions as NH_3 and $MgOH^+$. Thus, there is no **pH** at which this salt can dissolve mainly as the ions of the parent solid.

Further examination of the literature (tables of constants) reveals that $Mg(OH)_2$ and $Mg_3(PO_4)_2$ and even more complex products could precipitate at **pH** above 10, so that modification of this incomplete diagram is required. We see why the analytical method for Mg^{2+} asks for precipitation in ammonia buffer about **pH** 9, but approached from below by first adding $(NH_4)_2HPO_4$ to the sample followed by gradual addition of ammonia to cause precipitation through slow increase of $[PO_4^{3-}]$.

Problem of Completeness of the Equilibrium Model: Ag_3PO_4

Equilibrium pictures depend upon what species are chosen and the values of the K's available. In the previous examples, enough research has been done to suggest that we can hope to know the major species involved. Only in our first example, $RbClO_4$, did the single datum of its water solubility allow a valid calculation of the

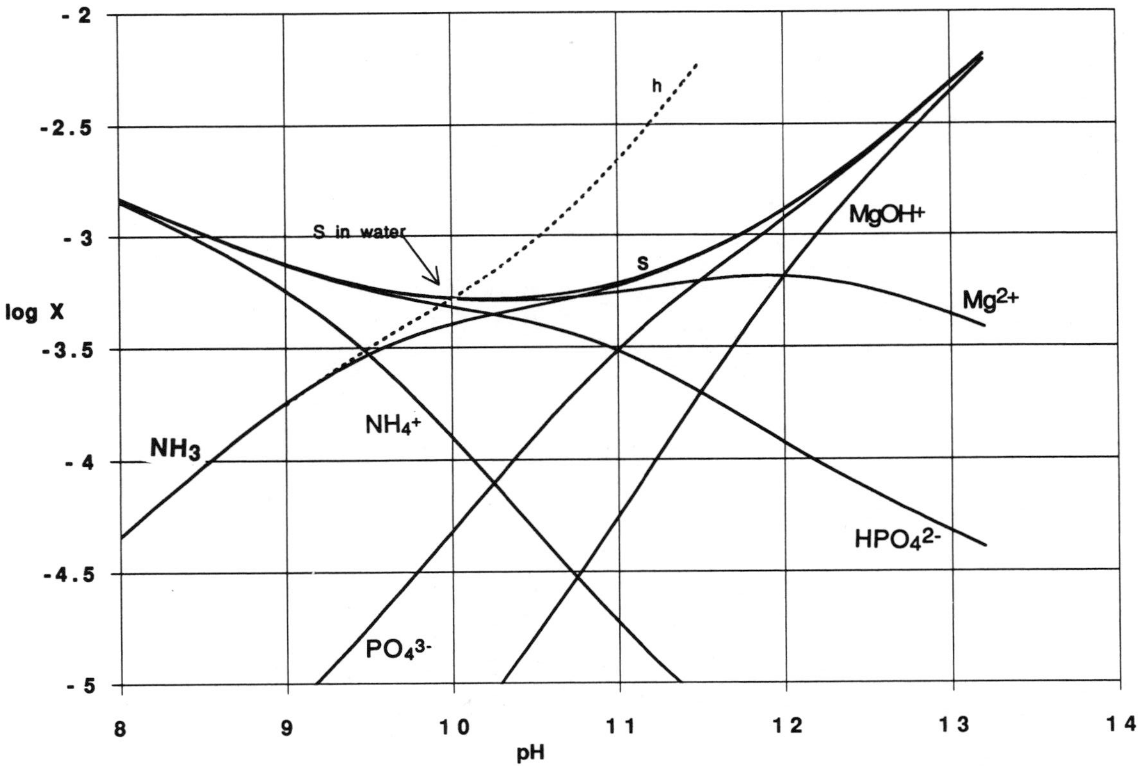

Fig. VI-15. Solubility of $MgNH_4PO_4$ vs. p**H**. Log (species) in saturated solutions. Approximate diagram near ionic strength 1 *M* at high p**H**. The dotted *h* line for the *S* in pure water is explained in the discussion.

K_{s0}. In other cases an erroneous value of K_{s0} results from the assumption that only the ions of the solid are formed in water. The error varies from small to enormous. Leussing[7] treats many facets of solubility investigation.

We are told that the K_{s0} of HgS is 10^{-54}. Beginning students often calculate, and alas are told in textbooks, that the solubility in pure water is 10^{-27} M. Now, this is about 0.001 ion per liter, effectively *zero*! This is a good place to show the advantage of thermodynamic determination of the constant from free energy data, since solubility measurement is not feasible in such a case. But HgS cannot form just Hg^{2+} and S^{2-}. The first is quite an acid, with pK_a about 4, and sulfide is close to a strong base; pK_{a2} of H_2S is about 13. The trace amount dissolving in water, pH about 7, must form largely such species as $HgOH^+$ and HS^-. The actual water solubility is found to be about 8 powers of 10 greater than 10^{-27}! The more complete picture includes several other species. (See the entertaining article by Butler,[8] who calculates the five Hg species and four sulfur species present.)

The water solubility of Ag_3PO_4 (Seidell, ref. 2) was obtained in 1903 as 0.00644 g/l, or 1.54×10^{-5} M. Several other Russian papers over the years seem to have confirmed this magnitude. One hint of a complex in one reference compilation is for $AgPO_4^{2-}$, with K_f about 10^{12}. The Ag(I) forms weak OH^- complexes in basic solutions. But phosphate forms HPO_4^{2-} extensively, with pK_{a3} 12. So, something like the HgS situation occurs. However, over the years K_{s0} tables and textbook problems have appeared using pK_{s0} from 15 to 21. Using the preceding solubility value, we get, by the naive method,

$$K_{s0} = [Ag^+]^3[PO_4^{3-}] = (3S)^3(S) = 27S^4 = 1.5 \times 10^{-18} \qquad pK_{s0} = 17.8$$

Values near this are often used. The true value can only be found after getting valid constants involving major species formed so that the actual ionic equilibrium species can be used. As usual (Eq. VI-1),

$$K_{s0} = (3S\alpha_{0M})^3(S\alpha_{0,PO_4}) = 27S^4\alpha_{0M}^3\alpha_{0,PO_4}$$

What can we do? Suppose the S does produce about 10^{-5} M phosphate ion, which largely gives HPO_4^{2-} and OH^-. This makes **pH** about 9, which leaves about 90% of the Ag^+ as the ion but reduces phosphate to about 10^{-8} M. This gives K_{s0} about 10^{-21}. Here, we omitted the possible $AgPO_4^{2-}$ complex. If it is formed with $K = 10^{12}$, the picture changes radically. The constant is so high that we try the picture

$$Ag_3PO_{4(s)} \rightleftharpoons AgPO_4^{2-} + 2Ag^+$$

as the major dissolved species. This gives $[Ag^+]$ of about 3×10^{-5} and $[AgPO_4^{2-}]$ of 1.5×10^{-5} *M*. In the *K* expression for the complex, we can then get free $[PO_4^{3-}]$ of about 5×10^{-13}, which gives K_{s0} of about 10^{-26}. We must conclude that sufficient data have not been presented (here) to deal with this system. Perhaps they exist in the original Russian papers.

It seems reasonable that the small silver and phosphate ions and their 1:3 charge attraction should produce strong complexing. The great basicity of phosphate and OH^- complexing of Ag^+ may make investigation of complexes in this system difficult to untangle. Other phosphate systems exhibit complexing by HPO_4^- and OH^-, as in the well-known apatite mineral $Ca_5OH(PO_4)_3$, so one may wonder how much remains unknown about the Ag(I) system. Measurement of $[Ag^+]$ using a silver electrode and of the **p**H might help. We assume there has been proof that the compound Ag_3PO_4 truly is formed (see ref. 7, Section B, p. 690). Clearly the water solubility alone gives no valid idea of K_{s0} or of the species present.

Tyrosine

Tyrosine [2-amino-3-(4-hydroxyphenyl) propanoic acid] is treated as an H_3L^+ acid. The uncharged form, H_2L, has low solubility, 0.0025 *M* (its K_{s2}). Thus, adding or losing protons increases the *S* of this ampholyte. The H_2L is constant at $\alpha_2 S$ in saturated solutions, just as is the intrinsic solubility of preceding cases. The other constants at 25°C and $\boldsymbol{I} = 0.1$ *M* are pK_{a1} 2.34, pK_{a2} 9.11, and pK_{a3} 10.61. The solubility function is obtained from $\alpha_2 S = 0.0025$:

$$S = 0.0025/\alpha_2 = 0.0025(\mathbf{H}/K_1 + 1 + K_2/\mathbf{H} + K_2K_3/\mathbf{H}^2)$$

This can also be obtained from the summation of all forms of tyrosine, taking their ratios to the constant form, H_2L, 0.0025 *M*. Note the effect of high and of low **H** on each term (**Fig. VI-16**). This is analogous to a K_{s0} using proton complexing to H_3L^+ and then loss to HL^- and L^{2-}, which increase *S* to the left and to the right of a minimum (cf. Figs. VI-10*a–c*). From the plot the reader can deduce a procedure for evaluating the acid constants from solubility data via the *S* equation. This has been used for EDTA (the H_4Y form) and other low solubility acids.

The reader might sketch the α curves of this triprotic acid to relate the **p**H to the solubility dependence. The log *C* lines relate exactly to a log α diagram of this acid, but bent to fit the α_2 to the changing total tyrosine. One of the least soluble amino acids, tyrosine is easily

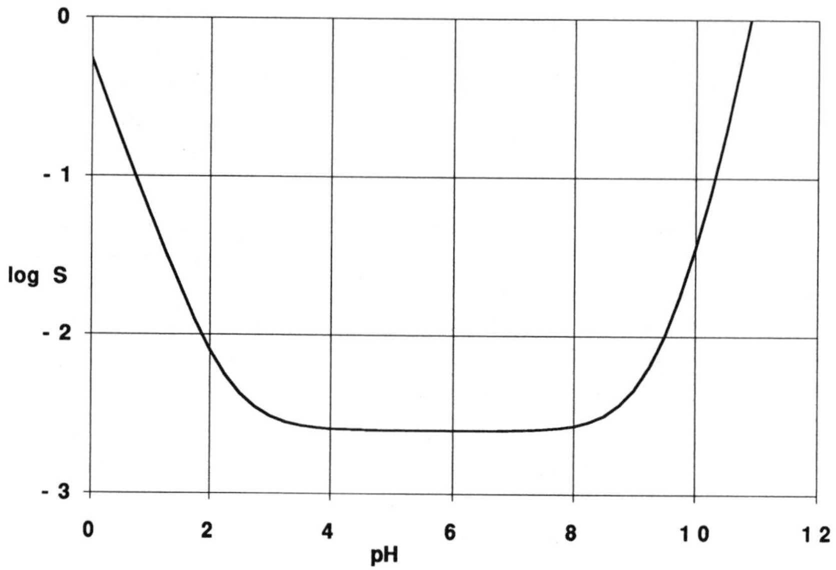

Fig. VI-16. Tyrosine: **pH** effect on solubility.

separated from protein hydrolysates by neutralization since the **pH** range of low solubility is large, about 3–9.

VI-8. POLYNUCLEAR HYDROXIDE SOLUBILITIES

Refer to Section V-9 for α, $\bar{\mathbf{n}}$ calculations in systems having mono- and polynuclear complexes. The Cu(II)–OH system is plotted first (**Fig. VI-17**). It may seem unexpected to find that the log species lines are straight also in these systems if saturation with respect to one species occurs.

In the *unsaturated* systems (Chapter V), polynuclear terms (here $Cu_2(OH)_2^{2+}$) contain α_0 terms to higher powers than 1 that do not cancel and make the α terms dependent upon C as well as upon **H**. However, in these saturated solutions, $[Cu^{2+}]$ depends only upon **H** so that the S equation is a function of **H** alone. See the next section. See the "Note on α Designations," p. 229.

Explanations

The constants for the formations of species[9] are $\log K_{11} = -9$; $\log K_{22} = -10.7$; $\log K_{12} = -18$; $\log K_{13} = -28$; and $\log K_{14} = -38.8$.

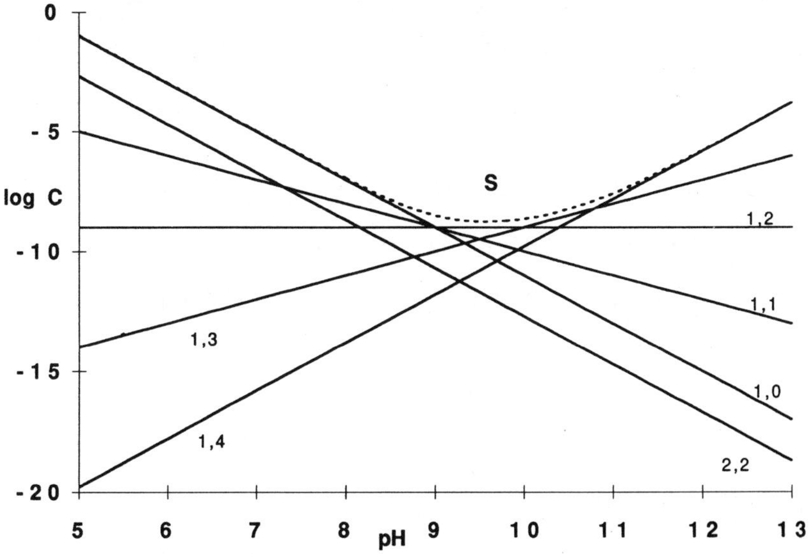

Fig. VI-17. Log of solubility and species vs. **pH** in the system $Cu(OH)_2$, $\boldsymbol{I} = 1\,M$, 25°C. The top line is the log of the sum of the species concentrations based on Cu. These are calculated values from the published constants.

In Chapter V, we showed how values of the K's operating at low **pH** are deduced from **pH** data. Now we see how solubility data of $Cu(OH)_2$ [or CuO] vs. **pH** leads to the K's of the more basic region:

Acid formation constant:

$$Cu^{2+}_{(aq)} + 2H_2O \rightleftharpoons Cu(OH)_{2(aq)} + 2H^+$$

$$K_{12} = 10^{-18} = \frac{[Cu(OH)_2]\mathbf{H}^2}{[Cu^{2+}]}$$

Acid solubility:

$$Cu(OH)_{2(s)} + 2H^+ \rightleftharpoons Cu^{2+} \qquad K_{sa10} = 10^9 = [Cu^{2+}]/\mathbf{H}^2$$

Alternatively, the solubility product K_{s0} can be used in these derivations. Results identical to those with the more convenient acid form obtain

$$K_{s0} = K_{sa10}K_w^2 = 10^{-18.6} = [Cu^{2+}][OH^-]^2$$

For ionic strength 1 M ($NaClO_4$), $pK_w \cong 13.80$ (ref. 9, p. 85].

Combine to $[Cu(OH)_{2(aq)}] = K_{12}K_{sa10} = 10^{-9}$. A solubility equilibrium adds a constraint to the system, requiring a constant value of the dissolved molecular species (if any) and/or dependence of $[Cu^{2+}]$, and thus all other Cu species, on **H** alone. Here, $[Cu^{2+}] = \alpha_0 S = K_{sa10}\mathbf{H}^2$.

Now, the sum of the six species (S in molarity of Cu) is

$$S = [Cu^{2+}] + [CuOH^+] + 2[Cu_2(OH)_2^{2+}] + [Cu(OH)_2] + [Cu(OH)_3^-] + [Cu(OH)_4^{2-}]$$
$$= [Cu^{2+}] + K_{11}[Cu^{2+}]/\mathbf{H} + 2K_{22}[Cu^{2+}]^2/\mathbf{H}^2 + K_{12}[Cu^{2+}]/\mathbf{H}^2 + K_{13}[Cu^{2+}]/\mathbf{H}^3 + K_{14}[Cu^{2+}]/\mathbf{H}^4$$

Using $[Cu^{2+}] = K_{sa10}\mathbf{H}^2$ gives

$$S = K_{sa10}\mathbf{H}^2[1 + K_{11}/\mathbf{H} + 2K_{22}K_{sa10} + K_{12}/\mathbf{H}^2 + K_{13}/\mathbf{H}^3 + K_{14}/\mathbf{H}^4]$$

Thus, **H** determines S, which is also the total Cu(II) dissolved. Then inserting $10^9\mathbf{H}^2 = [Cu^{2+}]$ serves to evaluate species. The logs of S and species are plotted in Fig. VI-17.

Determination of Species and Constants

Inspection of the S equation and the plot in Fig. VI-17 suggests means of K determination similar to those described in foregoing ML and MLX cases. When solubility is low, the polynuclear terms will be minor, as is the 2, 2 in Cu(II). Experimental data on solubility as a function of p**H** are plotted as log S vs. p**H** in Fig. VI-16. A curve like the top line of Cu(II) in Fig. VI-17 results. The slopes of log S vs. p**H** indicate the types of mononuclear species. In the Cu(II) example we see the following (try several readings from S, p**H** in each of these steps):

1. At the acid extremes (left side), a slope of log S of -2 indicates the $\mathbf{H}^2$ dependence of $[Cu^{2+}]$ in K_{sa10}, which is estimated as the observed $S/\mathbf{H}^2$ by assuming that the first term of the S equation predominates as $\mathbf{H} \rightarrow$ large.
2. At points of slope of -1, one may estimate S is $[CuOH^+]$ and the K_{11} as $S\mathbf{H}/[Cu^{2+}]$ using $K_{sa10}\mathbf{H}^2$ from step 1 for the $[Cu^{2+}]$ at this point. Thus,

$$K_{11} = S/K_{sa10}\mathbf{H}$$

3. At slope zero, we have the maximum possible portion S of the constant $[Cu(OH)_{2(aq)}] = K_{12}K_{sa10} = 10^{-9}$.
4. At slope +1, we have S is mainly $[Cu(OH)_3^-]$, and thus,

$$K_{13} = S\mathbf{H}^3/K_{sa10}\mathbf{H}^2 = S\mathbf{H}/K_{sa10} \quad \text{(see step 2)}$$

5. At slope +2, if $S \cong [Cu(OH)_4^{2-}]$, $K_{14} = S\mathbf{H}^4/K_{sa10}\mathbf{H}^2 = S\mathbf{H}^2/K_{sa10}$.

These first approximations are then refined for the overlap, subtracting neighboring species from S in each step. Combining these with the base titration results must then yield α and $\bar{\mathbf{n}}$ values in agreement with the experimental data. Computer iteration has been developed for this work.[9] Because of the overwhelming of minor species by major ones, even slight uncertainties in the data may give erroneous constants and even suggest species that are not present. That is, several choices of species and constants may give acceptable agreement with the data. The references point out many cases of unreliable conclusions from data in this difficult area of investigation.

Note on α Designations

Interpretation of α_n terms must be made with reference to the equilibrium chemical reaction involved (e.g., α_{0M} for a metal ion may refer to M–L complexing, or to Brønsted acid-base interactions). The first α_{0M} in Section VI-7 refers to silver ion not complexed by acetate ions. On p. 221, α_{1Mg} refers to the metal ion acting as an acid HA, and thus refers to the completely protonated (the aqua) ion, Mg^{2+}. In other contexts it might refer to $MgOH^+$ or to $Mg(PO_4)^-$ complexes. The species formulas or the chemical equations must be given in each case. More complex cases like $\alpha_{1,2,1}$ must be clearly defined in each system. See various Cu(II)-OH^--NH_3 systems discussed on pp. 175, 181, and 226.

SUMMARY

Solubility of slightly soluble ion-forming compounds and the effects of protonic and ML complexing with these ions are described and illustrated. Applications to varied cases are shown: $CaCO_3$ solubility as a function of CO_2 gas pressure; tyrosine and $MgNH_4HPO_4$ solubility vs. p**H**; and $Cu(OH)_2$ solubility vs. p**H** involving polynuclear complexes.

EXERCISES

1. Searching for ideal simple solubility like $RbClO_4$, one finds that, although NH_4ClO_4 is too soluble (about 1 M), tetramethylammonium perchlorate does have solubility close to that of $RbClO_4$. (Strangely, tetraethyl and higher homologues are much more soluble.) Examine the data that follows and determine K_{s0} and then K_{s0}^0 by graphical [vs. $\sqrt{I}/(1+\sqrt{I})$] and by numerical ($f_\pm$) methods. Use M for the $(CH_3)_4N^+$ ion. Temperature is 25°C. (*Ans.*: pK_{so}^0 of about 2.49.)

Medium	S of $MClO_4$, M	I	K_{s0}	f	K_{s0}^0
Water	0.0731	0.0731			
0.0398 $HClO_4$	0.0581	0.0979			
0.0497 $HClO_4$	0.0551				
0.01003 NaCl	0.0739				
0.02508 NaCl	0.0754				
0.05016 NaCl	0.0772				

2. (a) By analogy with the AgCl example treated in Fig. VI-11, write the solubility equations for AgBr and AgI as functions of free $[NH_3]$. Calculate and sketch their log S curves in 0–10 M ammonia. (b) Consider the problems in trying to precipitate selectively first AgI, then AgBr, and finally AgCl in soluble mixtures of the three halides. Can one find β values for other ligands for Ag(I) that might solve some of the difficulties caused by ammonia evaporation?

3. By examining the slopes and intersections of log species lines in AgSCN and Ag acetate examples (Figs. VI-6 and VI-9), deduce where to sketch the log species lines in the AgX cases of Figs. VI-10*a–c*. Compare these relatively simple (straight lines) with the curvature caused when an additional ligand or **pH** effect is added in the $AgCl–NH_3$, $CaCO_3–CO_2$, and later examples.

4. (a) Can EDTA prevent the precipitation of $BaSO_4$ at reasonable concentrations and **pH**? You need K_{s0} and the $Ba(edta)^{2-}$ formation constants. (b) Can **pH** conditions be devised for separation of Ba^{2+} and Pb^{2+} by sulfate precipitation? You need α_0 vs. **pH** for EDTA. See Figs. III-18 and V-15. (*Note*: The *American Chemical Society* abbreviation list uses EDTA for the name of the substance or ligand species and edta in formulas as a ligand where confusion with capitals for elements is avoided.)

REFERENCES

1. Guenther, W. B., *J. Am. Chem. Soc.*, **91**, 7619 (1969).
2. Compilations of solubility and complexing data with references to the primary literature should be consulted before assuming lack of complications: Sillen, L. G. and Martell, A. E., Eds., *Stability Constants of Metal-Ion Complexes*, Special Publications, No. 17, 25, The Chemical Society, London, 1964, 1971.
   Martell, A. E. and Smith, R. M., Eds., *Critical Stability Constants*, Plenum, New York (6 vols.), 1974–1989.
   Seidell, A., *Solubilities*, W. F. Linke, Ed., D. Van Nostrand, Princeton NJ, 1958, 2 vols.
3. Russo, S. O. and Hanania, G., *J. Chem. Educ.*, **66**, 148 (1989). Describes degrees of ion pairing for several cases.
4. Guenther, W. B., *Chemical Equilibrium*, Plenum, New York, 1975. pp. 173–176. Gives details of the calcium sulfate problem.
5. Yeatts, L. B. and Marshall, W. L., *J. Phys. Chem.*, **73**, 81 (1969).
6. Cave, G. C. B. and Hume, D. N., *J. Am. Chem. Soc.*, **75**, 2893 (1953).
7. *Solubility*, D.L. Leussing, in Kolthoff, I. M. and Elving, P. J., Eds. *Treatise on Analytical Chemistry*, Wiley-Interscience, New York, 1959. Part I. Vol. 1, Ch. 17. pp. 675–730.
8. Butler, J. N., *J. Chem. Educ.*, **38**, 460 (1961).
9. Baes, C. F. and Mesmer, R. E., *The Hydrolysis of Cations*, Wiley, New York, 1976. See also Martell, A. E. and Motekaitis, R. J., *Determination and Use of Stability Constants*, VCH Publishers, New York, 1988.

APPENDIX A

ASPECTS OF SOLUTION CHEMISTRY

A-I. CHEMISTRY BACKGROUND NEEDED FOR STUDY OF EQUILIBRIUM

The nature of aqueous solutions and types of reactions occurring in them as described in general and analytical chemistry texts is needed for the study of equilibrium. Books on these topics are described in the Annotated Bibliography.

Protonic (Brønsted) acid–base systems, H_nA, have n acidic protons on the central base group A (mononuclear cases). Weak bases are treated largely as conjugates of acids through their K_a, as in recent international compilations. Some polynuclear examples ($H_4P_2O_7$, $Cu_2(OH)_2^{2+}$) are mentioned. (The system H_nL is also used, in this book often with the implication that ligand behavior toward metal ions is in view and that the acid may be strong or weak. Here, H_nA is used for weak acids considered alone.)

Most acids have the protons attached to oxgyen (HOCl, formic acid, HOOCH, H_3PO_4 or $OP(OH)_3$, $Cu^{2+}_{(aq)}$ or $Cu(HOH)_6^{2+}$, phenols). Quite a few have protons attached to nitrogen, as in protonated ammonia and amines. Few have protons attached to other elements, e.g., HCl, H_2S, and HCN. Almost all of the thousands of known acids are weak: they react only to a small extent with water ("dissociate") to produce the fundamental water acid species, the aquated proton, $H^+_{(aq)}$, H_3O^+, etc. The major form, however, is tetraaquated $H_9O_4^+$. For ease in reading complex formulas with many proton terms, we choose a common symbol for the sum of aquated proton molarity, **H**, as universally used in the symbol pH and in recent tables of critical stability constant (see Martell and Smith in the Annotated Bibliography). We use boldface to indicate molarity, since pH often refers to the *activity* of aquated protons, which we write pa_H in this book.

The few common strong acids are HCl, HBr, HI, $HClO_4$, HNO_3, and H_2SO_4 (only the first proton is strong). Rare ones will turn up when one cannot find a K_a value for them. Examples are $H_4Fe(CN)_6$, $H_3Fe(CN)_6$, $H_3Co(CN)_6$, and some sulfonic acids. These acids are treated as 100% in ionic form in dilute aqueous solutions, say, below 1 *M*. Clearly, 7 *M* HCl, not all ionized: one can smell the gas escaping. Often, HSCN is treated as strong, though it seems to have a K_a^0 of about 0.1. This makes a 0.1 *M* solution about 62% ionic. All these acids are *weak* acids in most other solvents. Acid strength is dependent on the relative proton affinity of the solvent and the A group. Acids do not really "dissociate". Their protons are removed by water, less so by alcohols, but not by benzene, etc. Many organic texts have extensive discussion of acid-base strength with examples and *K* values. One example is given by Sykes (Ch. 3)[1].

Unfortunately, there is no way to predict acid–base behavior. Texts and compilations must be consulted to obtain information on the number and *K* values for the acidic protons in a species. For example, adenine ($C_5H_5N_5$) has only two basic nitrogen atoms and is part of the system of the parent acid $C_5H_6N_5^+$, having pK_a^0's 4.20 and 9.85.

A conjugate base differs from its acid by one proton. The system H_nA has *n* conjugate pairs. In H_3PO_4 these are H_3PO_4–$H_2PO_4^-$, $H_2PO_4^-$–HPO_4^{2-}, and HPO_4^{2-}–PO_4^{3-}. The ratios in the pairs are uniquely determined by the single equilibrium **H** value of a specific mixture.

The ratio $\mathbf{H}/K_{a1} = [H_3PO_4]/[H_2PO_4^-]$, and so on for $\mathbf{H}/K_{a2}$ and $\mathbf{H}/K_{a3}$, the other two ratios. This is the basis of the $\bar{\mathbf{n}}$ and α relations that unify acid–base mathematics (including Lewis ML types) as explained in this book.

Modern practice is to use the proton variable and K_a consistently with Brønsted acid–base systems. There is rarely a need to use older K_b or K_h forms with **OH**. The "hydrolysis" concept for proton transfers is extraneous to the Brønsted view of aqueous reactions.

Tables of Constants. An example from *Critical Stability Constants* (see Martell and Smith in the Annotated Bibliography) will help explain the information that can be obtained. The systematic name shows the structure of the common named species, citric acid. The H_3L designation indicates that three acid protons are available on the completely acid form, which here is the uncharged molecule. This is quite important since there is also an OH group, and some authors have used the H_4L convention for this acid. Since the acid constants are given in IUPAC form in this compilation, K_1 refers to the *formation* of the first protonated species of L, the citrate 3-ion.

$$H^+_{(aq)} + L^{3-} HL^{2-} \qquad K_1 = HL/H{\cdot}L = 1/K_{a3}$$

In the H_4L formulation, L^{4-} would be used here, making the K_1 value quite different (about 10^{16}). Numbers are not attached to K's since the expressions make the species involved completely clear.

ABBREVIATED EXAMPLE TABLE FOR WEAK ACID–METAL LIGAND COMPLEXING: $C_6H_8O_7$, 2-HYDROXYPROPANE-1,2,3-TRICARBOXYLIC ACID (CITRIC ACID), H_3L

Metal Ion	Equilibrium	log *K*, 25°C 0.1	 0	
H^+	$HL/H{\cdot}L$	5.69 ± 0.05	$6.396 + 0.004$	
	$H_2L/H{\cdot}HL$	4.35 ± 0.05	4.761–0.002	
	$H_3L/H{\cdot}H_2L$	2.87 ± 0.08	3.128–0.07	
Li^+	$ML/M{\cdot}L$	0.83		
Ni^{2+}	$ML/M{\cdot}L$	5.430[h]		
	$MHL/M{\cdot}HL$	3.30[h]		
	$MH_2L/M{\cdot}H_2L$	1.75[h]		[h]20°C, 0.1 *M*

Note that the second column specifically states the species involved. (A system of notations like $K_{1,2}$, $K_{1,1,0}$, $K_{2,1,1}$, . . . can be used and must be defined within each context as seen in metal ion–OH–ligand complexes in Chapter V. The preceding system avoids the problems of a number system.) The next columns state the temperature and the ionic strength for which the log K's are given with their estimated uncertainties (or likely range direction). The last column gives the very precise results of Bates and Pinching at NBS in 1949.[2] Nickel(II) complexing is shown to occur with citrate^{3-} as well as with $HCit^{2-}$ and H_2Cit^-. These are the results considered reliable by the editors of the tables. They do not preclude the formation of other species that have not been well investigated. The introduction to the tables gives detailed explanations of these and other features. The superscript h indicates conditions different from the column heading. Note that use of the formation direction allows including **H** in these listings as a Lewis acid combining with the L group.

Metal ion–ligand equilibria must be assumed to exist between any positive and negatives ions (and any Lewis acids and bases) until experiment shows it to be negligible. Understanding of chemistry in seawater, for example, was meager until conditional complexing constants were measured for such species as $CaSO_4^0$, $CaHCO_3^-$, KSO_4^-, $NaCO_3^-$, $MgCl^+$, and many others. Earlier workers felt forced to postulate gross oversaturation of $CaCO_3$ in the oceans based on assumptions that major ions were all strong electrolytes having negligible associations. The problems are complex and are treated in detail by Pytkowicz (Vol. II; see Annotated Bibliography). Note that temperature and pressure gradients greatly affect speciation and prevent achievement of equilibrium throughout the solution.

A-2. ACTIVITY COEFFICIENTS: VALUES AND USES

Extensive discussions of quantitative treatments of electrolyte solutions are found in most of the reference texts in the Annotated Bibliography. Equations and tables are summarized here. This is not a substitute for rigorous thermodynamic treatment of activities of electrolytes and ions.[3]

The basic Debye–Hückel (DH) relation for the estimated ionic activity coefficient f is

$$\log f = -0.509z^2 \frac{\sqrt{\boldsymbol{I}}}{1 + A\sqrt{\boldsymbol{I}}}$$

where A is a size factor and z is the ionic charge. Kielland[4] estimated the A's for the actual hydrated ions in 1937. The agreement with pure electrolytes is good up to ionic strength 0.1 M so that his tables have found wide use. These estimates seem accurate to within 0.5% below $\boldsymbol{I} = 0.02\ M$ and within about 2% at 0.1 M for singly charged ions. Higher individual deviations are seen for multiply charged ions, especially if the opposite ions are not singly charged. See data and discussion by Harned and Owen.[3]

The A factor is 0.328$\mathring{a}$, where $\mathring{a}$ is the *effective* diameter in angstroms. Table A-1 that follows shows that many common +1 and

-1 ions are near $3\mathring{a}$ so that A is close to 1. So, the simple DH relation (called the Guntelberg equation) for small ions is

$$\log f = -0.509z^2 \frac{\sqrt{I}}{1+\sqrt{I}} \quad \text{(DH)}$$

This function continues down as I increases, as seen in **Fig. A-1**.

Davies used another correction term that helps somewhat but does not require a size estimate. It has a subtractive term in I:

$$\log f = -0.509z^2\left[\frac{\sqrt{I}}{1+\sqrt{I}} - 0.3I\right] \quad \text{(Davies)}$$

The Davies function is calibrated to $NaClO_4$. It goes through a broad minimum (vs. I) in the I region of 0.2–0.7 M. Some, but not all, electrolytes do this. Furthermore, the question of what happens to the ion of interest in mixtures with other ions is unanswered. The ease of calculation and good results for many materials makes this

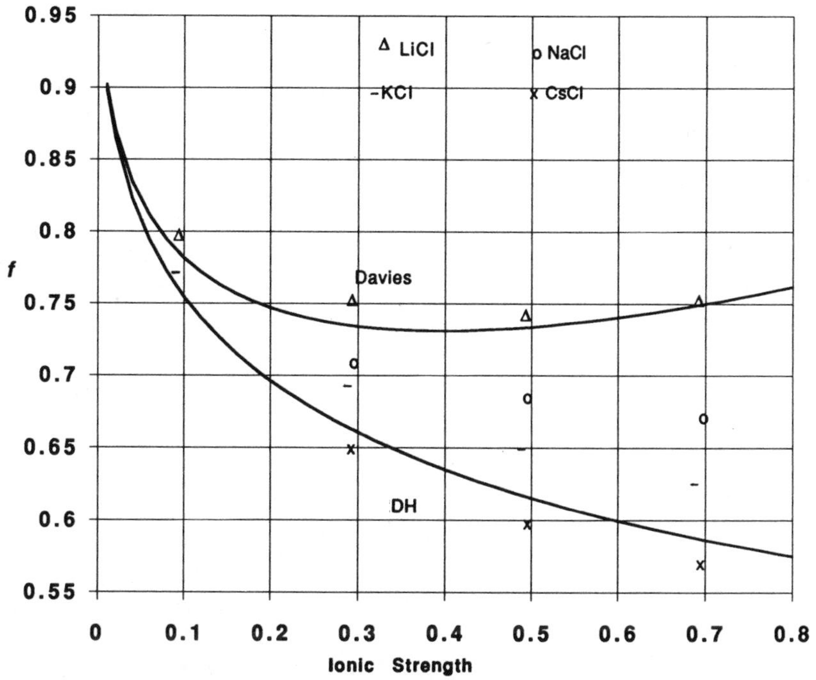

Fig. A-1. Experimental values of four halides compared with two approximation equations for *mean ionic activity coefficients*, the simple Debye–Hückel and the Davies plots of activity coefficients for singly charged ions.

the preferred estimate in the I region of up to about 0.2 M. The plots of these two functions are shown in **Figs. A-1**, **A-2*a*,*b*** and **A-3**.

Note that these plots are for the pure electrolytes and may be quite different in mixed electrolytes of the same I. One sees the need for data below $I = 0.05\ M$ if one needs to extrapolate to zero ionic strength with precision better than 1%. The impossibility of arranging such mixtures often dictates reporting conditional constants for high, constant I.

For media of approximately constant ionic strength, 0.4 M is a good choice since many ions change f very slightly with I in this region compared with 0.1 M.

For higher ionic strengths and mixed electrolytes, two or more constants adjustable for each ion and foreign electrolyte have recently been developed to give highly accurate f values. But this is at the price of having to know the *interaction* coefficients that must be found from experimental data for each case. Their use is detailed in a series of papers by Pitzer and co-workers.[5]

Table A-1 gives Kielland's ionic activity coefficients.

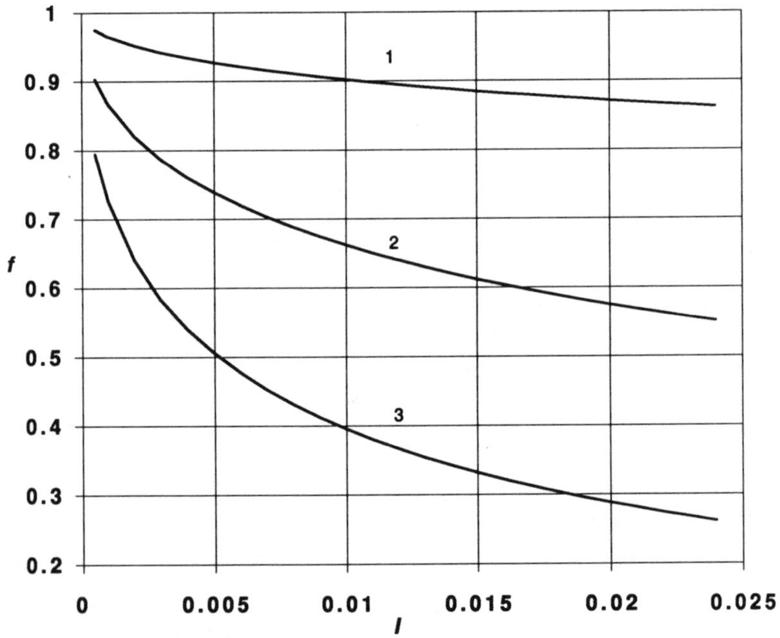

Fig. A-2*a*. Davies equation ionic activity coefficient f values for lower ionic strengths I. Ion charges 1, 2, 3 of either sign.

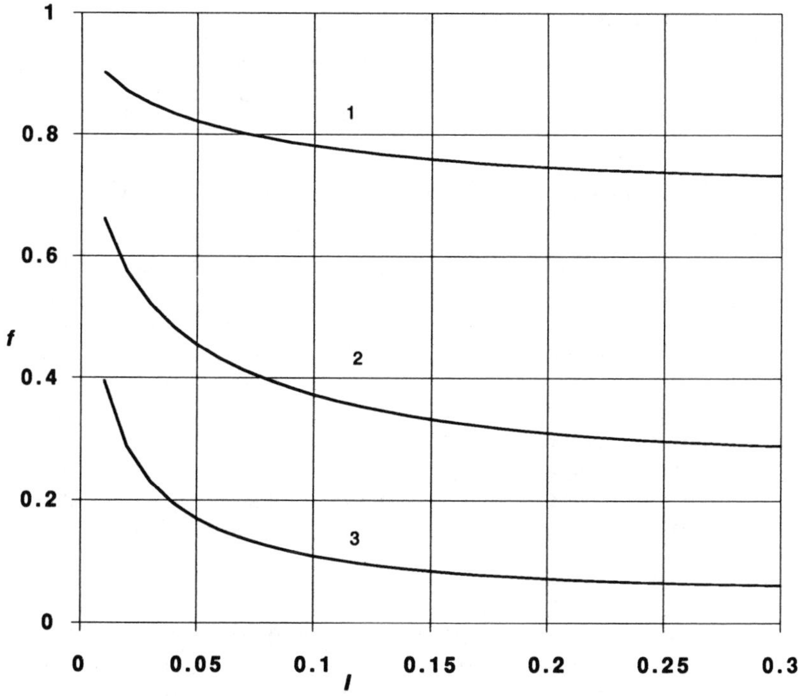

Fig. A-2*b*. Davies ionic activity coefficients for higher ionic strengths.

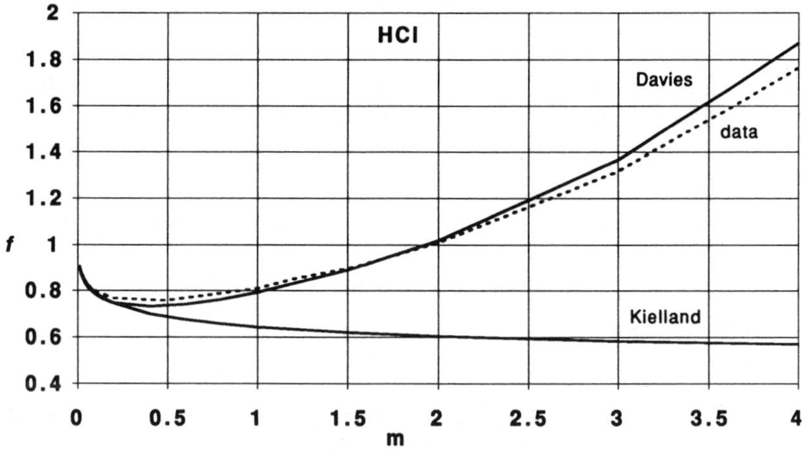

Fig. A-3. Experimental activity coefficients of HCl to high concentration compared with Davies equation and Kielland table functions.

TABLE A-1. Ionic Activity Coefficients and Values of å

Parameter	Ionic strength μ							
å	0.0005	0.001	0.0025	0.005	0.01	0.025	0.05	0.1
			Individual Activity Coefficients of Ions in Water at 25°C[a]					
ION CHARGE 1								
9	0.975	0.967	0.950	0.933	0.914	0.88	0.86	0.83
8	0.975	0.966	0.949	0.931	0.912	0.88	0.85	0.82
7	0.975	0.965	0.948	0.930	0.909	0.875	0.845	0.81
6	0.975	0.965	0.948	0.929	0.907	0.87	0.835	0.80
5	0.975	0.964	0.947	0.928	0.904	0.865	0.83	0.79
4.5	0.975	0.964	0.947	0.928	0.902	0.86	0.82	0.775
4	0.975	0.964	0.947	0.927	0.901	0.855	0.815	0.77
3.5	0.975	0.964	0.946	0.926	0.900	0.855	0.81	0.76
3	0.975	0.964	0.945	0.925	0.899	0.85	0.805	0.755
2.5	0.975	0.964	0.945	0.924	0.898	0.85	0.80	0.75
ION CHARGE 2								
8	0.906	0.872	0.813	0.755	0.69	0.595	0.52	0.45
7	0.906	0.872	0.812	0.755	0.685	0.58	0.50	0.425
6	0.905	0.870	0.809	0.749	0.675	0.57	0.485	0.405
5	0.903	0.868	0.805	0.744	0.67	0.555	0.465	0.38
4.5	0.903	0.868	0.805	0.742	0.665	0.55	0.455	0.37
4	0.903	0.867	0.803	0.740	0.660	0.545	0.445	0.355
ION CHARGE 3								
9	0.802	0.738	0.632	0.54	0.445	0.325	0.245	0.18
6	0.798	0.731	0.620	0.52	0.415	0.28	0.195	0.13
5	0.796	0.728	0.616	0.51	0.405	0.27	0.18	0.115
4	0.796	0.725	0.612	0.505	0.395	0.25	0.16	0.095
ION CHARGE 4								
11	0.678	0.588	0.455	0.35	0.255	0.155	0.10	0.065
6	0.670	0.575	0.43	0.315	0.21	0.105	0.055	0.027
5	0.668	0.57	0.425	0.31	0.20	0.10	0.048	0.021
ION CHARGE 5								
9	0.542	0.43	0.28	0.18	0.105	0.045	0.020	0.009

Values of the Parameter å for Selected Ions

INORGANIC IONS, CHARGE 1

9	H^+
6	Li^+
4–4.5	Na^+, $CdCl^+$, ClO_2^-, IO_3^-, HCO_3^-, $H_2PO_4^-$, HSO_3^-, $H_2AsO_4^-$, $Co(NH_3)_4(NO_2)_2^+$
3.5	OH^-, F^-, NCS^-, NCO^-, HS^-, ClO_3^-, ClO_4^-, BrO_3^-, IO_4^-, MnO_4^-
3	K^+, Cl^-, Br^-, I^-, CN^-, NO_2^-, NO_3^-
2.5	Rb^+, Cs^+, NH_4^+, Tl^+, Ag^+

TABLE A-1 (*Continued*)

INORGANIC IONS, CHARGE 2	
8	Mg^{2+}, Be^{2+}
6	Ca^{2+}, Cu^{2+}, Zn^{2+}, Sn^{2+}, Mn^{2+}, Fe^{2+}, Ni^{2+}, Co^{2+}
5	Sr^{2+}, Ba^{2+}, Ra^{2+}, Cd^{2+}, Hg^{2+}, S^{2-}, $S_2O_4{}^{2-}$, $WO_4{}^{2-}$
4.5	Pb^{2+}, $CO_3{}^{2-}$, $SO_3{}^{2-}$, $MoO_4{}^{2-}$, $Co(NH_3)_5Cl^{2+}$, $Fe(CN)_5NO^{2-}$
4	$Hg_2{}^{2+}$, $SO_4{}^{2-}$, $S_2O_3{}^{2-}$, $S_2O_8{}^{2-}$, $SeO_4{}^{2-}$, $CrO_4{}^{2-}$, $HPO_4{}^{2-}$, $S_2O_6{}^{2-}$
INORGANIC IONS, CHARGE 3	
9	Al^{+3}, Fe^{+3}, Cr^{+3}, Sc^{+3}, Y^{+3}, La^{+3}, In^{+3}, Ce^{+3}, Pr^{+3}, Nd^{+3}, Sm^{+3}
6	$Co(ethylenediamine)_3{}^{3+}$
4	$PO_4{}^{3-}$, $Fe(CN)_6{}^{3-}$, $Cr(NH_3)_6{}^{3+}$, $Co(NH_3)_6{}^{3+}$, $Co(NH_3)_5H_2O^{3+}$
INORGANIC IONS, CHARGE 4	
11	Th^{4+}, Zr^{4+}, Ce^{4+}, Sn^{4+}
6	$Co(S_2O_3)(CN)_5{}^{4-}$
5	$Fe(CN)_6{}^{4-}$
INORGANIC IONS, CHARGE 5	
9	$Co(S_2O_3)_2(CN)_4{}^{5-}$
ORGANIC IONS, CHARGE 1	
8	$(C_6H_5)_2CHCOO^-$, $(C_3H_7)_4N^+$
7	$OC_6H_2(NO_3)_3{}^-$, $(C_3H_7)_3NH^+$, $CH_3OC_6H_4COO^-$
6	$C_6H_5COO^-$, $C_6H_4OHCOO^-$, $C_6H_4ClCOO^-$, $C_6H_5CH_2COO^-$, $CH_2{=}CHCH_2COO^-$, $(CH_3)_2CCHCOO^-$, $(C_2H_5)_4N^+$, $(C_3H_7)_2NH_2{}^+$
5	$CHCl_2COO^-$, CCl_3COO^-, $(C_2H_5)_3NH^+$, $(C_3H_7)NH_3{}^+$
4.5	CH_3COO^-, CH_2ClCOO^-, $(CH_3)_4N^+$, $(C_2H_5)_2NH_2{}^+$, $NH_2CH_2COO^-$
4	${}^+NH_3CH_2COOH$, $(CH_3)_3NH^+$, $C_2H_5NH_3{}^+$
3.5	$HCOO^-$, H_2 citrate$^-$, $CH_3NH_3{}^+$, $(CH_3)_2NH_2{}^+$
ORGANIC IONS, CHARGE 2	
7	$OOC(CH_2)_5COO^{2-}$, $OOC(CH_2)_6COO^{2-}$, Congo red anion^{2-}
6	$C_6H_4(COO)_2{}^{2-}$, $H_2C(CH_2COO)_2{}^{2-}$, $(CH_2CH_2COO)_2{}^{2-}$
5	$H_2C(COO)_2{}^{2-}$, $(CH_2COO)_2{}^{2-}$, $(CHOHCOO)_2{}^{2-}$
4.5	$(COO)_2{}^{2-}$, H citrate^{2-}
ORGANIC IONS, CHARGE 3	
5	Citrate^{3-}

Source: Ref. 4, reprinted with permission, *Journal of the American Chemical Society*, **59**, 1937.

Davies Ionic Activity Coefficients. The values of the Davies ionic activity coefficients can be read from Figs. A-2*a*,*b* and A-3 with precision justified by the method.

The agreement of Davies and experimental HCl is not a general thing. However, it is true that the Kielland table is not valid above the ***I*** of approximately 0.1 *M* given since this function continues to drop while many ions rise.

The NBS system of standard pa_H references is given in Appendix C.

REFERENCES

1. Sykes, P., *A Guidebook to Mechanism in Organic Chemistry*, 5th ed., Longman, New York, 1981.
2. Bates, R. G. and Pinching, G. D., *J. Am. Chem. Soc.*, **71**, 1274 (1949).
3. Harned, H. S. and Owen, B. B., *The Physical Chemistry of Electrolyte Solutions*, 3rd ed., Reinhold, New York, 1958. Activity data for many electrolyte mixtures are given in Chs. 12–14 and in the Appendices.
4. Kielland, J., *J. Am. Chem. Soc.*, **59**, 1675 (1937).
5. Pitzer, K. S. et al., *J. Am. Chem. Soc.*, **96**, 5701 (1974) *and J. Phys. Chem.*, **77**, 2300 (1973). *Other papers in the series are listed.*

APPENDIX B

MATHEMATICAL TACTICS

This appendix presents more detailed amplifications and alternatives for some of the calculations of chemical equilibrium shown in the chapters.

B-1. SPREADSHEET CALCULATIONS FOR THE $\bar{\mathbf{n}}, \bar{\mathbf{n}}'$ METHOD

The computer does not make decisions about the chemical situation. However, one can write a variety of programs to use given acid and base forms to derive C_{AB}, C_A values and use them with the $\bar{\mathbf{n}}$ and $\bar{\mathbf{n}}'$ functions to determine **H**, or K_a, and also in the reverse sense. From given **H** one can calculate the C's required. Example spreadsheet methods follow.

Spreadsheet software provides assisted programming that is easy to learn and apply for many kinds of calculations in science, especially for tables of values for which one parameter is varied. The spreadsheet commands often combine a sequence of steps so that much time is saved. Consult the directions for the specific machines and applications at hand. The following is presented in terms related to Machintosh Plus with Microsoft Excel spreadsheet methods. Mathematica, also mentioned, requires larger memory systems than the usual Machintosh Plus.

Functions. For a general equation treating up to six complexes, H_6L or ML_6, we may relate the ratios expressed with acid ("dissociation") constants K_a's, which run inversely from 6 to 1, and the ML formation constants K's, which run from 1 to 6, to fit in one form of the $\bar{\mathbf{n}}$ equation. (For any number of complexes, n, use $1, \ldots, n$ for ML ranges and $n, n-1, \ldots, 1$ for the acid.)

ML	HL
$b = K_1\mathbf{L} = [ML]/[M] = R_{1-0}$	$b = \mathbf{H}/K_{a6} = [HL]/[L] = R_{1-0}$
$c = K_2\mathbf{L}$	$c = \mathbf{H}/K_{a5}$
$d = K_3\mathbf{L}$	$d = \mathbf{H}/K_{a4}$
$e = K_4\mathbf{L}$	$e = \mathbf{H}/K_{a3}$
$f = K_5\mathbf{L}$	$f = \mathbf{H}/K_{a2}$
$g = K_6\mathbf{L} = [ML_6]/[ML_5] = R_{6-5}$	$g = \mathbf{H}/K_{a1} = [H_6L]/[H_5L] = R_{6-5}$

(B-1)

Note that both types use the same ratio form, e.g., R_{1-0}. A triprotic acid would start with $b = \mathbf{H}/K_{a3}$ and continue to the K_{a1}, third term. That is, formation constants of the proton complexes are reciprocals of the "dissociation" constants in reverse order ($K_{f1} = 1/K_{a3}$). Next (see Chapters I and III)

$$\bar{\mathbf{n}} = \frac{b[1 + c(2 + d\{3 + e[4 + f(5 + 6g)]\})]}{1 + b[1 + c(1 + d\{1 + e[1 + f(1 + g)]\})]} \qquad \text{(B-2)}$$

The computer uses only one form of parenthesis. For the triprotic (or ML_3) case, e, f, and g are zero. Brief application to the simpler functions for monoprotic calculations are given in Section II-5.

Refer to instructions with software for meanings and uses of specific commands in spreadsheets. Manual rather than automatic calculation modes are desirable when frequent adjustments to a table are to be made. A general explanation of the secant method is presented in Section B-2. Those new to spreadsheet methods might wish to practice with that first.

(a) Simplest Method: Survey to Find pH Unit Range of Intersection. Find the p**H** of a 1:1 buffer having 0.0100 *M* each of citric acid and NaH_2 citrate.

Step 1. Manually search by p**H** units for a delta sign change. Set the calculation menu to manual. Enter formulas for calculation of the three columns (2–4) using **H** values entered in the first column. Choose CALCULATE NOW. Here, $\bar{\mathbf{n}}' = 2.5 - 50\mathbf{D}$ and $\Delta = \bar{\mathbf{n}} - \bar{\mathbf{n}}'$

[K_a values are given in part (b)]:

H	$\bar{n}$	$\bar{n}'$	Δ
0.1	2.98838	−2.5	5.4884
0.01	2.89400	2.000	0.8940
0.001	2.42606	2.450	−0.0239
0.0001	1.75988	2.495	−0.735

Since the answer lies between p**H** 2 and 3 (Δ sign change), we may continue the solution program at p**H** 3 in the spreadsheet as follows.

Step 2. Trial-and-error **H** steps can be entered until a desired precision result is obtained. (Try **H** = 0.002, 0.003, ...) A reader who is new to spreadsheets might try this before proceeding. See also Section II-5.

(b) A Simple Spreadsheet Convergence Method. The **H** choices can be programmed to move the Δ toward zero. A first choice **H** is entered and calculations made as before for the first line. The second **H** is chosen to increase the Δ if the first was negative, meaning the first **H** was too small ($\bar{n} < \bar{n}'$), or the opposite if $\bar{n} > \bar{n}'$. The command IF(Δ<0, H_1*2, H_1/2) will achieve this. [In words this means that if Δ_1 is negative, **H** is doubled; if not, it is halved. These are *true–false* choices for the IF clause.] The factor 2 is arbitrary.

The second line is calculated and its Δ is used with the first to extrapolate (or interpolate) to zero. Using the two points to establish two straight lines (two points and slopes) is called the secant method (see Section B-2) and does not require differentiation to find instantaneous slopes as in the Newton method. From the graphs that follow the reader may verify that the secants intersect at

$$\mathbf{H}_x = \mathbf{H}_1 + (\mathbf{H}_2 - \mathbf{H}_1)/(1 - \Delta_2/\Delta_1) \quad \text{(B-3)}$$

where $\mathbf{H}_x$ is the first approximate intersection value of **H** (column 5). The $\mathbf{H}_x$ is used as $\mathbf{H}_3$ in the third step and this continues until constant **H** is found, Δ is satisfactorily close to zero. An example is shown for the preceding problem, where 0.1 *M* ionic strength constants are given.

Citric acid: at $\boldsymbol{I} = 0.1\ M$, 25°C
$K_1 = 0.0011749$ Case: 0.01 *M* H_3Cit, 0.01 *M* H_2Cit^-
$K_2 = 4.3652 - 05$
$K_3 = 1.8197E - 06$ 2.5 − 50**D**

H	$\bar{\mathbf{n}}$	$\bar{\mathbf{n}}'$	Δ	$\mathbf{H}_x$	pH
5.000E − 04	2.2231	2.4750	−2.52E − 01		3.301
1.000E − 03	2.4261	2.4500	−2.39E − 02	0.0010525	3.000
1.053E − 03	2.4409	2.4474	−6.46E − 03	0.0010719	2.978
1.072E − 03	2.4462	2.4464	−2.04E − 04	0.0010726	2.970
0.0010726	2.4464	2.4464	−1.78E − 06	0.0010726	2.970
0.0010726	2.4464	2.4464	−4.92E − 10	0.0010726	2.970

Note the progression of **H** values, Δ, and p**H**. The final Δ's reflect the changes beyond the digits shown in the $\bar{\mathbf{n}}$ columns, which were shortened from those actually on the spreadsheet. A check can be made by restarting with the first **H** choice ($\mathbf{H}_1$) too large instead of too small, say, 0.002 *M*. No other changes need be entered, just press CALCULATE NOW. Slower convergence with too high $\mathbf{H}_1$ is demonstrated in the next example.

The reader might reproduce this and then change mixtures. (There is a table of citrate mixtures with answers given in the exercises at the end of Chapter III for practice.) Put the new $\bar{\mathbf{n}}'$ formula in the whole of column 3 (FILL DOWN). Change the first **H** if it seems needed, and press CALCULATE NOW. If convergence is not obtained, try other **H** values. This simple program does well for **H** much too small, but high **H** must be close: about twofold high. Higher **H** can produce negative **H** in the following steps, as can be seen on the graphs that follow. The method in (a) can be used to find a good starting **H**.

The following program seems to handle all cases.

(c) A More Flexible and Subtle Spreadsheet Program. Citric acid is used. Enter the six titles for the sheet calculation in columns B–G and the formulas for them in each cell below. The $\bar{\mathbf{n}}$ will use the three K_a's for any system as entered for a specific problem in cells B1–3. Here we use the K_a's for 25°C and 0.1 *M* ionic strength. [If one has chosen to use the standard equations in formation directions, be sure to change K_a's to reciprocal K_f's as shown in Eqs. B-1.

Example 1. Find the rigorous p**H** of the buffer having 0.0100 *M* citric acid and 0.00800 *M* NaH_2 citrate. Thus, $C_{AB} = 0.0460$ *M* and $C_A =$ 0.0180 *M*, so that $\bar{\mathbf{n}}' = 2.311 - 55.6\mathbf{D}$, the formula to put into D5, and FILL DOWN that column, say, 10 lines to start. Equation B-2 or an equivalent form is used in column C. Their difference (Δ) is in column E. Column G is −log 10 (column B) if desired. Columns B and F are described next.

A	B	C	D	E	F	G
$K_{a1} = 0.0011749$					Case: Buffer	
$K_{a2} = 4.3652E-05$			Citric acid		0.01 M H_3Cit	
$K_{a3} = 1.8197E-06$					0.008 M H_2Cit^-	
4	**H**	$\bar{\mathbf{n}}$	$\bar{\mathbf{n}}'$	$\Delta = C - D$	$\mathbf{H}_x$	pH
5	1.000E − 02	2.8940	1.7550	1.14E + 00		2.000
6	5.000E − 03	2.8067	2.0330	7.74E − 01	−5.591E − 03	2.301
7	2.500E − 03	2.6710	2.1720	4.99E − 01	−2.040E − 03	2.602
8	1.250E − 03	2.4902	2.2415	2.49E − 01	7.653E − 06	2.903
9	6.250E − 04	2.2882	2.2763	1.20E − 02	5.934E − 04	3.204
10	5.934E − 04	2.2730	2.2780	−4.97E − 03	6.027E − 04	3.227
11	6.027E − 04	2.2776	2.2775	7.90E − 05	6.026E − 04	3.220
12	6.026E − 04	2.2775	2.2775	5.17E − 07	6.026E − 04	3.220

Column F, starting in line 6, is the interpolation (extrapolation) formula as explained in (b) and illustrated graphically in Chapter III and in Section B-2. It is Eq. B-3:

$$\mathbf{H}_x = \mathbf{H}_1 + (\mathbf{H}_2 - \mathbf{H}_1)/(1 - \Delta_2/\Delta_1)$$

where subscripts 1 and 2 refer to the previous line **H** [column B, line 5 (B5)] and the present (B6) and their corresponding deltas. This is just the straight line intersection method and may refer to the zero crossing of delta or the intersection of the two $\bar{\mathbf{n}}$'s. (See Chapter III.) Near the true **H**, it will be closer to the true **H** than either $\mathbf{H}_2$ or $\mathbf{H}_1$. Over a small range of **H**, the lines are close to the straight approximation being used.

In this case with the starting $\mathbf{H}_1$ so high, the F ($\mathbf{H}_x$) values fall off the graph (negative) until we get closer in rows 8 and 9. Figure III-7*b* shows how this happens with lower $\bar{\mathbf{n}}'$ mixtures at high **H**. The $\bar{\mathbf{n}}$ line secants are too high to meet $\bar{\mathbf{n}}'$. The B formula, halving **H** each time, will eventually lower the secants enough to give positive $\mathbf{H}_x$ values and convergence. Thus, one may surmise that a low first estimate of **H** is desirable for faster convergence. Look back at the previous example, which did start with **H** too low.

We must contrive to select $\mathbf{H}_x$ values for the next line trial so that convergence occurs.

Because of the slopes, Δ will always be positive if the **H** tried is too large and negative if it is too small. This leads to a strategy for column B formulas: (A single formula is not feasible here.)

B5: Choose a first trial **H**. Any guess will do. Since if it is far off, the F lines will go toward the correct value and a better choice can be made in a rerun. In this case we were over 10-fold off, but convergence occurred in just seven lines. This may not happen in the method of part (b) in the preceding.

B6: = IF(E5)>0,B5/2,B5*2). This cuts the first guess in half if it was too high and doubles it if too low. The commas set off true–false choices in the IF statement.

B7: = IF(E5*E6<0,F6,$B6^2$/B5). If Δ's change sign, their product is negative and we take the interpolated $\mathbf{H}_x$ (F6); if not, we adjust **H** in the same direction as the previous line hoping to get still closer. (Note how this happens through line 9, then line 10 crosses zero Δ, and we go to the interpolated $\mathbf{H}_x$ estimate next.)

B8: = IF(E6*E7<0,F7,IF((1−E7/E6)>0.6,F7,$B7^2$/B6)). This is similar to the previous except that we are taking F7 (though it is an extrapolation) if the fractional decrease in Δ is great (over 60%), showing that $\mathbf{H}_x$ has become much closer. The succeeding B cells use this same formula (FILL DOWN).

Note that lines 10–12 check that the $\bar{\mathbf{n}}$'s become equal and Δ approaches zero relative to those above it. Even the first Δ may be very small in basic cases (Na_3Cit), but the relative change will be as great in column E. A rerun or further extension of the columns downward is needed if no sign change in Δ is found. Here we see two changes of sign, and the interpolations agree well so close to the root.

The change factor 2 in the B6 formula and the 0.6 in the B8 formula are arbitrary: good convergence occurs with a range of such choices. Factor 10 and 0.1 worked. The factor 10 might be desirable if one has no idea whether the result is acidic or basic. Convergence has been obtained in 12 lines (5–17) even if the first guess is up to 4 p**H** units off. This results from the regular change in the slopes for our functions ($\mathbf{H}^5$ power for an H_3A case). However, for a general cubic–quadratic pair, the intersections might be multiple and the curves too far from straight to allow so easy a formulation. Remember that our master function has but one positive real root. The plots in Chapters III and V illustrate this.

(d) Bisection Method. A related method of broader application to the solving of polynomials is the bisection method, Norris,[1] p. 74. It will be shown here although it offers little advantage with the favorable $\bar{\mathbf{n}}$ functions we are discussing. It can deal with multiple roots. However it requires the determination of two trial values that bracket the true root being sought. These are easily found by the (**a**) method preceeding—one gives a negative Δ, or, $\bar{\mathbf{n}} - \bar{\mathbf{n}}'$ value and the other a positive Δ value. Then the mean of these bracketing trials will be closer to the true root than one of the starters. The Δ produced by their mean is determined and then the mean is combined with the starter having a Δ of opposite sign to form a new bracketing pair. Continued bisections of bracketing pairs leads to a numerical value of the root to the desired precision. We shall see that this method converges slowly via oscillations, but cannot diverge.

The spreadsheet shows how this can be arranged for the citric acid system example: 0.01 *M* NaH_2Cit. The polynomial here is a fifth power expression in **H**, that is, $\bar{\mathbf{n}} - \bar{\mathbf{n}}' = 0$ for a triprotic system. This passes through zero when the true root, **H**, is used giving equal values of the $\bar{\mathbf{n}}$'s. It is positive for **H** too large and negative for **H** too small. Rows 9 to 11 show the needed portion of a type (a) determination of bracketing **H**'s by a preliminary calculation. Putting these **H** values into the bisection table, rows 16–32, leads to p**H** 3.674 as found in example 2 at the end of Chapter 3. To show the steps in the calculation, we have broken this into more columns than necessary. Formulas here are consistent for $\mathbf{H}_1$ larger than $\mathbf{H}_2$.

In row 16 we place the two bracketing **H** values from the first portion and their mean formula. The next six columns have $\bar{\mathbf{n}}$, $\bar{\mathbf{n}}'$, and their difference, Δ, formulas using the mean value of $\mathbf{H}_{(m)}$ and then using the $\mathbf{H}_1$ value. In column J, we get the product, *p*, of the Δ values. If this is negative, the pair of **H** values used are bracketing, while if this is positive both Δ's must be of the same sign, thus both lying to one side of the true root. This feature is used in the formulas for $\mathbf{H}_1$ and $\mathbf{H}_2$ in rows 17 to 32. For example, for cell A17 we use $\mathbf{H}_1$ = IF(J16<0,A16,C16), and next, B17, $\mathbf{H}_2$ = IF(J16<0,C16,B16). (No spaces are used in these Excel formulas.) The first is read: if *p* is negative, use the previous $\mathbf{H}_1$, and if positive use the mean in C16. Following down the table shows the changes in values that are found from these formulas. Note how the **H** values change, one at a time, the first down and the second up, while the means and resultant p**H** oscillate with decreasing amplitude.

The reader might experiment with both values of **H** to one side to see that convergence to the lower, incorrect, value occurs. Also, wider separation of **H**'s still converge if they bracket the root. For example, one might be convinced that **H** must lie between 0.01 and 0.00001 M and thus omit the first (**a**) method calculation. Convergence is obtained to p**H** 3.674. If these were not bracketing, convergence to p**H** 5 would occur.

	A	B	C	D	E	F
1	Citric acid, K1	0.001174898				
2	k2	4.36516E-05				
3	k3	1.8197E-06				
4						
5						
6						
7	Bracketing H	for 0.01 M	H2Cit-	2-100*D		
8						
9		H	n-bar	n-bar/	delta	pH
10	trials 1	0.001	2.426060	1.900000	0.52605983	3
11	2	0.0001	1.759875	1.990000	-0.2301249	4
12						
13	Bisections	using the two	trials as the	H1 and H2		
14						
15	H1	H2	mean H, m	n-bar (m)	n-bar/ (m)	delta (m)
16	1.0000E-03	1.0000E-04	5.5000E-04	2.250837	1.945000	0.305837
17	5.5000E-04	1.0000E-04	3.2500E-04	2.099741	1.967500	0.132241
18	3.2500E-04	1.0000E-04	2.1250E-04	1.979777	1.978750	0.001027
19	2.1250E-04	1.0000E-04	1.5625E-04	1.892000	1.984375	-0.092375
20	2.1250E-04	1.5625E-04	1.8438E-04	1.939471	1.981563	-0.042091
21	2.1250E-04	1.8438E-04	1.9844E-04	1.960376	1.980156	-0.019781
22	2.1250E-04	1.9844E-04	2.0547E-04	1.970250	1.979453	-0.009203
23	2.1250E-04	2.0547E-04	2.0898E-04	1.975055	1.979102	-0.004047
24	2.1250E-04	2.0898E-04	2.1074E-04	1.977426	1.978926	-0.0015
25	2.1250E-04	2.1074E-04	2.1162E-04	1.978604	1.978838	-0.000234
26	2.1250E-04	2.1162E-04	2.1206E-04	1.979191	1.978794	0.000397
27	2.1206E-04	2.1162E-04	2.1184E-04	1.978898	1.978816	8.19E-05
28	2.1184E-04	2.1162E-04	2.1173E-04	1.978751	1.978827	-7.59E-05
29	2.1184E-04	2.1173E-04	2.1179E-04	1.978824	1.978821	2.99E-06
30	2.1179E-04	2.1173E-04	2.1176E-04	1.978788	1.978824	-3.65E-05
31	2.1179E-04	2.1176E-04	2.1177E-04	1.978806	1.978823	-1.67E-05
32	2.1179E-04	2.1177E-04	2.1178E-04	1.978815	1.978822	-6.88E-06

A briefer one-row method using the ITERATE function can be devised. After getting the row values with the first choice of **H**'s, simply replace them with the formulas by calculating the new **H** values in two added columns, L and M, with the IF formulas just given, and then making A and B equal to these. Then conduct iterations to constancy. The cubic example (**a**) in Section B-3 can be treated in this way for practice, compare Figs. B-2*a*, *b*, *c*. When one chooses an interval having two or three roots as found in this example, convergence is obtained only to one of them. Without further investigation one would not know whether this represents three equal roots or one of three different roots to the cubic.

	G	H	I	J	K
1					
2					
3					
4					
5					
6					
7					
8					
9					
10					
11					
12					
13					
14					
15	n-bar (H1)	n-bar/ (H1)	delta (H1)	p=F*I	pH= log10(C)
16	2.426060	1.900000	0.52605983	0.16088856	3.25964
17	2.250837	1.945000	0.305837	0.04044428	3.48812
18	2.099741	1.967500	0.1322413	0.0001358	3.67264
19	1.979777	1.978750	0.00102694	-9.486E-05	3.80618
20	1.979777	1.978750	0.00102694	-4.322E-05	3.73430
21	1.979777	1.978750	0.00102694	-2.031E-05	3.70238
22	1.979777	1.978750	0.00102694	-9.451E-06	3.68725
23	1.979777	1.978750	0.00102694	-4.156E-06	3.67989
24	1.979777	1.978750	0.00102694	-1.54E-06	3.67625
25	1.979777	1.978750	0.00102694	-2.401E-07	3.67444
26	1.979777	1.978750	0.00102694	4.0791E-07	3.67354
27	1.979191	1.978794	0.00039721	3.2519E-08	3.67399
28	1.978898	1.978816	8.1869E-05	-6.215E-09	3.67422
29	1.978898	1.978816	8.1869E-05	2.4442E-10	3.67410
30	1.978824	1.978821	2.9855E-06	-1.089E-10	3.67416
31	1.978824	1.978821	2.9855E-06	-4.997E-11	3.67413
32	1.978824	1.978821	2.9855E-06	-2.053E-11	3.67412

Spreadsheet and Formulas for the pH Calculation: Mixture of 0.02 *M* Formic, Acetic, and Propionic Acids in Section II-6

	A	B	C
1	H	$(\Sigma\ \alpha_1 C) + \mathbf{D} - 0.06$	pH
2	0.005	0.00376	2.30103
3	0.004	0.00247	2.39794
4	0.002090	−0.0007	2.67979
5	0.002513	0.00015	2.59988
6	0.002438	9.93E − 06	2.61303
7	0.002432	−1.4E − 07	2.61397
8	0.002432	1.3E − 10	2.61396

See the setup of this problem, Section II-6. Line 1 gives the titles for each column. The formula bar for B2 in Excel terms is (C times the sum of the three α's minus 0.06) and always starts with the equals sign, =0.02/(1+(10^−3.56)/A2) + 0.02/(1+(10^−4.56)/A2)+0.02/(1+(10^−4.67)/A2)+A2−(10^−13.8)/A2−0.06. (Note: no spaces are used in actual Excel formulas.) The first **H** is in A2, while the second formula for A3 is =IF(B2>0,0.8*A2,1.2*A2). The succeeding **H** values are obtained by the interpolation formula, aiming at zero for the Δ in B. The formula for A4 is =A3+(A2−A3)/(1−B2/B3) (line 4 and fill down).

Example 2. A six-K calculation follows, the 0.100 M Ni(II), 0.600 M NH_3 mixture, in Chapter V (Fig. V-1). Here $\bar{\mathbf{n}}' = 6 - 10[NH_3]$ and $L = NH_3$. These are formation constants,

A	B	C	D	E	F
1 K_1 =	630.0	K_4 =	15.9		
2 K_2 =	173.8	K_5 =	5.62	Ni–NH3, 6:1 case	
3 K_3 =	53.7	K_6 =	1.07		

4	**L**	$\bar{\mathbf{n}}$	$\bar{\mathbf{n}}'$	Δ	**H** (or **L**) int.
5	0.0100	1.9824	5.9000	−3.918	
6	0.0200	2.5522	5.8000	−3.248	0.06849
7	0.0400	3.1342	5.6000	−2.466	0.10306
8	0.0800	3.7207	5.2000	−1.479	0.13998
9	0.1600	4.2816	4.4000	−1.184E − 01	0.16696
10	0.1670	4.3141	4.3304	−1.623E − 02	0.16807
11	0.1681	4.3192	4.3193	−1.501E − 04	0.16808
12	0.1681	4.3192	4.3192	−1.875E − 07	0.16808
13	0.1681	4.3192	4.3192	−2.160E − 12	0.16808
14	0.1681	4.3192	4.3192	0.000E + 00	#DIV/0!

Here we started much too low, $\mathbf{L} = 0.01$. Even though Δ did not change sign, it seems to be approaching zero to the number of figures shown. A rerun with the first guess as 0.1681 will check this convergence. Note how far from the maximum, 6, the $\bar{\mathbf{n}}$ is with these rather weak complexing constants. (The K's given are not logs.) The log **L** and the α values could be added if desired. They are omitted here for space considerations.

Details of the spreadsheet for a complex solubility function ($CaCO_3$ in water as a function of pressure of CO_2 gas) are given in Chapter VI.

B-2. GRAPHICAL ILLUSTRATIONS

Graphical illustrations of the functions discussed in the preceeding with K and C parameters chosen to bring out the slopes in crucial ranges are shown in Section III-7. Here we show a sample with curves chosen to make the secant method visible. It is harder to see in our usual **H** calculations because the $\bar{\mathbf{n}}, \bar{\mathbf{n}}'$ curves are so nearly straight in the narrow **H** range used that the secants are almost on the $\bar{\mathbf{n}}$ lines. Thus, convergence is rapid.

Example. The two functions are

$$\text{(i)} \quad 0.1x^3 + 2x^2 + 3x + 4 \qquad \text{(ii)} \quad 30 - 2x^2 - x$$

These are like the $\bar{\mathbf{n}}$ pair in that the first has positive slope and the second negative slope for positive x. Thus, they should have only one simultaneous positive solution. In **Fig. B-1** we plot their values (heavy lines) for $x = 0, \ldots, 3$ at intervals of 0.1. Then we put straight lines through the line values for $x = 0.5$, 1.5 to represent the secant lines intersecting at X_a (2.365), and then the secant lines for points at 2.5 and 3 intersecting at X_b (2.1264). We see that both have a positive error but are much closer to the true intersection (2.056) for

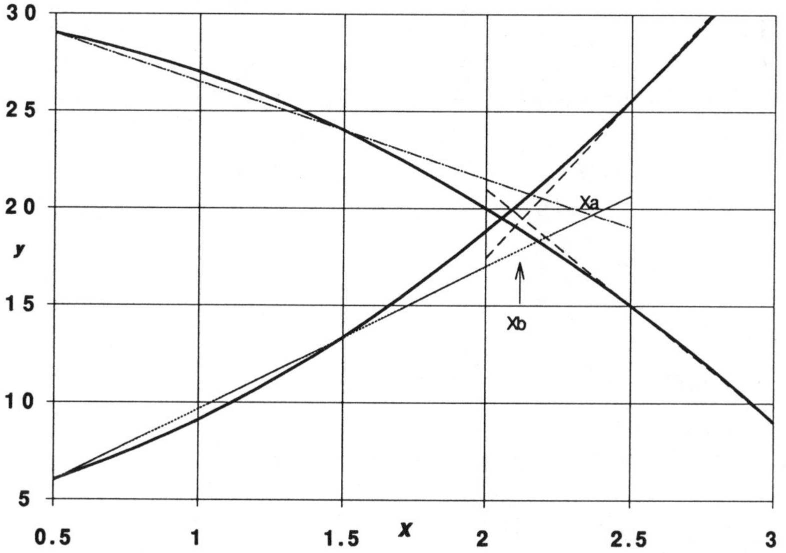

Fig. B-1. Two pairs of secant lines on two curves: bold lines for functions (i) and (ii). X_a secant lines are dotted, X_b lines are dashed.

this fifth-power system in x. Putting these two new x values into the equations and calculating the new secant intersection gives 2.0602. A trial with 2.05 crosses the Δ sign and allows interpolation to 2.056, which then gives 19.49 as the value of each function (columns B and C). Such trials can be added at the end of a spreadsheet by filling down the formulas and inserting any desired x values as done in the last row, x_{sim} for a check of the simultaneous value of x.

The intersections are calculated by the formula for a straight line given by two points B and C as described by Eq. B-3. That formula (given below) is in column E for that row and the previous one, showing how well it converges toward the true intersection for this pair of curves. (A less favorable case is described in the next section.) Early rows are not shown, but that for $x = 1.4$ was used to find the intersection in row 2.

The portion of the spreadsheet that follows is for x values 1.5, . . . , 2.5, which include the Δ sign change between 2.0 and 2.1.

SPREADSHEET FOR FUNCTIONS (i) AND (ii) (PARTIAL)

	A	B	C	D	E	
1	x	$f_{(i)}$	$f_{(ii)}$	Δ	Intersection	
2	1.5	13.3375	24	10.6625	2.1569	
3	1.6	14.3296	23.28	8.9504	2.1228	
4	1.7	15.3713	22.52	7.1487	2.0968	
5	1.8	16.4632	21.72	5.2568	2.0779	
6	1.9	17.6059	20.88	3.2741	2.0651	
7	2	18.8	20	1.2	2.0579	
8	2.1	20.0461	19.08	−0.9661	2.0554	
9	2.2	21.3448	18.12	−3.2248	2.0573	
10	2.3	22.6967	17.12	−5.5767	2.0629	
11	2.4	24.1024	16.08	−8.0224	2.0720	
12	2.5	25.5625	15	−10.5625	2.0842	
r_{sim}	2.056	19.49	19.49	−0.00164	2.0559	Check

The formulas in terms of cells are as follows:

Column A cells: choices of x for plotting.
First line of B, B2: $f_{(i)} = 0.1*A2\hat{}3+2*A2\hat{}2+3*A2+4$.
C: $f_{(ii)}=30-2*A2\hat{}2-A2$.

D: Δ=C2−B2.
Second E: Intersection =A2+(A3−A2)/(1−D3/D2).

B-3. ITERATION METHODS

Spreadsheets can handle certain kinds of cross references to uncalculated terms (loops). Examples of such cases occur in the first example on solubility in Chapter VI and the sulfuric acid problem in Chapter II. In both, the ionic strength is iterated with the unknown ion concentrations to determine a corrected K value for use in the equilibrium equations. In Chapter V, with Bjerrum's data for Cu(II)–NH_3 complexing, one starts with approximate constants deduced from the data and continues using selected averages of newly calculated K's. See these sections for examples of spreadsheet methods.

Solutions of Higher Order Equations

Decartes's rule of signs states that the maximum number (there may still be none) of real positive roots of equations equals the number of sign changes. The equation must be in order of powers of x. For example,

(a) $$x^3 - 2x^2 - 6x + 3 = 0$$

This has two changes of sign and thus may have two positive real roots. What these are will be shown by several methods. Clearly, if there are no sign changes, the polynomial can have no zero value for any positive x. The general complete relation for single monoprotic acid–base systems (Chapter II) can be organized by powers as

(b) $$x^3 + (C_0 + K_a)x^2 - (K_a C_1 + K_w)x - K_a K_w = 0 \qquad x = \mathbf{H}$$

or

(b′) $$x^3 + (C_0 - C_{sa} + C_{sb} + K_a)x^2 - (K_a C_{AB} + K_w)x - K_a K_w = 0$$

Relation (b′) is the broader one including a strong acid and base (see Eq. II-8). Since concentrations (C's) and constants (K's) are positive, this equation has at most one positive real root. It has one *change* of sign: the negative sign, once arrived at in either the second or third term, remains. The other two roots may be negative and/or imaginary. This accords with our chemical intuition that a specified weak acid–base mixture has only one meaningful p**H**.

The equation for the ampholyte (0.01 M H_2A^-) of the triprotic acid (pK's 2, 3, 4) was shown in Chapter III to have the standard form

$$\text{(c)} \qquad 10^9\mathbf{H}^5 + 2\times 10^7\mathbf{H}^4 + 10^4\mathbf{H}^3 - 99\mathbf{H}^2 - 0.02\mathbf{H} - 10^{-13.80} = 0$$

Again, the single variance of signs tells that there is not more than one positive real root. With the constants closely spaced and in this acidic solution (approximate p**H** 2.5 for this ampholyte; see the α $\bar{\mathbf{n}}$ diagram, Figs. III-1 and III-2), one sees that the final K_w term is negligible, the others being on the order of 1–10^{-4}. Let $\mathbf{H} = x/100$ and multiply through by $10/x$ to obtain a simpler form to evaluate:

$$\text{(c}'\text{)} \qquad x^4 + 2x^3 + 0.1x^2 - 0.099x - 0.002 = 0$$

The single positive root of either form will be the same. Sometimes a physically impossible extraneous root may be introduced by simplifications of complete equations. Reduction of relation (b) to a quadratic can, for some unusual conditions, give two positive roots, but one will be higher than is chemically possible (more acid than is put in).

Returning to relation (a), let us solve it in several ways and demonstrate why solution of the standard power equation is often much more difficult than use of $\bar{\mathbf{n}}$ forms arising naturally in chemistry to provide easy iterating functions or spreadsheet computations.

Plotting the polynomial of relation (a) ($=y$) from $x = -4$ to $x = +4$ shows a simple cubic crossing $y = 0$ three times, near -2, $+0.5$, and $+3.5$ (**Fig. B-2*a***). Rearranging the expression may be done in many ways to obtain pairs of useful functions, for example:

$$x^3 - 2x^2 = 6x - 3 \quad \text{(a cubic cut by a line; Fig. B-2}b\text{)}$$

$$x^2 - 2x - 6 = -3/x \quad \text{(a parabola and a hyperbola)}$$

$$x^3 = 2x^2 + 6x - 3 \quad \text{(a cubic cut by a parabola; Fig. B-2}c\text{)}$$

The reader might plot other arrangements of this equation to show the same results with different curves. Construct a spreadsheet to calculate each function for given small increments of x. (The roots are -1.925423, 0.4480706, and 3.477352.)

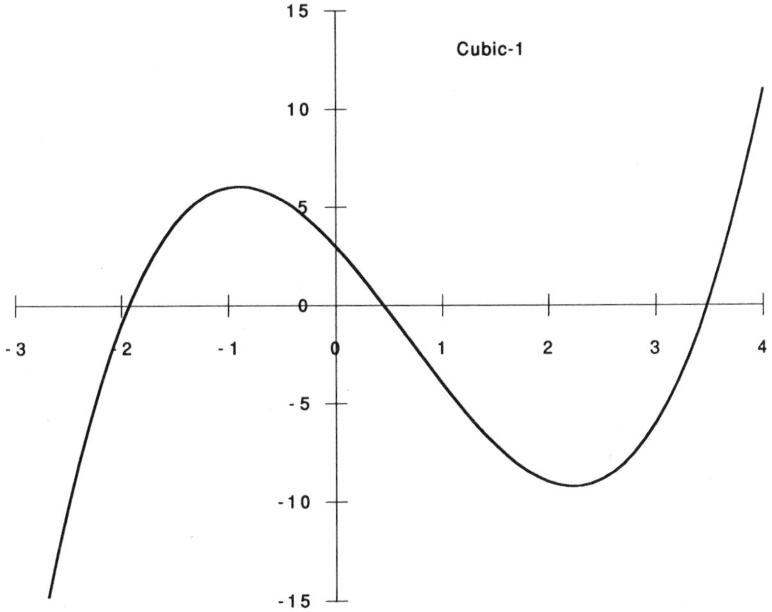

Fig. B-2*a*. Plot of the full cubic $f(x)$, relation (a).

Compare the simpler spreadsheet secant line method for solving these. Since there are now three roots, we find that the method converges to one root depending on the relative slopes at the starting points chosen. This problem does not arise with the preceding equilibrium functions. Their lines have no change in the sign of their slopes and thus no more than one possible intersection. Investigate the secant method near $x = 1, 2, \ldots$ in the two-line forms of the preceding cubic.

The *Newton–Raphson method* is based upon the idea that a tangent line to the curve at a point near a root will intersect zero at a point closer to that root. For $f(x) = 0$ and $f'(x)$ the derivative, simple triangulation gives the improved root x_{r+1} in terms of the first approximation, x_r, as

$$x_{r+1} = x_r - f(x)/f'(x)$$

Trials may be inserted until satisfactory precision is obtained. For relation (a) we have $f'(x) = 3x^2 - 4x - 6$. Trying $x = -2$ gives $f(x) = -1$ and $f'(x) = 14$. Thus,

$$x_{r+1} = -2 - (-1)/14 = -1.9286$$

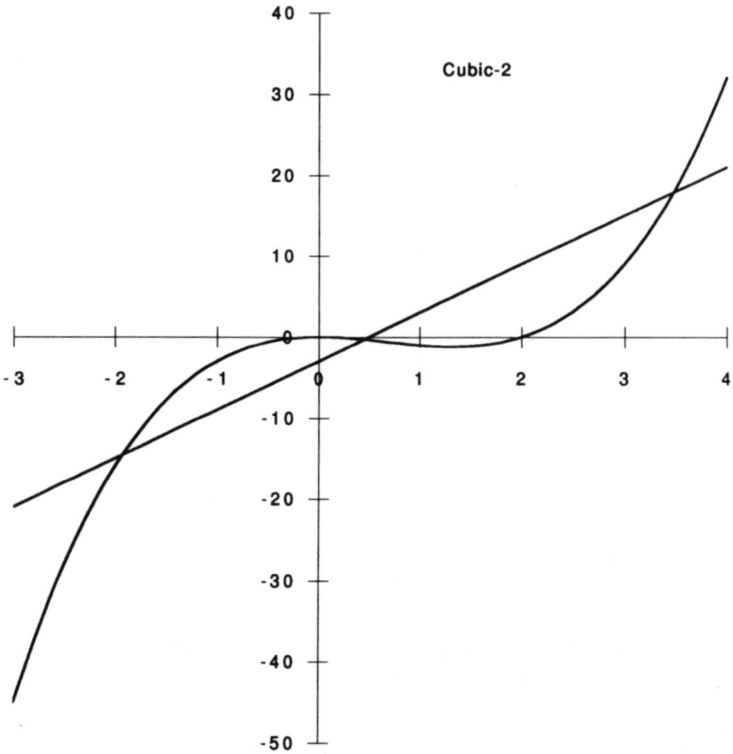

Fig. B-2*b*. Plot of relation (a) arranged as a line and a cubic.

Continuing with this as the new trial root gives the root −1.9254, which remains unchanged in further trials. Then, trying the other roots near 0.5 and 3.5 quickly converges to 0.44807 and 3.37735. Continuous functions of the type we find in equilibrium problems converge easily. This may be done in a spreadsheet on which each successive line uses the x_{r+1} found in the previous line or a circular automatic iteration may be used. If the first guess is close, only a few iterations are needed and a calculator serves. For higher power cases like relation (c), a computer is quite desirable. The Newton–Raphson iteration for relation (c′) yields $x = 0.2009$, so that $\mathbf{H} = 0.002009$ and **pH** is 2.697. The plot of the x^4 polynomial for relation (c′) shows that widely extended values of x in both directions have no other zero crossing for positive x. Examine relation (c′) to see why this is so: As $x \to 0$, $f(x) \to -0.002$, and as x becomes large (left on graph), $f(x)$ rises rapidly above $x = 1$ ($\mathbf{H} = 0.01\ M$). (This is the reasoning behind the Cartesian rule.) When such a narrow range is used and the function varies gradually, the preceding secant line (a straight line

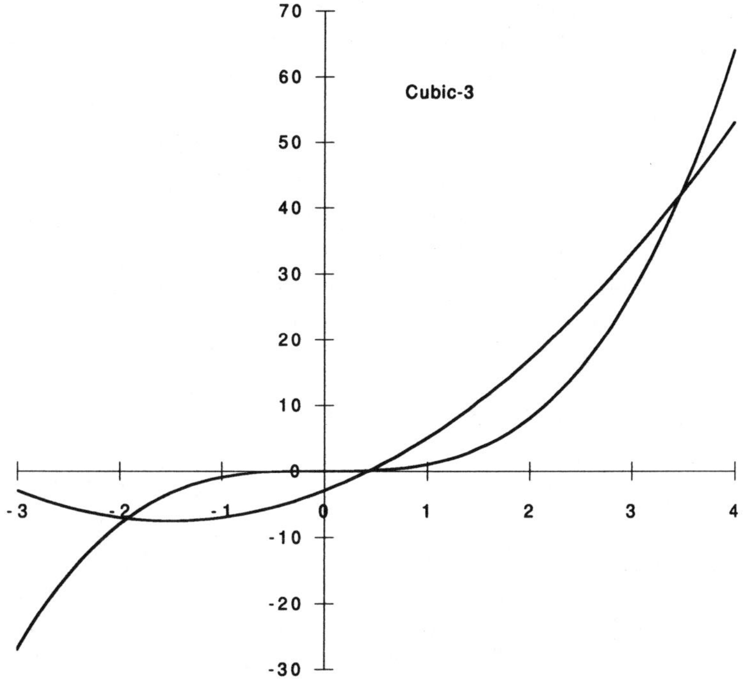

Fig. B-2*c*. Plot of relation (a) arranged as a cubic and a parabola.

approximation) works as well and saves differentiation and evaluations.

The point of these exercises is that many high power equilibrium equations are naturally broken into two functions one of which is the same for every mixture involving that system. That function is $\bar{\mathbf{n}}$, which always increases with **H** (or **L**). The other is the quadratic $\bar{\mathbf{n}}'$, having slope everywhere opposite in sign to that of $\bar{\mathbf{n}}$. With a microcomputer, or even a programmable calculator, these $\bar{\mathbf{n}}$ terms can be evaluated for use in many situations. This method is exactly based on the chemistry involved and makes a securely founded approach to equilibrium that this text tries to demonstrate. The Newton–Raphson method works well, but getting the standard power form of **H** equations and their derivative is tedious and unnecessary for such favorable functions. The $\bar{\mathbf{n}} = \bar{\mathbf{n}}'$ (secant rather than tangent) method described in the preceding is faster even with simple computation facilities.

With more sophisticated computer systems, the intersection (or root of $\bar{\mathbf{n}} - \bar{\mathbf{n}}' = 0$) can be found in seconds. For example, using the software, Mathematica*, one may use the command N[Solve[a==0,x]] to obtain the three numerical roots. Repeating with relation (c) gives the five roots, three negative, one zero, and one positive, 0.002009327. Relation (c′) gave four roots; only one was positive, the one obtained in relation (c). Alternatively, the command FindRoot[c==0,(x,0.01)] will find the positive root starting with 0.01 using Newton's method. Systems of simultaneous equations and matrices can be solved numerically to deal with the evaluation of equilibrium constants starting with data, $\bar{\mathbf{n}}$ or F_0 vs. **H** or **L** as described in the chapters and in what follows.

B-4. SIMULTANEOUS LINEAR EQUATIONS

The $\bar{\mathbf{n}}$, **L** (or **H**) data used in Eqs. IV-8, IV-9, and V-5 to V-7 give sets of simultaneous equations that are linear in the unknown equilibrium constants. Bjerrum's data for Cu(II) ammonia are solved for K values in Chapter V by an iterative process (called the Gauss–Seidel, successive displacement). Brute force Gaussian substitution to eliminate down to one K followed by back substitution to get the other constants is shown here. It is longer to program and very tedious to do by hand owing to the many sign changes. We chose four data with $\bar{\mathbf{n}}$ near "half-integral" values, which gives a good diagonal system that is well conditioned and thus easy to work with preserving significant digits. These properties also make the iterative procedure work well. (See the book by Norris[1] for a clear and detailed explanation of these methods with numerical examples.) The equation used is

$$\bar{\mathbf{n}} = (1 - \bar{\mathbf{n}})\beta_1\mathbf{L} + (2 - \bar{\mathbf{n}})\beta_2\mathbf{L}^2 + (3 - \bar{\mathbf{n}})\beta_3\mathbf{L}^3 + (4 - \bar{\mathbf{n}})\beta_4\mathbf{L}^4$$

The first row, R1 in the table that follows, is made by using the **L**, $\bar{\mathbf{n}}$ data given to get the coefficients $(1 - \bar{\mathbf{n}})\mathbf{L}, \ldots$ shown.

The 4×4 matrix of coefficients (of the unknown β's), block 1 of the table, is treated as shown to scale them and eliminate the six values below the diagonal (Block 4). The last row (4) is then just

$$(\text{the } \beta_4 \text{ coefficient})\ \beta_4 = \text{the } \bar{\mathbf{n}} \text{ term left}$$

$$\beta_4 = 28{,}433/5.765E - 9 = 4.932E + 12$$

*Mathematica is the trademark of Wolfram Research Inc. The book *Mathematica*, Addison-Wesley, Reading, MA, 1988, describes its uses; solving as described here is found on pp. 74 ff and 476–478.

This is substituted in the final row, R3, to get β_3, and so on.

Method. Many approaches will solve such a set, but the following process helps keep magnitudes and rounding errors low. (This is a shortcut of the method of elimination as taught in elementary algebra.) In the fourth block, we eliminate using the equation having the largest coefficient in the column being worked on, i.e., move row 4 (R4 of block 3) to be the new R2*. Multiply this row by the factor $-8.702/-142.559 = 0.061041$ and subtract each term from the one below it in the new R3*. (Move R2 of block 3 to R4.) Watch signs carefully! Then repeat with factor $2.1113/-142.559 = -0.014808$ and subtract from the new R4, the old R2. Next, to get zero in column 3 of R4, multiply R3 by the factor $-1.6799/9.124 = -0.18412$ and subtract from R4 to get the final R4*. Working through this or Norris's examples will clarify the method for those wishing to use it. The process can be programmed, and Norris gives a simple program. Mathematica solves such sets.

BLOCK 1

Bjerrum Data		Coefficients of β's			
$[NH_3] = L$	$\bar{n}$	$(1-\bar{n})L$	$(2-\bar{n})L^2$	$(3-\bar{n})L^3$	$(4-\bar{n})L^4$
0.0000462	0.486	2.3747E − 05	3.2315E − 09	2.4791E − 13	1.6009E − 17
0.000535	1.877	−0.0004692	3.5206E − 08	1.7197E − 10	1.7393E − 13
0.00229	2.784	−0.0040854	−4.111E − 06	2.5939E − 09	3.3441E − 11
0.02265	3.743	−0.062129	−0.0008942	−8.634E − 06	6.764E − 08

BLOCK 2

Scale: Divide each row by absolute value of largest coefficient.

R1 20466	+1	0.0001361	1.044E − 08	6.742E − 13
R2 4000.42	−1	7.5034E − 05	3.665E − 07	3.707E − 10
R3 681.5	−1	−0.0010063	6.3492E − 07	8.1855E − 09
R4 60.244	−1	−0.014392	−0.000139	1.0887E − 06

BLOCK 3

Add R1 from Block 2 to the other 3.

20466	1	0.0001361	1.044E − 08	6.742E − 13
24466.43	0	0.00021113	3.7694E − 07	3.7137E − 10
21147.5	0	−0.0008702	6.4536E − 07	8.1862E − 09
20526.24	0	−0.0142559	−0.000139	1.0887E − 06

BLOCK 4

	R1* 20466	1	0.0001361	1.044E − 08	6.742E − 13
	R2* 20526.24	0	−0.014256	−0.000139	1.0887E − 0
	R3* 19894.56	0	0	9.124E − 06	−5.827E − 08
	R4′ 24770.4	0	0	−1.68E − 06	1.6493E − 0

Change R4′ to R4* below.

Final	R4* 28433.3	0	0	0	5.7646E − 0

Results by back substitution		Log beta	
β_4	4.932E + 12	12.693	from the final R4*
β_3	3.368E + 10	10.527	from R3*
β_2	4.689E + 07	7.671	from R2*
β_1	1.373E + 04	4.138	from R1*

These results are practically identical to those obtained in Chapter V from the full set of Bjerrum's data. The log K values can be obtained simply by subtracting the log β values. The same Bjerrum data are used to illustrate a variety of calculation methods in a recent text by Connors. See the references at the end of Chapter V.

EXERCISES

1. Differentiate the Davies equation for ionic activity coefficients and solve. Set equal to zero to find the $\boldsymbol{I}$ at which the minimum f occurs. Show that ionic charges have no effect on this. See Appendix A.
2. Program a spreadsheet table for calculation of α's, $\bar{\mathbf{n}}$, and $\bar{\mathbf{n}}'$ for a triprotic acid, H_3A, having all pK_a's 2.000, for various mixtures of low concentrations, 0.01–0.05 M, of the pure acid, its ampholytes, and 2:1 buffers. See Fig. III-17. Generalize about deviations of usual approximations related to magnitude and closeness of K_a's and to concentrations.

REFERENCE

1. Norris, A. C., *Computational Chemistry*, Wiley, New York, 1981, p. 72 and Ch. 5.

APPENDIX C

EXPERIMENTAL AIDS AND PREPARATIONS

Standard reference works of analytical chemistry supply directions for quantitative laboratory work. Here are gathered a few methods revised or originated by the author that may be of help to others in the field of precision wet chemical work.

C-1. NBS BUFFER STANDARDS

Table C-1 shows the $pa_H(S)$ values of the buffers made from their standard materials. These are on an activity scale [$pa_H(S)$] defined by Bates.[1] [The National Bureau of Standards (NBS) is now called the National Institute of Standards and Technology.]

Measurements

For high precision pH work, one needs several buffers including one or more primary standard grade solutions covering the range of pH in use. Using temperature control to 25°C, run through the series of standards setting the instrument slope with two of them. Repeat with two others and compare results. Then read values for the unknown solutions and repeat the standard series, at least the buffers nearest the unknowns. With large volumes and/or covered containers and inert atmosphere, reproducibility may be within about twice the readability of the instrument. Accuracy may be the same within the

TABLE C-1. pH(S) of NBS Primary Standards from 0 to 95°C

Temperature (°C)	KH Tartrate (Saturated, at 25°C)	KH_2 Citrate ($m = 0.05$)	KH Phthalate ($m = 0.05$)	KH_2PO_4 ($m = 0.025$), Na_2HPO_4 ($m = 0.025$)	KH_2PO_4 ($m = 0.008695$), Na_2HPO_4 ($m = 0.03043$)	Borax ($m = 0.01$)	$NaHCO_3$ ($m = 0.025$), Na_2CO_3 ($m = 0.025$)
0	—	3.863	4.003	6.984	7.534	9.464	10.317
5	—	3.840	3.999	6.951	7.500	9.395	10.245
10	—	3.820	3.998	6.923	7.472	9.332	10.179
15	—	3.802	3.999	6.900	7.448	9.276	10.118
20	—	3.788	4.002	6.881	7.429	9.225	10.062
25	3.557	3.776	4.008	6.865	7.413	9.180	10.012
30	3.552	3.766	4.015	6.853	7.400	9.139	9.966
35	3.549	3.759	4.024	6.844	7.389	9.102	9.925
38	3.548	3.755	4.030	6.840	7.384	9.081	9.903
40	3.547	3.753	4.035	6.838	7.380	9.068	9.889
45	3.547	3.750	4.047	6.834	7.373	9.038	9.856
50	3.549	3.749	4.060	6.833	7.367	9.011	9.828
55	3.554	—	4.075	6.834	—	8.985	—
60	3.560	—	4.091	6.836	—	8.962	—
70	3.580	—	4.126	6.845	—	8.921	—
80	3.609	—	4.164	6.859	—	8.885	—
90	3.650	—	4.205	6.877	—	8.850	—
95	3.674	—	4.227	6.886	—	8.833	—

pH range of about 4–10. This is about ±0.002 at best for recent research instruments reading to 0.001 pH unit. Unknown pH values will then have reasonably reliable significance as $-\log a_H$ *if* ionic strength is close to the standard series (0.05–0.1 *M*) and the inorganic ions accounting for most of this are K^+, Cl^-, and perhaps a few other small ions. Researchers please should consult the book by Bates.[1] pH measurement and interpretation are not simple. Much reported pH data have no significance beyond 0.1 unit.

C-2. LABORATORY PREPARATION OF BUFFERS

Quite accurate secondary buffer standards can be prepared from reagent grade chemicals by the precision ratio method. A few examples are given here from Ref. 2

Rapid titration methods for preparation of solutions of precise buffer ratio are described. These avoid errors caused by incorrect quantities of hygroscopic solids and the frequently uncertain stoichiometry of the solid salts of polyprotic acid systems. Only one solution of known concentration is required in each case and the pH (NBS pa_H scale is intended in this section) of the standard is quite insentitive to the concentration of the starting acid. Large quantities can be made, reserving expensive primary standards for occasional checking of these working solutions.

The method consists in titrating a weak acid with NaOH to a suitable end point and adding another known aliquot of the original acid solution to this conjugate base to obtain a buffer of accurately known ratio. This has advantages over weighing solid conjugates that may not be available pure and dry and thus do not yield the proper buffer ratio. This method requires only one moderately accurate stable stock solution, the acid. The titration end point and duplicate pipetting lock in the buffer ratio to better than one part per thousand (or the limit of a repeat pipetting). A more extended leapfrog method in the phosphate buffer scheme makes each addition precisely determine the next, starting with a measured sample of KH_2PO_4, adding NaOH and then H_3PO_4 to replicate the NBS 6.86 pH standard. Only the first need be precisely known since it will determine the amounts of the others at the successive end points.

Preparations

For even more precise results, the acid may be made a little high, titrated with standard base, and then it, or the final buffer, diluted to the correct volume. In any case, the buffer ratio is correct. The NBS buffers are defined for molality, a temperature-independent mass basis. The densities and dilution coefficients of the buffers are given by Bates[1] (p. 95). The densities are within three parts per thousand of 1.000 g/ml for the buffers in this work. Since the buffer ratios are unaffected, it is unlikely that the measurable pH will differ detectably if molarity is used. However, the molarity solutions that follow may be changed to molality by adding the volume of water given after the density in each case. This was calculated from the density and mass of salts. Buffers may be kept sterile by addition of a small (1-mm) crystal of thymol to the storage bottle. Storage in many small screw-capped bottles filled to the top preserves them from air contact if they are used serially.

Acetate. 0.1000 *M* acetic acid–sodium acetate, pa_H 4.652, from 20 to 30°C, is defined for molarity in this case.

Stock solutions: 0.400 *M* acetic acid. Weigh 24.00 g (in air) of reagent glacial acetic acid, 99.7%, into a liter volumetric flask and make up to the mark. [Glacial acetic acid can be easily purified by partial crystallization, melting point 16.60°C (discarding the liquid). Then if the remainder has the correct freezing point on a calibrated thermometer, it is a reliable standard acid. Its purity is given by

$$\text{wt}\ \% = 100 - 0.53(16.60 - t_f)$$

where t_f is the freezing point of the sample in question. For example, a sample that freezes at 16.30°C is 99.84% acetic acid if the impurity is assumed to be water.] 0.5 *M* NaOH. Dilute 5 ml of 19 *M* (50%) to 200 ml just before use. It need not be standardized but should be fresh so that carbonate is low.

Pipet 25 ml of the acetic acid into a 100-ml volumetric flask. Put in a drop of phenolphthalein and add NaOH to the pink end point. Pipet in another 25 ml of the acetic acid and dilute to the mark. This buffer has proved stable and reliable over years of use, checking against other standards.

Phosphate. 0.0250 *m* in KH_2PO_4 and in Na_2HPO_4. pa_H6.881 at 20°C and 6.863 at 25°C. −0.003/degree. $d = 1.0028$ g/ml (4.1 ml water/l).

Stock solutions: 0.500 *M* KH_2PO_4 (dry the solid reagent at 110°C and weigh 17.01 g/250.0 ml). 0.500 *M* phosphoric acid, 34 ml of the 85% acid per liter. 1 *M* NaOH, 52 ml 50% (19 *M* NaOH) per liter.

A. Place the NaOH in a buret and read precisely to 0.01 ml. Pipet 50.00 ml of the KH_2PO_4 into a beaker with about 100 ml water and titrate carefully with the NaOH until pH 9.4 is reached. Record the volume of NaOH, V_b. Reserve this beaker. Into another beaker with about 100 ml water measure precisely V_b milliliters of the NaOH. Place the H_3PO_4 in a buret, read, and run in about 23 ml. Stir, enter the pH electrodes into the solution, and continue with the acid until pH 9.4 is reached. Record V_a milliliters of acid used. Measure in precisely this volume additional of H_3PO_4. Transfer, with rinsing, all the contents of both beakers to a volumetric flask and make up to 1 l. Add 4.1 ml of water to convert this solution to molality at 25°C.

B. This brief all-sodium method is also reliable. Pipet 75 ml of 0.500 *M* H_3PO_4 into a 500-ml beaker and add about 200 ml water and 73 ml NaOH or until the pH nears 8 using a pH meter. Then go by drops until pH 9.4 is reached. Pipet in 25 ml more of the acid and make up to 1 l in a volumetric flask. This makes the correct buffer ratio and ionic strength, but all cations are Na^+. One-third of the cations are K^+ in the NBS standard. We find that the all-sodium ion buffer is indistinguishable from the NBS standard, within 0.002 unit, when used on meters reading to 0.001 unit.

The NBS solids must be properly dried to obtain agreement. The dried solid reagent grade materials may not give reliable results since the stoichiometry may be incorrect and the Na_2HPO_4 is wet and must be thoroughly dried for NBS grade use. The KH_2PO_4 is the more reliable reagent so that the method given here largely avoids the problem.

Citrate. 0.0500 *m* KH_2 citrate. pH 3.788 at 20°C, 3.776 at 25°C. 0.002/degree. $d = 1.0029$ g/ml (8.5 ml water/l).

A stock solution of 0.333 *M* citric acid was prepared by dissolving 32.0 g anhydrous citric acid, or 35.0 g of the monohydrate, in water and diluting to 500 ml. Hydrated citric acid loses water easily so it is best to prepare the anhydrous form. Dry the reagent over sulfuric acid (preferably in vacuum) for several days, then place it over Drierite. The acid titrates close to 100.0% after a week. (If the hydrate is heated, it melts to give a solution that dries to form a cake.) Approximately 1 *M* KOH or NaOH is made as in the preceding. The NaOH may not give quite the same pH owing to possible specific ion activity effects and junction potential. The 50% NaOH reagent solution usually contains about one part per thousand carbonate, while the 45% KOH reagent solution contains about 1%

carbonate, which may introduce measurable positive pH error if the CO_2 is not evolved.

Pipet 50 ml of the 0.333 *M* citric acid solution into a 500-ml beaker and add the base until the pH is 8–9 using a pH meter or phenolphthalein. It takes about 50 ml of 1 *M* base. Pipet in 100 ml more of the citric acid and dilute to 1.000 l in a volumetric flask. Add 8.5 ml water to convert to a molal solution.

To avoid the carbonate problem and have the potassium buffer, one can make the solid reagent by the method of Kolthoff.[3]

Potassium Hydrogen Phthalate. Primary standard grade KHP is usually a reliable substitute for the NBS 4.004 pH_s standard. Prepare 0.05 *m* solution at 25°C.

C-3. PREPARATION OF RELIABLE CHEMICAL STANDARDS

High purity primary standards are readily available, but at high prices. Some of these are quite easily produced in the laboratory from reagent grade materials. Directions for making them exist in older analytical treatises and in the primary literature. This section presents the author's experiences in the preparation, preservation, and use of a few common standards together with brief instructions for preparation and references to the literature. Standards for acid–base, redox, complexometric, and a few other methods are given.

Preparation of pure materials can serve as student exercises allied to careful work in the analytical course. Time may prevent very many purifications, but some experience in clean chemical work is desirable for instilling self-confidence and initiative in those who want a career in chemical research. Time as well as funds may be saved by making some materials rather than ordering them. Tales abound of chemistry graduates waiting for gas cylinders or special reagents when a liter of dry CO_2, HCl, or NH_3 or a few grams of recrystallized borax or $K_2Cr_2O_7$ will suffice. Each can be made from common materials in the time it takes to place an order. One may not know just how to make them, but a visit to the inorganic section of the library, *Inorganic Syntheses* journals, nearby preparation books, and analytical treatises will yield the needed information.

(*Note*: In this section, *weight* means a balance reading, while *mass* implies a buoyancy-corrected value for the amount of matter.)

Stock Solutions: NaOH. Reagent grade 50% NaOH (19 *M*) is highly desirable since it contains under 0.1% carbonate. (Fisher Co., 1988, claims 0.05%). If this reagent is not available, make a stock of it from solid NaOH and store it in a plastic bottle. It requires many days for the low solubility Na_2CO_3 to settle out in this viscous medium. Centrifuging or suction filtering, not through paper, has been used. (Approximately 52 mL of the 50% solution contains 1 mol NaOH.) Dilute it with warm, freshly distilled water or deionized water that has been bubbled with N_2 or air passed through an ascarite tube to remove CO_2. For large volumes of 0.1–0.5 *M* NaOH, it may be more convenient to make a smaller stock of 1–5 *M* NaOH (10 times the final concentration desired) stored in plastic with ascarite protection. Quantitative dilution followed by standardization establishes the molarity of the parent stock for future uses. Two liters of diluted solution is better stored in full 250- or 500-ml plastic screw-cap bottles used one at a time to minimize air contact. For frequent use, an ascarite protected autofilling buret saves much time. Standardization of a weighed portion to get the mass molarity (moles per kilogram weighed in air) allows rapid and very precise preparation of dilute solutions. (See Section C-4 on mass titrations.)

Solutions of *common acids* can be made by dilution of more concentrated stocks of precisely known density. Tables of these densities are in ref. 4. Note that they are absolute so that buoyancy corrections are required. Temperature must be known. For example, 20% HCl is 1.09947 g/ml; 21% is 1.10465 g/ml, a difference of 0.00518 per percentage point. (These are at 15°C.) The change for 0.02% is thus 0.00010_4, which is 1/1000 and is easily measurable. One could weigh a 25-ml volumetric flask (to 0.1 mg) empty and filled with the acid to deduce its precise molarity from the density table.

Bottling. Solutions of acids, redox materials, and metal ions are best preserved in the $2\frac{1}{2}$-l bottles in which acids are shipped. The glass has been well leached and the screw cap can make them airtight. Glass-stoppered containers breathe and lose water unless grease or plastic wrapping is used. For this reason, storing solutions in the volumetric flask in which they are made is not a reliable practice. Bases need thick plastic (thin plastic is said to pass CO_2), but testing of the seal is needed since many plastic screw caps leak. In any case, storage for many months is not so reliable for bases as for acids. Some hydrated solids and CB HCl can advantageously be stored in the refrigerator but must be warmed to room temperature to avoid moisture condensation before opening.

Acid–Base Standards

A standard is a material of known quantitative reactivity toward some other materials. It should be easily preserved and measured (usually weighed) without change. Note that high purity, while most common, is not required. A chief example is azeotropic HCl at about 20% HCl.

Early chemists made good to bad guesses about materials useful for standards. Tests by many others decided which were useful. Modern development of new standards and reproducible preparations are illustrated by collective reports in *The Analyst* (London) and from the National Bureau of Standards in Washington.[5,6] These should be consulted for details, and they serve as easily read instruction for students of analytical chemistry. The details of preparation and assay by independent laboratories and the troubles encountered are illuminating. We begin with the most easily prepared and reliable standards. In general, larger or smaller amounts may be handled in proportion. High precision ($1/10^5$ parts) mass titration methods are described in Section C-4. Modern balances greatly facilitate such work. ["Rational weights" are buoyancy corrected, the weight of 1 mol in air ($d = 0.0012$ g/ml) vs. standard masses of density 8.0. Thus, a balance reading divided by this rational value yields the correct number of moles in a sample weighed under the condition mentioned. For very different conditions, calculate the buoyancy correction.]

In all operations, products must be covered to keep dust and other material out, whether in the hood, the oven, or the open air. Wet wash desktops before preparative work and banish linty paper towels to a distant washup sink.

Bases for Use with Strong Acids

Borax ($Na_2B_4O_7 \cdot 10H_2O$). Formula mass is 381.37. Density is 1.73 g/ml. Rational equivalent weight is 190.60. This is a 1:1 compound of boric acid and sodium borate, and thus it is a buffer as well as a base standard. The reagent is almost always quite low in water content and does not readily absorb water from wet air. Therefore, it *must* be recrystallized for use. It is then reliable as long as it is kept in a container at 50+% humidity. The amount below is adjusted so that crystals form only below 55°C giving the 10-hydrate. A 5-hydrate forms at higher temperatures.

Dissolve 30 g borax in 100 ml water at about 80°C with vigorous stirring to prevent overheating on the bottom and decomposition

(B_2O_3 deposits?). Filter while hot through a warmed buchner or fritted glass funnel. Cool while stirring it in an ice bath. Suction filter, washing the product twice with each of the following: (1) a little ice-cold water, (2) ethanol, and (3) ether. Spread thinly in a large dish, dry in air until ether is gone, and store it over saturated NaBr or over a wet mixture of NaCl and sucrose. The last is about 70% humidity and is recommended in the literature, although the 50–60% NaBr seems to work as well. However, we have noted serious loss of water after allowing the humidifying agent to dry out over several months. Note that loss of water increases the base reactivity per gram so that an acid titrated against the borax will calculate too low.

It requires no further treatment for use. Weigh out a sample (1.9 g for 50 ml of 0.2 *N* HCl) to 0.1 mg. Dissolve in the strong acid sample (heat if slow) and titrate to the green end point of bromocresol green (or a mixed indicator of it and methyl red). A color match of 50 ml of 0.1 *M* boric acid and NaCl with the same quantity of indicator* (or a pH meter to pH 5.0–5.2) must be used for accuracy better than two parts per thousand. For example, taking 1% more borax than expected for the pipetted HCl sample, one can finish the titration with 0.05 *M* HCl to get a five-figure result limited mainly by the precision of the end point determination. Property stored, this standard is ready for immediate use. We have had reliable results with it for many years.

THAM [Tris(hydroxymethyl)aminomethane]. Formula mass is 121.14. Rational equivalent weight is 121.05. Density is 1.3 g/ml. Reagent grade is close to 100.0% and is available at low cost. Recrystallization directions (water plus methanol) for lower grades are found in Belcher (pp. 15–16).[7] Dry at 100°C for 1 hr. Decomposition may occur above 103°C. Titrate a weighed sample vs. strong acids to pH 4.7: bromocresol green–alizarin red-S 0.05% each, using a matching solution adjusted to pH 4.7, say, 0.1 *M* acetic acid–acetate. The solid seems indefinitly stable and gives results in agreement with other standards.

Sodium Carbonate (Na_2CO_3). Formula mass is 105.99. Rational equivalent weight is 52.98. Density is 2.53 g/ml. This material is cheap and available in high purity except that it absorbs water once opened. It has an equivalent weight too low and absorbs water too

*This unbuffered standard in the older literature seems questionable. A buffer of 0.1 *M* acetic acid and 0.2 *M* sodium acetate should be more reliable. For best work, check it with a pH meter.

fast for accurate use by weighing individual samples for titration. In spite of these drawbacks, it is widely recommended since it can be prepared in very reliable purity.[5] It is best to dry just the amount needed in a weighing bottle so that it need be opened only to pour out the sample after the first weighing (weighing by difference). It should be poured into a flask containing water and swirled immediately to prevent formation of a hard mass of the 10-hydrate, which dissolves very slowly. Then the empty container is weighed.

A. *Moderate Accuracy Method.* Sodium bicarbonate, reagent grade, may contain a little water and Na_2CO_3 but is converted quantitatively to the anhydrous carbonate by heating above 150°C for 2 h. Cover tightly and store in a desiccator over a strong drying agent such as Dryerite, a specially dehydrated $CaSO_4$, not the reagent powder. USP baking soda may contain about 0.2% NaCl if produced by the Solvay process. However, most soda in the United States is now produced from natural (Trona) deposits and is of quite high purity. It might do for work at the 0.2% level. If recrystallized as follows, it should be satisfactory.

B. *Preparation of High Purity* Na_2CO_3. Add 256 g $NaHCO_3$ to 1 l water at 86 ± 1°C with constant stirring. (Much CO_2 is evolved.) Hold the temperature until all is dissolved. Stir off heat until 75° is reached (some solid forms) and immediately filter through fine paper or glass sinter in a warm funnel, discarding solids. Stir filtrate and cool in an ice bath until about 18°C is reached. Filter through Whatman no. 541 paper, washing the solids with a small amount of ice water and suck as dry as possible. Spread out in a dish and dry at about 100°C, breaking up lumps that form. [This may be stored and portions converted as needed.) Convert to carbonate by crushing lumps and heating at 270°C to constant weight. This is the method of ref. 5, which I have followed.

Simply raising the oven gradually to about 250°C seems to make the conversion smoothly. One batch was heated at about 160°C for three days and seemed pure $NaCO_3$. (See method A just preceding.) The material must be dried again after storage of more than a few hours. It takes up water even in a good desiccator. The wet precipitate made is a tronalike substance, a mixed carbonate–bicarbonate hydrate, but has no precise composition. Trona is $Na_2CO_3 \cdot NaHCO_3 \cdot 2H_2O$, sodium sesquicarbonate. Reference 5 shows results on two preparations done in two laboratories. Thirteen assays averaged 99.996 ± 0.008% (standard deviation σ) vs. HCl standardized gravimetrically vs. pure Ag.

The usual carbonate titration by strong acid with a boilout before finishing the titration gives satisfactory results. However, the small carbonate sample size is a problem for high precision work. A weighed stock solution with, say, 10 aliquots of Na_2CO_3 can be advantageous. Then a weighing of 25 or 50 ml of this gives a mass of standard known to five or six figures. A large (1 l) titration flask fitted with a funnel to cut spray loss is needed for the reaction of large samples. A 5-min boiling time to remove CO_2 should not be shortened.

Acids

The less commonly recommended HCl and H_2SO_4 are included since they are inexpensive and might sometimes be the only high grade materials available. Sulfamic acid[5,8] is the material of choice if one is buying reagents. The weak acids oxalic and potassium acid phthalate give problems with carbonate-containing bases (Section IV-3).

Sulfamic Acid (HSO_3NH_2). Formula mass is 97.09. Density is 2.13 g/ml. Rational equivalent weight is 97.05. It is a strong acid. (Actually it may have pK_a of about 1.) The reagent grade now often has a stated assay on the container close to 100.0%. It can be stored in a desiccator over Dryerite ready for use. Purification of this, or especially of lower grades, follows.

In 300 ml water, 125 g of reagent grade acid is dissolved, giving a saturated solution at about 75°C. This is filtered by suction three times in rapid succession, removing some solid (impure, not saved) each time and cooling to about 50°C. The final filtrate is now cooled rapidly in an ice-salt bath with stirring and left in the cold bath for $\frac{1}{2}$ hr. The crystals are collected on the suction filter and washed twice each with ice water, ethanol, and ether. After air drying, it is ground and stored in a desiccator ready for use.

When dry, it is stable for a short time at 100°C. But if moisture is present, heat causes rearrangement to ammonium bisulfate, NH_4HSO_4, with weight gain. Oven drying is not recommended. This method is from ref. 9 (Vol. II, p. 99). For very high accuracy, in ref. 5b four filtrations were used. They assayed it vs. their standard Na_2CO_3, back titrating with 0.01 M NaOH. Twenty-six results in four laboratories gave $100.000_5 \pm 0.009\%$ (σ).

Potassium Hydrogen Phthalate ($KC_8H_5O_4$). Formula mass is 204.23. Rational equivalent weight is 204.11. Density is 1.64 g/ml. Inexpensive and available in high purity, this is a good standard for low carbonate strong bases. Lower grade materials are easily recrystallized one or more times from water by standard methods. Solubility is about 56 g/100 g water at 100°C. Drying up to about 130°C is possible. An indicator changing at pH 8–9 is needed. (See ref. 9, Vol. II.)

Oxalic Acid Dihydrate ($H_2C_2O_4 \cdot 2H_2O$). Formula mass is 126.07. Rational equivalent weight is 63.00. Density is 1.65 g/ml. In contrast with borax, this material carries too much water when recrystallized from water. Dehydration followed by water vapor absorption gives a pure material. Reagent grade material may be recrystallized from water but is often so pure that readjustment of the hydration is all it needs for moderate precision use. The product is first extensively dehydrated at 60°C or in a vacuum desiccator. It need not be completely anhydrous. (It should lose about 20% of its weight. The hydrate is about 29% water.) It is then placed in a 50% humidistat (see before under Borax) until constant weight is achieved (several days) (ref. 9, Vol. II). Once made, this remains ready for immediate use if kept in the humidistat. Titrate with low carbonate strong base to pH 8–9.

Azeotropic Hydrochloric Acid. Water and HCl form an azeotrope (constant boiling, CB) with about 20% (6.1 *M*) HCl. This is quite definite in composition but varies slightly with atmospheric pressure at the time of distillation. Committees of analysts have not recommended this standard owing to problems sometimes found in making and storing it reliably. My experience is that it is quite reliable. This agrees with very careful coulometric verification at NBS,[6] confirming work early in this century that established its composition by gravimetric Ag (to AgCl) analysis. Another coulometric report[10] claims that the very fast distillation of some of the early workers gives erratic high results. I have had some similar experiences, but slower (equilibrium?) distillation gives reliable acid consistent with the other references. However, Eckfeldt and Shaffer[10] claim that slow distillation gives results 0.05% higher (rather than lower) than that in the usual tabes. This seems a contradiction. The NBS work tested a number of acid standards by coulometry and found constant boiling HCl was 99.999% of the historically determined value. In any case, this material can be a valuable independent check of acid–base standardizations. The distilled acid has been preserved in well-closed hard glass bottles without change for years. A laboratory with borax, oxalic acid, and the azeotrope on hand can check strong acid and base titrating solutions very quickly. Reliance on only a single standard is unwise.

To distill azeotropes, erect an all-glass still with a 40-cm empty fractionating column before the condenser. The delivery adapter should enter a small-check storage bottle (after collecting early discard acid). All should be clean but need not be dried. The column acts as a spray trap and allows slight fractionation to give a steady concentration of azeotropic acid.

The starting acid is 6.2 *M* (5.59–5.60 mmol/g′). The symbol g′ refers to an apparent gram of solution weighed in air in this work. The azeotrope is 5.569 mmol/g′ (a weight molarity; this is not molality) when distilled at 730 torr. In Sewanee at common pressure of 715 torr, the acid produced is 5.5791 mmol/g′. Reference 9 (Vol. II, p. 67–68) gives a range of values vs. pressure. However, linear interpolation is quite good: −0.00067 per torr. The apparent molarity values can simply be multiplied by the balance reading of acid to obtain the moles of acid. These are the reciprocals of the usual table entries: apparent weight of solution containing 1 mol HCl. Thus, at 740 torr the table gives 179.77 g′/mol (rational molarity), which gives 5.5627 mmol/g′. Make temperature, latitude, and altitude corrections found in handbooks (about 2–4 torr here) on the barometer readings to get corrected pressure to 0.5 torr. (Do *not* correct to sea level. You need station pressure.) Even an error of 2 torr gives but 2/10,000 error in the acid. Note that the pressure varies greatly and must be read during the distillation. At altitude about 2000 ft our pressure varies from about 690 (in hurricane) to 718 in summer highs.

It is worthwhile to titrate the starting acid (weigh 1 g–0.1 mg and titrate with standard 0.2 *M* NaOH) and adjust it to the correct value. The azeotrope is reached much sooner from slightly high acid than from low. It can be made approximately by diluting 520 ml of reagent acid (12 *M*) to 1 l or adding 930 ml water per liter of the 12 *M* HCl. These vary with brand and time since opening the bottle. From the rough (3–4 significant figure) titration, calculate the water or acid needed to correct the acid. Distill 1 l (in a 2-l pot) with two boiling chips. When acid begins to distill over, collect it in a 10-ml graduated cylinder and adjust the flame so that 2–3 ml min are obtained. After 150 ml (larger graduated cylinder) have distilled, increase the rate to about 4 ml/min and collect about 150 ml more. (This clean acid may be saved for lab use as 6 *M* HCl.) Then collect two portions of 300 ml each in 500 ml plastic screw-cap glass bottles sitting in ice and protected from moist air with clean tissue. Read the barometer. If the distillation must be interrupted before completed, an initial discard may be needed before collecting good azeotrope again. It is best to start in the morning and do this without breaks. The process may be monitored by weighing 1-g samples and titrating as described before for the acid. If an initial titration near the start shows acid below the azeotrope value, stop and add 12 *M* HCl to get above it. If this is not done, at least one-half of the acid must be discarded before good azeotrope is reached. (A curve of boiling point vs. temperature in this region is so flat that temperature is not a guide to constant boiling unless a thermometer reading to 0.01°C is used.) It seems best not to allow any interruption of heating (drafts) and acid throughput, thus disturbing the column distribution. If the starting acid is correct, the two good portions are indistinguishable within 2/10,000. Their assay can be precisely checked as follows.

Weigh (to 0.1 mg) a dry 50-ml conical glass-stoppered weighing bottle (or plastic-stoppered Erlenmeyer). Use a dry pipet to add about 2 ml of the CB HCl. Immediately stopper and reweigh. Rapidly add water and transfer quantitatively to a 500-ml titration flask with much wash water. Multiply the CB HCl weight by 1.08 to get the amount of pure borax needed to give about 1% excess. Weigh the borax by difference into the flask and swirl to dissolve in about 50 ml total water solution. Back titrate with standard 0.05 *M* HCl as described before under Borax. Calculate net equivalents of borax that reacted per g′ HCl. Similar checks can be made using other standard bases.

Absolute Sulfuric Acid. Kunzler[11] describes vacuum distillation and other methods to make pure 100% sulfuric acid. One fairly simple way is to mix some fuming (excess SO_3) acid with reagent (98%) acid. A fog of H_2SO_4 forms over the surface if even slight excess SO_3 is present when moist air is blown over it. 98% acid is added by drops until the fog just stops forming. This is called the fair and foggy method and has to be seen to be believed. It works quite well. The pure acid is immediately weighed and diluted. I have made several preparations that agreed with other standards to within several parts per ten thousand as Kunzler claims. The fuming acid is quite hazardous, and unsupervised students should not handle it. This is a possibility to keep in mind for unusual situations in which sulfuric is wanted or is the only available material. Sulfuric is effectively a strong diprotic acid in dilute solutions for titration purposes since the pK_{a2} is 2.0.

Other Acids. Reagent grades of benzoic, succinic, citric, furoic, and a few other acids are usually of high purity and require only drying several days over strong desiccant to serve as secondary checks of strong bases. See ref. 9, Vol. II.

Precipitation Materials. Modern reagent grade NaCl, KCl, and KSCN are usually close to 100.0% purity after drying at 150°C (ref. 9, Vol. II, p. 252). Silver nitrate reagent contains only some water and can be dried to a quite reliable standard by melting (melting point about 210°C). It should not be heated hotter because decomposition to $AgNO_2$ and O_2 occurs (ref. 9, Vol. II). Weigh a clean 20-ml beaker and add the approximate amount of $AgNO_3$ needed (17 g for 1 l of 0.1 *M*). Just melt it with a small flame and keep it liquid for 5 min. Cool in a desiccator and weigh. Dissolve it in 10 ml water and transfer to a volumetric flask. Dust causes blackening. Dirty starting material should be recrystallized from water. (See ref. 9, Vol. II.) Filtered water may be needed to keep out organic matter. Silver solutions must not touch paper, plastic, or other organic matter. Sintered glass filters must be used for reliably stable solutions. Do not store in plastic containers. They turn black.

Refer to standard analytical works for titration methods. We find the absorption indicator methods (Fajans') the most precise, although Volhard and Mohr's are useful for certain conditions.

Complexometric Titrations. EDTA, commonly disodium dihydrogen $EDTA \cdot 2H_2O$ (formula mass, 372.24; density, 1.73; rational equivalent weight, 372.05) is now available in high purity. For work more precise than 1%, it is standardized against metal ion solutions. These can be made by dissolving high purity Zn, Pb, Cu, Bi, or others in dilute nitric acid. Zinc oxide heated to red heat to dry it and remove carbonate and $CuSO_4 \cdot 5H_2O$ preserved in a desiccator over the same material wetted with drops of water are also reliable standards.

Redox Materials. The materials $K_2Cr_2O_7$ and As_2O_3 are very reliable materials (ref. 9, Vol. III). They can be compared with each other as described in what follows and with most other oxidizing and reducing reagents. Standard treatises give extensive detail and applications to large numbers of inorganic and organic analyses. Here are presented details on the interrelations of a few materials involved in a study of the purity and stability of ferrous ammonium sulfate (FAS).[12] FAS is compared with dichromate by means of five-to-six-figure weighings of the solids. The dichromate and As_2O_3 can be similarly directly compared. These can then be used to standardize most other redox reagent solutions, I_2, $KMnO_4$, $Na_2S_2O_3$, etc.

Reference Ferrous Ammonium Sulfate

Purification of reagent grade FAS, which is usually 98–99.5% pure as Fe(II), can produce material 99.9+%.[12] An acidic recrystallization in the presence of Fe powder to reduce Fe(III) followed by rapid drying effects the improvement. Brief directions are given here. Half to four times the quantities can be managed reliably. Small portions are less successful. The product is ready for use in about 1 hr.

Dissolve 100 g FAS in 165 ml of 0.2 *M* sulfuric acid with heating to 60°C and stirring with a glass rod. Add about 0.2 g pure Fe powder and stir hot about 3 min. (Do not use a magnetic bar while Fe powder is present!) Filter hot through a heated suction filter. Cover it and stir magnetically in a bath of room temperature water until it is about 30°C. Shift to cold water and add ice to the bath slowly so that it cools to about 10°C in about 15 min. Strong stirring and not too fast precipitation make small separated crystals that do not trap liquid. Filter with suction on fritted glass if possible and wash once with 60% acetone or ethanol. Then wash twice with reagent grade acetone. Spread the solid 2–3 mm deep in rounded wall evaporating dishes and dry in a vacuum dessicator by evacuating briefly and admitting air about 10 times. Warm slightly on a hot plate or in a 55°C oven for a few minutes and repeat the evacuations. The solid should now be well separated, free flowing, and nearly white. Store over wet NaBr (50% humidity) or in a tightly closed brown jar.

About half of the original weight is obtained. Adding the wash solvents to the filtrate will recover the rest, which can be recrystallized. Adding acetone to the original FAS solution precipitates all the FAS, but it is not pure. Syntheses of FAS are described elsewhere.[12]

Recrystallization of Potassium Dichromate

The reagent grade is usually close to 100.0%. Traces of CrO_3 and sulfates may be present. These are easily removed with water. Dissolve 50 g in 85 ml water by heating and stirring. Cool with rapid stirring to about 20°C. Filter, suck dry, and dry at 160°C for about 2 hr. Several recrystallizations can be made if one starts with lower grade material.

C-4. RAPID, HIGH PRECISION MASS TITRATIONS

This modification of the mass titration method of the NBS[6] yields results with average deviations less than 0.5 part per thousand. It can now be rapidly used with acid–base and other titrations for high precision work much less tedious than former "weight titration" techniques. Here we take a mass of standard dichromate solution and a 2-g FAS sample in slight excess so that a small (<1%) final titration with diluted dichromate identifies the equivalence point to within 0.2 part per thousand. For the reaction with Fe(II), 1 *M* Cr(VI) is 0.5 *M* dichromate and is 3 *N*. Use of normality, when defined, simplifies repeated calculations.

Dry and weigh to 0.1 mg about 5 g of primary standard potassium dichromate into a dry, weighed 500-ml volumetric flask. Fill to the mark and weigh the solution to within 0.01 g. Correct the dichromate for buoyancy and calculate the air normality (or molarity if desired) of the solution, that is, the milliequivalents (millimoles) of Cr(VI) per apparent gram of solution. Dilute a portion of this 10-fold and fill a 10- or 25-ml buret with this solution. Calculate the mass of FAS needed for 25 g of the 0.2 *N* (0.07 *M*) Cr(VI) solution and increase it by 0.5%. Into a dry, weighed flask pipet (approximately) 25 ml of the first dichromate solution and weigh to within 0.1 mg. In a 300-ml or larger flask prepare the acid indicator solution as follows. Mix 6 ml each of 6 *M* sulfuric and 85% phosphoric acids, 30 ml water, and 3 drops of indicator [0.2% diphenylaminesulfonate (DPS) or 0.01 *M* 4,7-dimethylferroin]. Add 2 drops 0.02 *M* Fe(II) and titrate with the diluted Cr(VI) until the indicator is just oxidized [purple (DPS) or red to blue (ferroin)]. Transfer, with washing, the weighed dichromate and the sample of FAS into this flsk and titrate to the end point with the diluted dichromate. The volume normality of the dichromate can be used for the diluted solution in calculating the approximately 0.02 meq (0.007 mmol) added in the end point determination. Add this to the weighed aliquot and calculate the mass of FAS titrated and compare it with that taken. (See the example calculation that follows.) The preoxidation of the indicators had almost no effect on the results but should be performed for results approaching the high precision of the NBS report[6] where the end points were determined potentiometrically.

Sample weighings were made on a Mettler AE 160 balance checked against a set of weights calibrated at NBS. A secondary set of weights was used to check the balance operation at 10 and 100 g each day of use. This balance is internally calibrated to read in terms of weights of density 8.0 g/ml. (Note that chemists widely use 7.7 or 7.8 for steel weights. This was abandoned by balance makers and NBS about 1980.) This and the local air density of 0.00113 g/ml were used for buoyancy corrections. In the following example, *weight* or *air g* indicates the balance reading and *mass* the quantity of matter estimated through buoyancy correction. This correction may be made on the weight or, inversely, on the molar mass making a *rational molar weight*. The latter is simpler for repeated calculations and is illustrated next.

A typical assay illustrates the process and calculation. The standard solution contained 4.9500 air g potassium dichromate (equivalent mass 49.032) in 501.50 air g of solution. Using the apparent (rational) equivalent weight 49.018 gives an air-N of 0.20136 meq/air g of solution (0.033560 mmol/air g) (N_v = 0.20197 meq/ml, or 0.033662 M)). An air weight of solution of 25.0679 g was mixed with 1.9944 air g of purified FAS. Diluted dichromate (1.88 ml) was used to complete the reaction. This is 0.0380 meq, for a total of 5.0857 meq. Using the local rational molar weight of FAS, 391.96 yields 1.9934 g in air, a purity of 99.95%.

Fisher Scientific Co. primary standard As(III) oxide, 99.98%, was used to check the dichromate by the reaction in the NBS study.[6] They agreed within experimental uncertainty (0.02%). A preparation of recrystallized Baker Co. reagent grade potassium dichromate was indistinguishable (within 0.01%) from the G. F. Smith Co. primary standard of 100.00% used in most of this work.

The As(III) oxide was used to standardize 0.04 M potassium permanganate solution (HCl–KIO_3 catalyst method). Two batches of FAS were checked with permanganate, which gave results about 0.04% higher than the Cr method. Considering the HCl method of permanganate standardization, the agreement is probably satisfactory. We regard the Fe(II)–Cr(VI) chemistry as more reliable.

REFERENCES

1. Bates, R. G., *Determination of pH*, 2nd ed., Wiley, New York, 1973. This is the authoritative work on measurements and standards by a longtime researcher at NBS.
2. Guenther, W. B., *Analyst*, **113**, 683 (1988).
3. Kolthoff, I. M., *Acid-Base Indicators*, Macmillan, New York, 1937, p. 253.
4. Kolthoff, I. M. and Elving, P. J., *Treatise on Analytical Chemistry*, Interscience, New York.
5. (a) *Analyst*, **90**, 251 (1965); a report on sodium carbonate, committee headed by R. Belcher reviews acid–base standards and methods of their use. (b) *Analyst*, **92**, 587 (1967); a report of the committee on sulfamic acid.

6. (a) Taylor, J. K. and Smith, S. W., *J. Res. Nat. Bur. Stand.*, **63A**, 153 (1959); azeotropic HCl and other standards via coulometry, references to earlier work. (b) Sappenfield, K. M., Marinenko, G., and Hague, J. L., "Standard Reference Materials: Redox Standards," NBS Special Publication 260-24, U.S. Government Printing Office, Washington DC, January 1972.
7. Belcher, R. and Wilson, C. L., *New Methods of Analytical Chemistry*, 2nd ed., Chapman-Hall, London, 1964.
8. Sisler, H. H. et al., *Inorganic Syntheses*, Vol. II, McGraw-Hill, New York, 1946, pp. 176–179.
9. Kolthof, I. M. and Stenger, V. A. et al., *Volumetric Analysis*, Vol. II, Interscience, New York, 1947; Vol. III, 1957.
10. Eckfeldt, E. L. and Shaffer, E. W., Jr., *Anal. Chem.*, **37**, 1534 (1965).
11. Kunzler, J. E., *Anal. Chem.* **25**, 93 (1953).
12. Guenther, W. B., *Analyst*, **112**, 1743 (1987).

APPENDIX D

TABLES OF VALUES OF EQUILIBRIUM FUNCTIONS

Several systems are printed here to furnish values for use away from the computer.

TABLE D-1. α Values for Monoprotic Acids Centered on pL_a 6

pH	α_1	α_0	pH	α_1	α_0
0.0	0.99999901	1E − 06	6.2	0.38686318	0.61313682
0.2	0.99999842	1.5849E − 06	6.4	0.28474725	0.71525275
0.4	0.99999749	2.5119E − 06	6.6	0.20076001	0.79923999
0.6	0.99999602	3.9811E − 06	6.8	0.13680689	0.86319311
0.8	0.99999369	6.3095E − 06	7.0	0.09090909	0.90909091
1.0	0.99999	9.9999E − 06	7.2	0.05935094	0.94064906
1.2	0.99998415	1.5849E − 05	7.4	0.0382865	0.9617135
1.4	0.99997488	2.5118E − 05	7.6	0.02450337	0.97549663
1.6	0.99996019	3.9809E − 05	7.8	0.01560166	0.98439834
1.8	0.99993691	6.3092E − 05	8.0	0.00990099	0.99009901
2.0	0.99990001	9.999E − 05	8.2	0.00627001	0.99372999
2.2	0.99984154	0.00015846	8.4	0.00396529	0.99603471
2.4	0.99974887	0.00025113	8.6	0.00250559	0.99749441
2.6	0.99960205	0.00039795	8.8	0.00158239	0.99841761
2.8	0.99936944	0.00063056	9.0	0.000999	0.999001
3.0	0.999001	0.000999	9.2	0.00063056	0.99936944
3.2	0.99841761	0.00158239	9.4	0.00039795	0.99960205
3.4	0.99749441	0.00250559	9.6	0.00025113	0.99974887
3.6	0.99603471	0.00396529	9.8	0.00015846	0.99984154
3.8	0.99372999	0.00627001	10.0	9.999E − 05	0.99990001
4.0	0.99009901	0.00990099	10.2	6.3092E − 05	0.99993691

TABLE D-1 (*Continued*)

4.2	0.98439834	0.01560166	10.4	3.9809E − 05	0.99996019
4.4	0.97549663	0.02450337	10.6	2.5118E − 05	0.99997488
4.6	0.9617135	0.0382865	10.8	1.5849E − 05	0.99998415
4.8	0.94064906	0.05935094	11.0	9.9999E − 06	0.99999
5.0	0.90909091	0.09090909	11.2	6.3095E − 06	0.99999369
5.2	0.86319311	0.13680689	11.4	3.9811E − 06	0.99999602
5.4	0.79923999	0.20076001	11.6	2.5119E − 06	0.99999749
5.6	0.71525275	0.28474725	11.8	1.5849E − 06	0.99999842
5.8	0.61313682	0.38686318	12.0	1E − 06	0.999999
6.0	0.5	0.5			

TABLE D-2. Citric Acid. I 0.1 M, 25°C. pK_a's: 2.93, 4.36, 5.74

pH	α_3	α_2	α_1	α_0	$\bar{n}$
0	0.9988	1.174E-03	5.123E-08	9.322E-14	2.99883
0.1	0.9985	1.477E-03	8.116E-08	1.859E-13	2.99852
0.2	0.9981	1.859E-03	1.286E-07	3.708E-13	2.99814
0.3	0.9977	2.339E-03	2.037E-07	7.396E-13	2.99766
0.4	0.9971	2.943E-03	3.226E-07	1.475E-12	2.99706
0.5	0.9963	3.702E-03	5.110E-07	2.940E-12	2.99630
0.6	0.9953	4.656E-03	8.090E-07	5.861E-12	2.99534
0.7	0.9941	5.854E-03	1.281E-06	1.168E-11	2.99414
0.8	0.9926	7.359E-03	2.027E-06	2.327E-11	2.99264
0.9	0.9908	9.246E-03	3.206E-06	4.634E-11	2.99075
1	0.9884	1.161E-02	5.069E-06	9.224E-11	2.98838
1.1	0.9854	1.458E-02	8.010E-06	1.835E-10	2.98541
1.2	0.9817	1.828E-02	1.265E-05	3.647E-10	2.98169
1.3	0.9771	2.290E-02	1.995E-05	7.243E-10	2.97706
1.4	0.9713	2.867E-02	3.143E-05	1.437E-09	2.97127
1.5	0.9641	3.582E-02	4.945E-05	2.845E-09	2.96408
1.6	0.9552	4.468E-02	7.765E-05	5.625E-09	2.95516
1.7	0.9443	5.560E-02	1.216E-04	1.109E-08	2.94415
1.8	0.9308	6.900E-02	1.900E-04	2.182E-08	2.93062
1.9	0.9144	8.533E-02	2.959E-04	4.277E-08	2.91407
2	0.8945	1.051E-01	4.587E-04	8.348E-08	2.89399
2.1	0.8705	0.1288	7.076E-04	1.621E-07	2.86982
2.2	0.8421	0.1568	1.085E-03	3.129E-07	2.84102
2.3	0.8088	0.1896	1.651E-03	5.995E-07	2.80710
2.4	0.7702	0.2273	2.492E-03	1.139E-06	2.76771
2.5	0.7264	0.2699	3.725E-03	2.144E-06	2.72266
2.6	0.6776	0.3169	5.507E-03	3.990E-06	2.67205
2.7	0.6243	0.3676	8.043E-03	7.335E-06	2.61626
2.8	0.5676	0.4208	1.159E-02	1.331E-05	2.55600
2.9	0.5087	0.4748	1.646E-02	2.380E-05	2.49222
3	0.4492	0.5277	2.304E-02	4.192E-05	2.42606
3.1	0.3905	0.5776	3.174E-02	7.272E-05	2.35865

TABLE D-2 (*Continued*)

3.2	0.3343	0.6225	4.307E-02	1.242E-04	2.29099
3.3	0.2818	0.6605	5.753E-02	2.089E-04	2.22381
3.4	0.2338	0.6901	7.567E-02	3.459E-04	2.15748
3.5	0.1912	0.7102	9.804E-02	5.642E-04	2.09199
3.6	0.1539	0.7200	1.251E-01	9.065E-04	2.02700
3.7	0.1221	0.7191	0.1573	0.0014	1.96193
3.8	0.0954	0.7075	0.1949	0.0022	1.89611
3.9	0.0734	0.6854	0.2377	0.0034	1.82891
4	0.0556	0.6538	0.2854	0.0052	1.75988
4.1	4.148E-02	0.6136	0.3372	0.0077	1.68884
4.2	3.042E-02	0.5664	0.3919	0.0113	1.61595
4.3	2.193E-02	0.5141	0.4477	0.0163	1.54168
4.4	1.554E-02	0.4586	0.5029	0.0230	1.46671
4.5	1.082E-02	0.4021	0.5551	0.0319	1.39184
4.6	7.410E-03	0.3466	0.6023	0.0436	1.31780
4.7	4.989E-03	0.2937	0.6427	0.0586	1.24512
4.8	3.303E-03	0.2449	0.6744	0.0774	1.17404
4.9	2.152E-03	0.2008	0.6964	0.1007	1.10448
5	1.380E-03	0.1621	0.7077	0.1288	1.03610
5.1	8.711E-04	0.1288	0.7081	0.1622	0.96838
5.2	5.414E-04	0.1008	0.6975	0.2012	0.90074
5.3	3.313E-04	0.0777	0.6764	0.2456	0.83273
5.4	1.996E-04	0.0589	0.6457	0.2952	0.76413
5.5	1.183E-04	0.0440	0.6068	0.3492	0.69503
5.6	6.903E-05	0.0323	0.5611	0.4065	0.62592
5.7	3.965E-05	0.0233	0.5108	0.4658	0.55759
5.8	2.242E-05	0.0166	0.4578	0.5256	0.49108
5.9	1.249E-05	1.166E-02	0.4042	0.5842	0.42750
6	6.859E-06	8.059E-03	0.3518	0.6401	0.36793
6.1	3.718E-06	5.499E-03	0.3022	0.6923	0.31321
6.2	1.991E-06	3.708E-03	0.2565	0.7398	0.26393
6.3	1.055E-06	2.473E-03	0.2154	0.7821	0.22036
6.4	5.538E-07	1.634E-03	0.1792	0.8192	0.18248
6.5	2.884E-07	1.071E-03	0.1479	0.8510	0.15004
6.6	1.491E-07	6.975E-04	0.1212	0.8781	0.12261
6.7	7.667E-08	4.515E-04	0.0988	0.9008	0.09967
6.8	3.923E-08	2.908E-04	0.0801	0.9196	0.08068
6.9	1.999E-08	1.866E-04	0.0647	0.9351	0.06507
7	1.016E-08	1.193E-04	0.0521	0.9478	0.05232
7.1	5.145E-09	7.610E-05	4.182E-02	0.9581	0.04197
7.2	2.601E-09	4.844E-05	3.351E-02	0.9664	0.03361
7.3	1.313E-09	3.077E-05	2.680E-02	0.9732	0.02686
7.4	6.616E-10	1.953E-05	2.141E-02	0.9786	0.02145
7.5	3.331E-10	1.237E-05	1.708E-02	0.9829	0.01711
7.6	1.675E-10	7.835E-06	1.362E-02	0.9864	0.01363
7.7	8.419E-11	4.957E-06	1.085E-02	0.9891	1.086E-02
7.8	4.229E-11	3.135E-06	8.634E-03	0.9914	8.641E-03

TABLE D-2. Citric Acid. *I* 0.1 M, 25°C. pK_a's: 2.93, 4.36, 5.74 (*Continued*)

7.9	2.123E-11	1.982E-06	6.871E-03	0.9931	6.875E-03
8	1.066E-11	1.252E-06	5.465E-03	0.9945	5.468E-03
8.1	5.347E-12	7.909E-07	4.346E-03	0.9957	4.348E-03
8.2	2.682E-12	4.995E-07	3.455E-03	0.9965	3.456E-03
8.3	1.345E-12	3.154E-07	2.747E-03	0.9973	2.747E-03
8.4	6.746E-13	1.991E-07	2.183E-03	0.9978	2.183E-03
8.5	3.383E-13	1.257E-07	1.735E-03	0.9983	1.735E-03
8.6	1.696E-13	7.932E-08	1.378E-03	0.9986	1.379E-03
8.7	8.502E-14	5.006E-08	1.095E-03	0.9989	1.095E-03
8.8	4.262E-14	3.160E-08	8.702E-04	0.9991	8.703E-04
8.9	2.136E-14	1.994E-08	6.914E-04	0.9993	6.914E-04
9	1.071E-14	1.258E-08	5.492E-04	0.9995	5.493E-04
9.1	5.368E-15	7.940E-09	4.363E-04	0.9996	4.363E-04
9.2	2.691E-15	5.010E-09	3.466E-04	0.9997	3.466E-04
9.3	1.349E-15	3.161E-09	2.753E-04	0.9997	2.754E-04
9.4	6.759E-16	1.995E-09	2.187E-04	0.9998	2.187E-04
9.5	3.388E-16	1.259E-09	1.737E-04	0.9998	1.738E-04
9.6	1.698E-16	7.942E-10	1.380E-04	0.9999	1.380E-04
9.7	8.510E-17	5.011E-10	1.096E-04	0.9999	1.096E-04
9.8	4.265E-17	3.162E-10	8.709E-05	0.9999	8.709E-05
9.9	2.138E-17	1.995E-10	6.918E-05	0.9999	6.918E-05
10	1.071E-17	1.259E-10	5.495E-05	0.9999	5.495E-05
11	1.072E-20	1.259E-12	5.495E-06	1.0000	5.495E-06

TABLE D-3. Lysine. *I* 0.1 M, 25°C. pK_a's: 2.19, 9.15, 10.68

pH	α_3	α_2	α_1	α_0	$\bar{n}$
1	0.9394	0.0606	4.294E-10	8.971E-20	2.939
1.1	0.9248	0.0752	6.700E-10	1.762E-19	2.925
1.2	0.9072	0.0928	1.042E-09	3.449E-19	2.907
1.3	0.8859	0.1141	1.612E-09	6.720E-19	2.886
1.4	0.8605	0.1395	2.482E-09	1.302E-18	2.860
1.5	0.8304	0.1696	3.796E-09	2.508E-18	2.830
1.6	0.7955	0.2045	5.763E-09	4.793E-18	2.796
1.7	0.7555	0.2445	8.675E-09	9.083E-18	2.756
1.8	0.7105	0.2895	1.293E-08	1.704E-17	2.711
1.9	0.6610	0.3390	1.906E-08	3.164E-17	2.661
2	0.6077	0.3923	2.778E-08	5.803E-17	2.608
2.1	0.5516	0.4484	3.996E-08	1.051E-16	2.552
2.2	0.4942	0.5058	5.675E-08	1.879E-16	2.494
2.3	0.4370	0.5630	7.952E-08	3.315E-16	2.437
2.4	0.3814	0.6186	1.100E-07	5.773E-16	2.381
2.5	0.3288	0.6712	1.503E-07	9.928E-16	2.329
2.6	0.2801	0.7199	2.029E-07	1.688E-15	2.280
2.7	0.2361	0.7639	2.711E-07	2.838E-15	2.236

TABLE D-3. (*Continued*)

2.8	0.1971	0.8029	3.586E-07	4.728E-15	2.197
2.9	0.1632	0.8368	4.706E-07	7.810E-15	2.163
3	0.1341	0.8659	6.130E-07	1.281E-14	2.134
5.1	1.229E-03	0.9987	8.901E-05	2.341E-10	2.001
5.2	9.762E-04	0.9989	1.121E-04	3.711E-10	2.001
5.3	7.755E-04	0.9991	1.411E-04	5.883E-10	2.001
5.4	6.161E-04	0.9992	1.777E-04	9.325E-10	2.000
5.5	4.894E-04	0.9993	2.237E-04	1.478E-09	2.000
5.6	3.888E-04	0.9993	2.816E-04	2.343E-09	2.000
5.7	3.088E-04	0.9993	3.546E-04	3.713E-09	2.000
5.8	2.453E-04	0.9993	4.464E-04	5.884E-09	2.000
5.9	1.948E-04	0.9992	5.619E-04	9.325E-09	2.000
6	1.547E-04	0.9991	7.073E-04	1.478E-08	1.999
6.1	1.229E-04	0.9990	8.903E-04	2.342E-08	1.999
6.2	9.760E-05	0.9988	1.121E-03	3.711E-08	1.999
6.3	7.751E-05	0.9985	1.410E-03	5.880E-08	1.999
6.4	6.155E-05	0.9982	1.775E-03	9.315E-08	1.998
6.5	4.887E-05	0.9977	2.234E-03	1.476E-07	1.998
6.6	3.879E-05	0.9972	2.810E-03	2.338E-07	1.997
6.7	3.079E-05	0.9964	3.535E-03	3.702E-07	1.996
6.8	2.444E-05	0.9955	4.447E-03	5.862E-07	1.996
6.9	1.939E-05	0.9944	5.592E-03	9.280E-07	1.994
7	1.538E-05	0.9930	7.030E-03	1.469E-06	1.993
7.1	1.219E-05	0.9912	8.834E-03	2.323E-06	1.991
7.2	9.664E-06	0.9889	1.110E-02	3.674E-06	1.989
7.3	7.654E-06	0.9861	1.393E-02	5.806E-06	1.986
7.4	6.058E-06	0.9825	1.747E-02	9.169E-06	1.983
7.5	4.790E-06	0.9781	2.190E-02	1.447E-05	1.978
7.6	3.784E-06	0.9726	2.741E-02	2.280E-05	1.973
7.7	2.984E-06	0.9657	3.426E-02	3.588E-05	1.966
7.8	2.350E-06	0.9572	4.276E-02	5.636E-05	1.957
7.9	1.846E-06	0.9467	5.324E-02	8.835E-05	1.947
8	1.446E-06	0.9338	6.610E-02	1.381E-04	1.934
8.1	1.129E-06	0.9180	0.0818	2.152E-04	1.918
8.2	8.784E-07	0.8988	0.1008	3.339E-04	1.898
8.3	6.798E-07	0.8758	0.1237	5.157E-04	1.875
8.4	5.231E-07	0.8483	0.1509	7.917E-04	1.848
8.5	3.997E-07	0.8161	0.1827	1.207E-03	1.815
8.6	3.030E-07	0.7787	0.2195	1.825E-03	1.777
8.7	2.275E-07	0.7361	0.2612	2.735E-03	1.733
8.8	1.690E-07	0.6884	0.3075	4.054E-03	1.684
8.9	1.241E-07	0.6363	0.3578	5.938E-03	1.630
9	8.990E-08	0.5805	0.4109	0.0086	1.572
9.1	6.425E-08	0.5223	0.4655	0.0122	1.510
9.2	4.526E-08	0.4631	0.5197	0.0172	1.446
9.3	3.141E-08	0.4046	0.5715	0.0238	1.381
9.4	2.147E-08	0.3482	0.6193	0.0325	1.316

TABLE D-3. Lysine. *I* 0.1 M, 25°C. pK_a's: 2.93, 4.36, 5.74 (*Continued*)

9.5	1.446E-08	0.2953	0.6610	0.0437	1.252
9.6	9.599E-09	0.2467	0.6954	0.0578	1.189
9.7	6.282E-09	0.2033	0.7212	0.0755	1.128
9.8	4.054E-09	0.1651	0.7376	0.0972	1.068
9.9	2.580E-09	0.1323	0.7442	0.1235	1.009
10	1.620E-09	0.1046	0.7406	0.1547	0.950
10.1	1.004E-09	8.159E-02	0.7272	0.1913	0.890
10.2	6.132E-10	6.275E-02	0.7041	0.2331	0.830
10.3	3.694E-10	4.759E-02	0.6722	0.2802	0.767
10.4	2.193E-10	3.557E-02	0.6325	0.3319	0.704
10.5	1.283E-10	2.619E-02	0.5864	0.3874	0.639
10.6	7.393E-11	1.900E-02	0.5355	0.4454	0.574
10.7	4.197E-11	1.358E-02	0.4819	0.5046	0.509
10.8	2.348E-11	9.565E-03	0.4272	0.5632	0.446
10.9	1.295E-11	6.642E-03	0.3735	0.6199	0.387
11	7.050E-12	4.552E-03	0.3222	0.6732	0.331
11.1	3.791E-12	3.081E-03	0.2746	0.7223	0.281
11.2	2.016E-12	2.063E-03	0.2315	0.7665	0.236
11.3	1.062E-12	1.368E-03	0.1932	0.8054	0.1959
11.4	5.545E-13	8.992E-04	0.1599	0.8392	0.1617
11.5	2.874E-13	5.869E-04	0.1314	0.8680	0.1326
11.6	1.481E-13	3.807E-04	0.1073	0.8923	0.1080
11.7	7.591E-14	2.456E-04	0.0872	0.9126	0.0876
11.8	3.874E-14	1.578E-04	0.0705	0.9293	0.0708
11.9	1.970E-14	1.011E-04	0.0568	0.9431	0.0570
12	9.992E-15	6.452E-05	0.0457	0.9543	0.0458

TABLE D-4. Phosphoric Acid. *I* 0.1 M, 25°C. pK_a's: 1.95, 6.80, 11.67

pH	α_3	α_2	α_1	α_0	$\bar{n}$
0	0.98890	1.110E-02	1.759E-09	3.760E-21	2.9889
0.2	0.98253	1.747E-02	4.389E-09	1.487E-20	2.9825
0.4	0.97259	2.741E-02	1.091E-08	5.860E-20	2.9726
0.6	0.95724	4.276E-02	2.698E-08	2.296E-19	2.9572
0.8	0.93389	6.611E-02	6.611E-08	8.919E-19	2.9339
1	0.89912	0.1009	1.599E-07	3.418E-18	2.8991
1.2	0.84902	0.1510	3.792E-07	1.285E-17	2.8490
1.4	0.78013	0.2199	8.753E-07	4.701E-17	2.7801
1.6	0.69123	0.3088	1.948E-06	1.658E-16	2.6912
1.8	0.58550	0.4145	4.145E-06	5.591E-16	2.5855
2	0.47125	0.5287	8.380E-06	1.792E-15	2.4712
2.2	0.35993	0.6401	1.608E-05	5.448E-15	2.3599
2.4	0.26188	0.7381	2.938E-05	1.578E-14	2.2619
2.6	0.18291	0.8170	5.155E-05	4.388E-14	2.1829
2.8	0.12376	0.8762	8.762E-05	1.182E-13	2.1237

TABLE D-4. (*Continued*)

3	8.182E-02	0.9180	1.455E-04	3.111E-13	2.0817
3.2	5.323E-02	0.9465	2.378E-04	8.056E-13	2.0530
3.4	3.425E-02	0.9654	3.843E-04	2.064E-12	2.0339
3.6	2.188E-02	0.9775	6.168E-04	5.249E-12	2.0213
3.8	1.391E-02	0.9851	9.851E-04	1.329E-11	2.0129
4	8.820E-03	0.9896	1.568E-03	3.353E-11	2.0073
4.2	5.578E-03	0.9919	2.492E-03	8.443E-11	2.0031
4.4	3.522E-03	0.9925	3.951E-03	2.122E-10	1.9996
4.6	2.220E-03	0.9915	6.256E-03	5.325E-10	1.9960
4.8	1.397E-03	0.9887	9.887E-03	1.334E-09	1.9915
5	8.766E-04	0.9835	1.559E-02	3.333E-09	1.9853
5.2	5.483E-04	0.9750	2.449E-02	8.298E-09	1.9761
5.4	3.411E-04	0.9614	3.827E-02	2.055E-08	1.9621
5.6	2.105E-04	0.9405	5.934E-02	5.051E-08	1.9409
5.8	1.284E-04	0.9090	9.090E-02	1.226E-07	1.9092
6	7.693E-05	0.8631	1.368E-01	2.925E-07	1.8633
6.2	4.494E-05	0.7992	0.2008	6.802E-07	1.7993
6.4	2.538E-05	0.7152	0.2847	1.529E-06	1.7153
6.6	1.373E-05	0.6131	0.3869	3.293E-06	1.6132
6.8	7.063E-06	0.5000	0.5000	6.745E-06	1.5000
7	3.448E-06	3.869E-01	0.6131	1.311E-05	1.3869
7.2	1.601E-06	2.847E-01	0.7152	2.424E-05	1.2847
7.4	7.123E-07	2.008E-01	0.7992	4.292E-05	1.2007
7.6	3.062E-07	1.368E-01	0.8631	7.346E-05	1.1367
7.8	1.284E-07	9.090E-02	0.9090	1.226E-04	1.0908
8	5.289E-08	5.934E-02	0.9405	2.011E-04	1.0591
8.2	2.152E-08	3.827E-02	0.9614	3.258E-04	1.0379
8.4	8.690E-09	2.449E-02	0.9750	5.236E-04	1.0240
8.6	3.490E-09	1.559E-02	0.9836	8.372E-04	1.0148
8.8	1.397E-09	9.888E-03	0.9888	1.334E-03	1.0086
9	5.576E-10	6.257E-03	0.9916	2.120E-03	1.0041
9.2	2.222E-10	3.952E-03	0.9927	3.364E-03	1.0006
9.4	8.843E-11	2.492E-03	0.9922	5.328E-03	0.9972
9.6	3.513E-11	1.569E-03	0.9900	8.426E-03	0.9931
9.8	1.392E-11	9.857E-04	0.9857	1.330E-02	0.9877
10	5.502E-12	6.174E-04	0.9785	2.092E-02	0.9797
10.2	2.165E-12	3.849E-04	0.9669	3.276E-02	0.9676
10.4	8.456E-13	2.383E-04	0.9488	5.095E-02	0.9493
10.6	3.269E-13	1.460E-04	0.9214	7.843E-02	0.9217
10.8	1.245E-13	8.811E-05	0.8811	0.1189	0.8812
11	4.633E-14	5.198E-05	0.8238	0.1761	0.8239
11.2	1.672E-14	2.973E-05	0.7469	0.2531	0.7470
11.4	5.798E-15	1.634E-05	0.6506	0.3494	0.6506
11.6	1.917E-15	8.562E-06	0.5402	0.4598	0.5402
11.8	6.013E-16	4.257E-06	0.4257	0.5743	0.4257
12	1.792E-16	2.011E-06	0.3187	0.6813	0.3187
12.2	5.101E-17	9.072E-07	0.2279	0.7721	0.2279

TABLE D-4 (*Continued*)

12.4	1.399E-17	3.943E-07	0.1570	0.8430	0.1570
12.6	3.730E-18	1.666E-07	0.1051	0.8949	0.1051
12.8	9.749E-19	6.901E-08	0.0690	0.9310	0.0690
13	2.513E-19	2.819E-08	0.0447	0.9553	0.0447
13.2	6.418E-20	1.141E-08	0.0287	0.9713	0.0287
13.4	1.629E-20	4.592E-09	0.0183	0.9817	0.0183
13.6	4.120E-21	1.840E-09	0.0116	0.9884	0.01161
13.8	1.039E-21	7.359E-10	0.0074	0.9926	0.00736
14	2.618E-22	2.937E-10	0.0047	0.9953	0.00466

TABLE D-5. EDTA. *I* 0.1 M, 25°C. pK_a's: 0, 1.5, 2.00, 2.68, 6.11, 10.17

pH	α_6	α_5	α_4	α_3	α_2	α_1	α_0	n
0	0.4921	0.4921	0.0156	1.555E-04	3.250E-07	2.522E-13	1.705E-23	5.4763
0.2	0.3752	0.5946	0.0298	4.720E-04	1.563E-06	1.923E-12	2.060E-22	5.3444
0.4	0.2691	0.6759	0.0537	1.348E-03	7.075E-06	1.379E-11	2.342E-21	5.2127
0.6	0.1818	0.7236	0.0910	3.624E-03	3.015E-05	9.315E-11	2.507E-20	5.0834
0.8	0.1156	0.7296	0.1455	9.178E-03	1.210E-04	5.926E-10	2.528E-19	4.9514
1	0.0690	0.6905	0.2182	2.182E-02	4.560E-04	3.539E-09	2.392E-18	4.8058
1.2	0.0383	0.6076	0.3043	4.823E-02	1.597E-03	1.965E-08	2.105E-17	4.6328
1.4	0.0195	0.4894	0.3884	9.757E-02	5.122E-03	9.985E-08	1.695E-16	4.4205
1.6	0.0089	0.3539	0.4452	1.772E-01	1.475E-02	4.556E-07	1.226E-15	4.1649
1.8	0.0036	0.2256	0.4497	0.2837	0.0374	1.832E-06	7.814E-15	3.8741
2	0.0013	0.1251	0.3955	0.3955	0.0827	6.414E-06	4.336E-14	3.5669
2.2	0.0004	0.0603	0.3020	0.4787	0.1586	1.950E-05	2.089E-13	3.2652
2.4	0.0001	0.0254	0.2017	0.5067	0.2660	5.185E-05	8.805E-13	2.9867
2.6	0.0000	0.0095	0.1194	0.4754	0.3956	1.222E-04	3.289E-12	2.7427
2.8	0.0000	0.0032	0.0638	0.4023	0.5305	2.597E-04	1.108E-11	2.5392
3	0.0000	0.0010	0.0313	0.3130	0.6542	5.077E-04	3.432E-11	2.3781
3.2	1.814E-07	2.876E-04	1.440E-02	0.2283	0.7561	9.299E-04	9.963E-11	2.2570
3.4	3.169E-08	7.961E-05	6.319E-03	0.1587	0.8333	1.624E-03	2.758E-10	2.1700
3.6	5.352E-09	2.131E-05	2.680E-03	0.1067	0.8878	2.743E-03	7.382E-10	2.1094
3.8	8.830E-10	5.571E-06	1.111E-03	7.009E-02	0.9243	4.525E-03	1.930E-09	2.0678
4	1.434E-10	1.434E-06	4.531E-04	4.531E-02	0.9469	7.348E-03	4.967E-09	2.0389
4.2	2.302E-11	3.648E-07	1.827E-04	2.895E-02	0.9591	1.180E-02	1.264E-08	2.0175
4.4	3.662E-12	9.198E-08	7.301E-05	1.834E-02	0.9628	1.877E-02	3.187E-08	1.9997
4.6	5.780E-13	2.301E-08	2.895E-05	1.152E-02	0.9588	2.962E-02	7.972E-08	1.9820
4.8	9.042E-14	5.705E-09	1.138E-05	7.177E-03	0.9465	4.634E-02	1.977E-07	1.9609
5	1.399E-14	1.399E-09	4.420E-06	4.420E-03	0.9239	7.169E-02	4.846E-07	1.9327
5.2	2.131E-15	3.378E-10	1.692E-06	2.681E-03	0.8881	0.1092	1.170E-06	1.8935
5.4	3.178E-16	7.982E-11	6.336E-07	1.592E-03	0.8355	0.1629	2.766E-06	1.8387
5.6	4.601E-17	1.832E-11	2.304E-07	9.174E-04	0.7633	0.2358	6.346E-06	1.7651

TABLE D-5. EDTA. I 0.1 M, 25°C. pK_a's: 0, 1.5, 2.00, 2.68, 6.11, 10.17 (*Continued*)

5.8	6.410E-18	4.044E-12	8.064E-08	5.088E-04	0.6710	0.3285	1.401E-05	1.6720
6	8.523E-19	8.523E-13	2.693E-08	2.693E-04	0.5629	0.4368	2.953E-05	1.5634
6.2	1.076E-19	1.705E-13	8.541E-09	1.354E-04	0.4484	0.5514	5.908E-05	1.4486
6.4	1.289E-20	3.239E-14	2.571E-09	6.458E-05	0.3390	0.6608	1.122E-04	1.3390
6.6	1.474E-21	5.867E-15	7.381E-10	2.938E-05	0.2445	0.7553	2.033E-04	1.2443
6.8	1.620E-22	1.022E-15	2.038E-10	1.286E-05	0.1695	0.8301	3.541E-04	1.1692
7	1.727E-23	1.727E-16	5.459E-11	5.459E-06	0.1141	0.8853	5.985E-04	1.1135
7.2	1.803E-24	2.857E-17	1.431E-11	2.268E-06	0.0751	0.9239	9.898E-04	1.0741
7.4	1.853E-25	4.655E-18	3.695E-12	9.280E-07	0.0487	0.9497	1.613E-03	1.0471
7.6	1.885E-26	7.505E-19	9.441E-13	3.759E-07	0.0313	0.9661	2.600E-03	1.0287
7.8	1.904E-27	1.201E-19	2.396E-13	1.511E-07	0.0199	0.9759	4.163E-03	1.0158
8	1.914E-28	1.914E-20	6.047E-14	6.047E-08	0.0126	0.9807	6.630E-03	1.0060
8.2	1.915E-29	3.035E-21	1.520E-14	2.409E-08	7.981E-03	0.9815	1.052E-02	0.9975
8.4	1.909E-30	4.795E-22	3.806E-15	9.561E-09	5.019E-03	0.9784	1.661E-02	0.9884
8.6	1.894E-31	7.541E-23	9.486E-16	3.777E-09	3.142E-03	0.9707	2.612E-02	0.9770
8.8	1.868E-32	1.178E-23	2.350E-16	1.483E-09	1.955E-03	0.9572	4.083E-02	0.9611
9	1.825E-33	1.825E-24	5.768E-17	5.768E-10	1.206E-03	0.9356	6.324E-02	0.9380
9.2	1.761E-34	2.791E-25	1.398E-17	2.215E-10	7.339E-04	0.9026	9.670E-02	0.9040
9.4	1.667E-35	4.188E-26	3.324E-18	8.350E-11	4.384E-04	0.8545	0.1451	0.8553
9.6	1.537E-36	6.119E-27	7.698E-19	3.065E-11	2.550E-04	0.7877	0.2120	0.7883
9.8	1.368E-37	8.629E-28	1.720E-19	1.086E-11	1.432E-04	0.7009	0.2990	0.7012
10	1.164E-38	1.164E-28	3.679E-20	3.679E-12	7.688E-05	0.5966	0.4033	0.5968
10.2	9.419E-40	1.493E-29	7.477E-21	1.185E-12	3.925E-05	0.4827	0.5172	0.4828
10.4	7.232E-41	1.817E-30	1.442E-21	3.622E-13	1.901E-05	0.3706	0.6293	0.3707
10.6	5.286E-42	2.104E-31	2.647E-22	1.054E-13	8.769E-06	0.2709	0.7291	0.2709
10.8	3.706E-43	2.338E-32	4.662E-23	2.942E-14	3.879E-06	0.1899	0.8101	0.1899
11	2.514E-44	2.514E-33	7.946E-24	7.946E-15	1.661E-06	0.1289	0.8711	0.1289
11.2	1.666E-45	2.640E-34	1.322E-24	2.096E-15	6.941E-07	0.0854	0.9146	0.0854
11.4	1.085E-46	2.726E-35	2.164E-25	5.435E-16	2.853E-07	0.0556	0.9444	0.0556
11.6	6.991E-48	2.783E-36	3.501E-26	1.394E-16	1.160E-07	0.0358	0.9642	0.0358
11.8	4.470E-49	2.820E-37	5.623E-27	3.548E-17	4.679E-08	0.0229	0.9771	0.0229
12	2.844E-50	2.844E-38	8.988E-28	8.988E-18	1.879E-08	0.0146	0.9854	0.0146

TABLE D-6. Carbonic Acid. I 0.1 M, 25°C. pK_a's: 6.16, 9.93

pH	α_2	α_1	α_0	$\bar{n}$
4	0.99313	0.00687	8.072E-09	1.9931
4.1	0.99137	0.00863	1.277E-08	1.9914
4.2	0.98915	0.01085	2.020E-08	1.9892
4.3	0.98638	0.01362	3.192E-08	1.9864
4.4	0.98292	0.01708	5.041E-08	1.9829
4.5	0.97859	0.02141	7.954E-08	1.9786
4.6	0.97320	0.02680	1.254E-07	1.9732
4.7	0.96649	0.03351	1.973E-07	1.9665
4.8	0.95817	0.04183	3.101E-07	1.9582
4.9	0.94791	0.05209	4.861E-07	1.9479
5	0.93529	0.06471	7.602E-07	1.9353
5.1	0.91988	0.08012	1.185E-06	1.9199
5.2	0.90119	0.09881	1.840E-06	1.9012
5.3	0.87870	0.12129	2.843E-06	1.8787
5.4	0.85194	0.14805	4.369E-06	1.8519
5.5	0.82049	0.17950	6.669E-06	1.8205
5.6	0.78405	0.21594	1.010E-05	1.7840
5.7	0.74252	0.25746	1.516E-05	1.7425
5.8	0.69611	0.30386	2.253E-05	1.6961
5.9	0.64533	0.35464	3.310E-05	1.6453
6	0.59105	0.40890	4.804E-05	1.5910
6.1	0.53445	0.46548	6.885E-05	1.5344
6.2	0.47694	0.52296	9.738E-05	1.4768
6.3	0.42004	0.57982	1.359E-04	1.4199
6.4	0.36519	0.63462	1.873E-04	1.3650
6.5	0.31362	0.68613	2.549E-04	1.3134
6.6	0.26627	0.73338	3.430E-04	1.2659
6.7	0.22374	0.77580	4.568E-04	1.2233
6.8	0.18628	0.81312	6.028E-04	1.1857
6.9	0.15383	0.84538	7.890E-04	1.1530
7	0.12616	0.87281	1.025E-03	1.1251
7.1	0.10285	0.89582	1.325E-03	1.1015
7.2	0.08344	0.91486	1.704E-03	1.0817
7.3	0.06740	0.93042	2.181E-03	1.0652
7.4	0.05426	0.94296	2.783E-03	1.0515
7.5	0.04356	0.95290	3.540E-03	1.0400
7.6	0.03488	0.96063	4.493E-03	1.0304
7.7	0.02787	0.96644	5.691E-03	1.0222
7.8	0.02223	0.97057	7.195E-03	1.0150
7.9	0.01771	0.97321	9.083E-03	1.0086
8	1.409E-02	0.97447	1.145E-02	1.0026

TABLE D-6. Carbonic Acid. *I* 0.1 M, 25°C. pK_a's: 6.16, 9.93 (*Continued*)

8.1	1.119E-02	0.97440	1.441E-02	0.9968
8.2	8.874E-03	0.97301	1.812E-02	0.9908
8.3	7.029E-03	0.97023	2.274E-02	0.9843
8.4	5.558E-03	0.96593	2.851E-02	0.9771
8.5	4.388E-03	0.95995	3.567E-02	0.9687
8.6	3.457E-03	0.95201	4.453E-02	0.9589
8.7	2.716E-03	0.94182	5.546E-02	0.9473
8.8	2.128E-03	0.92900	6.887E-02	0.9333
8.9	1.662E-03	0.91312	8.522E-02	0.9164
9	1.292E-03	0.89371	1.050E-01	0.8963
9.1	9.992E-04	0.87028	0.1287	0.8723
9.2	7.683E-04	0.84237	0.1569	0.8439
9.3	5.865E-04	0.80962	0.1898	0.8108
9.4	4.441E-04	0.77179	0.2278	0.7727
9.5	3.332E-04	0.72887	0.2708	0.7295
9.6	2.473E-04	0.68115	0.3186	0.6816
9.7	1.815E-04	0.62927	0.3705	0.6296
9.8	1.315E-04	0.57420	0.4257	0.5745
9.9	9.412E-05	0.51721	0.4827	0.5174
10	6.646E-05	0.45976	0.5402	0.45989
10.1	4.631E-05	0.40335	0.5966	0.40344
10.2	3.186E-05	0.34938	0.6506	0.34945
10.3	2.166E-05	0.29902	0.7010	0.29906
10.4	1.456E-05	0.25308	0.7469	0.25311
10.5	9.694E-06	0.21207	0.7879	0.21209
10.6	6.395E-06	0.17614	0.8239	0.17615
10.7	4.187E-06	0.14517	0.8548	0.14518
10.8	2.723E-06	0.11886	0.8811	0.11887
10.9	1.761E-06	9.678E-02	0.9032	0.09678
11	1.134E-06	7.844E-02	0.9216	0.07844
11.1	7.271E-07	6.333E-02	0.9367	0.06333
11.2	4.648E-07	5.097E-02	0.9490	0.05097
11.3	2.964E-07	4.091E-02	0.9591	0.04091
11.4	1.886E-07	3.277E-02	0.9672	0.03277
11.5	1.198E-07	2.621E-02	0.9738	0.02621
11.6	7.600E-08	2.093E-02	0.9791	0.02093
11.7	4.816E-08	1.670E-02	0.9833	0.01670
11.8	3.049E-08	1.331E-02	0.9867	0.01331
11.9	1.929E-08	1.060E-02	0.9894	0.01060
12	1.220E-08	8.440E-03	0.9916	0.00844

TABLE D-7. Ni(II)-NH_3, α_6 to α_0, and $\bar{n}$. Bjerrum K values in Ch. V. [NH_3], or L, values from 10^{-5}–10 M

log L	α_6	α_5	α_4	α_3	α_2	α_1	α_0	n
-5	5.442E-22	5.086E-17	9.081E-13	5.859E-09	1.091E-05	6.270E-03	9.937E-01	0.0063
-4.9	2.163E-21	1.606E-16	2.277E-12	1.167E-08	1.726E-05	7.881E-03	9.921E-01	0.0079
-4.8	8.593E-21	5.067E-16	5.709E-12	2.324E-08	2.731E-05	9.901E-03	9.901E-01	0.0100
-4.7	3.412E-20	1.598E-15	1.430E-11	4.625E-08	4.316E-05	1.243E-02	9.875E-01	0.0125
-4.6	1.354E-19	5.037E-15	3.581E-11	9.198E-08	6.819E-05	1.560E-02	9.843E-01	0.0157
-4.5	5.368E-19	1.587E-14	8.959E-11	1.828E-07	1.076E-04	1.956E-02	9.803E-01	0.0198
-4.4	2.126E-18	4.991E-14	2.239E-10	3.628E-07	1.697E-04	2.450E-02	9.753E-01	0.0248
-4.3	8.410E-18	1.568E-13	5.588E-10	7.193E-07	2.673E-04	3.065E-02	9.691E-01	0.0312
-4.2	3.321E-17	4.920E-13	1.392E-09	1.424E-06	4.202E-04	3.827E-02	9.613E-01	0.0391
-4.1	1.309E-16	1.540E-12	3.462E-09	2.812E-06	6.593E-04	4.770E-02	9.516E-01	0.0490
-4	5.146E-16	4.809E-12	8.587E-09	5.540E-06	1.032E-03	5.929E-02	9.397E-01	0.0614
-3.9	2.016E-15	1.497E-11	2.123E-08	1.088E-05	1.609E-03	7.347E-02	9.249E-01	0.0767
-3.8	7.870E-15	4.641E-11	5.229E-08	2.128E-05	2.501E-03	9.069E-02	9.068E-01	0.0958
-3.7	3.057E-14	1.432E-10	1.281E-07	4.143E-05	3.867E-03	1.114E-01	8.847E-01	0.1192
-3.6	1.180E-13	4.391E-10	3.122E-07	8.017E-05	5.944E-03	1.360E-01	8.580E-01	0.1481
-3.5	4.523E-13	1.337E-09	7.548E-07	1.540E-04	9.069E-03	1.648E-01	8.260E-01	0.1834
-3.4	1.718E-12	4.033E-09	1.809E-06	2.932E-04	1.371E-02	1.980E-01	7.880E-01	0.2263
-3.3	6.455E-12	1.204E-08	4.288E-06	5.520E-04	2.051E-02	2.352E-01	7.437E-01	0.2779
-3.2	2.394E-11	3.546E-08	1.003E-05	1.026E-03	3.028E-02	2.758E-01	6.928E-01	0.3395
-3.1	8.742E-11	1.029E-07	2.312E-05	1.878E-03	4.403E-02	3.185E-01	6.355E-01	0.4123
-3	3.135E-10	2.930E-07	5.232E-05	3.375E-03	6.286E-02	3.612E-01	5.725E-01	0.4973
-2.9	1.101E-09	8.172E-07	1.159E-04	5.940E-03	8.787E-02	4.011E-01	5.050E-01	0.5951
-2.8	3.774E-09	2.225E-06	2.507E-04	1.021E-02	1.199E-01	4.348E-01	4.348E-01	0.7063
-2.7	1.259E-08	5.897E-06	5.278E-04	1.706E-02	1.593E-01	4.588E-01	3.644E-01	0.8306
-2.6	4.076E-08	1.516E-05	1.078E-03	2.769E-02	2.053E-01	4.696E-01	2.963E-01	0.9676
-2.5	1.277E-07	3.773E-05	2.131E-03	4.347E-02	2.560E-01	4.652E-01	2.332E-01	1.1163
-2.4	3.860E-07	9.062E-05	4.065E-03	6.587E-02	3.081E-01	4.448E-01	1.771E-01	1.2753
-2.3	1.124E-06	2.096E-04	7.467E-03	9.612E-02	3.572E-01	4.095E-01	1.295E-01	1.4431
-2.2	3.146E-06	4.660E-04	1.319E-02	1.348E-01	3.980E-01	3.625E-01	9.105E-02	1.6180

TABLE D-7. Ni(II)-NH_3, α_6 to α_0, and $\bar{n}$. Bjerrum K values in Ch. V. [NH_3], or L, values from 10^{-5}–10 M (*Continued*)

-2.1	8.452E-06	9.944E-04	2.235E-02	1.816E-01	4.257E-01	3.080E-01	6.144E-02	1.7984
-2	2.177E-05	2.035E-03	3.634E-02	2.344E-01	4.365E-01	2.509E-01	3.976E-02	1.9829
-1.9	5.374E-05	3.989E-03	5.658E-02	2.900E-01	4.289E-01	1.958E-01	2.465E-02	2.1702
-1.8	1.270E-04	7.489E-03	8.438E-02	3.435E-01	4.036E-01	1.463E-01	1.463E-02	2.3596
-1.7	2.873E-04	1.346E-02	1.204E-01	3.894E-01	3.634E-01	1.047E-01	8.315E-03	2.5505
-1.6	6.218E-04	2.314E-02	1.645E-01	4.224E-01	3.132E-01	7.165E-02	4.521E-03	2.7426
-1.5	1.288E-03	3.805E-02	2.149E-01	4.384E-01	2.581E-01	4.692E-02	2.351E-03	2.9358
-1.4	2.550E-03	5.985E-02	2.685E-01	4.351E-01	2.035E-01	2.938E-02	1.170E-03	3.1300
-1.3	4.827E-03	9.002E-02	3.207E-01	4.129E-01	1.534E-01	1.759E-02	5.562E-04	3.3250
-1.2	8.739E-03	1.294E-01	3.663E-01	3.746E-01	1.106E-01	1.007E-02	2.529E-04	3.5200
-1.1	1.513E-02	1.780E-01	4.001E-01	3.250E-01	7.619E-02	5.512E-03	1.100E-04	3.7139
-1	2.505E-02	2.341E-01	4.180E-01	2.697E-01	5.022E-02	2.886E-03	4.574E-05	3.9052
-0.9	3.971E-02	2.948E-01	4.181E-01	2.143E-01	3.169E-02	1.447E-03	1.821E-05	4.0921
-0.8	6.033E-02	3.558E-01	4.008E-01	1.632E-01	1.917E-02	6.952E-04	6.952E-06	4.2728
-0.7	8.801E-02	4.123E-01	3.690E-01	1.193E-01	1.113E-02	3.207E-04	2.547E-06	4.4457
-0.6	1.235E-01	4.595E-01	3.267E-01	8.391E-02	6.220E-03	1.423E-04	8.979E-07	4.6098
-0.5	1.671E-01	4.938E-01	2.788E-01	5.689E-02	3.350E-03	6.088E-05	3.051E-07	4.7642
-0.4	2.184E-01	5.126E-01	2.299E-01	3.726E-02	1.743E-03	2.516E-05	1.002E-07	4.9086
-0.3	2.764E-01	5.154E-01	1.836E-01	2.364E-02	8.784E-04	1.007E-05	3.185E-08	5.0428
-0.2	3.396E-01	5.030E-01	1.424E-01	1.456E-02	4.296E-04	3.913E-06	9.829E-09	5.1668
-0.1	4.060E-01	4.777E-01	1.074E-01	8.722E-03	2.045E-04	1.479E-06	2.952E-09	5.2805
0	4.734E-01	4.424E-01	7.900E-02	5.097E-03	9.492E-05	5.455E-07	8.645E-10	5.3839
0.1	5.396E-01	4.006E-01	5.682E-02	2.912E-03	4.307E-05	1.966E-07	2.475E-10	5.4768
0.2	6.028E-01	3.555E-01	4.005E-02	1.630E-03	1.916E-05	6.946E-08	6.946E-11	5.5595
0.3	6.615E-01	3.099E-01	2.773E-02	8.967E-04	8.369E-06	2.411E-08	1.915E-11	5.6320
0.4	7.147E-01	2.659E-01	1.890E-02	4.855E-04	3.600E-06	8.236E-09	5.196E-12	5.6948
0.5	7.619E-01	2.252E-01	1.271E-02	2.594E-04	1.528E-06	2.776E-09	1.391E-12	5.7486
0.6	8.029E-01	1.885E-01	8.455E-03	1.370E-04	6.409E-07	9.252E-10	3.683E-13	5.7942
0.7	8.381E-01	1.563E-01	5.568E-03	7.168E-05	2.663E-07	3.054E-10	9.657E-14	5.8324
0.8	8.678E-01	1.285E-01	3.638E-03	3.720E-05	1.098E-07	1.000E-10	2.512E-14	5.8641
0.9	8.926E-01	1.050E-01	2.361E-03	1.918E-05	4.496E-08	3.253E-11	6.489E-15	5.8902
1	9.131E-01	8.534E-02	1.524E-03	9.832E-06	1.831E-08	1.052E-11	1.668E-15	5.9116

TABLE D-8. Tartaric Acid. *I* 0.1 M, 25°C. pK_a's: 2.82, 3.97

pH	α_2	α_1	α_0	$\bar{n}$
0	0.9985	0.0015	1.6194E-07	1.9985
0.2	0.9976	0.0024	4.0641E-07	1.9976
0.4	0.9962	0.0038	1.0194E-06	1.9962
0.6	0.9940	0.0060	2.5550E-06	1.9940
0.8	0.9905	0.0095	6.3954E-06	1.9905
1	0.9851	0.0149	1.5976E-05	1.9851
1.2	0.9765	0.0234	3.9782E-05	1.9765
1.4	0.9633	0.0366	9.8572E-05	1.9632
1.6	0.9429	0.0568	2.4237E-04	1.9427
1.8	0.9123	0.0871	5.8902E-04	1.9117
2	0.8673	0.1313	1.4066E-03	1.8659
2.2	0.8039	0.1928	3.2749E-03	1.8006
2.4	0.7192	0.2734	7.3596E-03	1.7118
2.6	0.6142	0.3701	1.5786E-02	1.5984
2.8	0.4952	0.4729	3.1970E-02	1.4632
3	0.3737	0.5657	6.0612E-02	1.3131
3.2	0.2627	0.6302	0.1070	1.1557
3.4	0.1717	0.6527	0.1757	0.9960
3.6	0.1042	0.6279	0.2679	0.8363
3.8	0.0588	0.5615	0.3797	0.6791
4	0.0309	0.4678	0.5013	0.5296
4.2	1.521E-02	0.3650	0.6198	0.3954
4.4	7.075E-03	0.2690	0.7240	0.2831
4.6	3.142E-03	0.1893	0.8076	0.1956
4.8	1.347E-03	0.1287	0.8700	0.1314
5	5.636E-04	0.0853	0.9141	0.0864
5.2	2.318E-04	0.0556	0.9442	5.606E-02
5.4	9.421E-05	0.0358	0.9641	3.601E-02
5.6	3.801E-05	0.0229	0.9771	2.298E-02
5.8	1.526E-05	0.0146	0.9854	1.461E-02
6	6.109E-06	0.0092	0.9907	9.258E-03
6.2	2.440E-06	5.854E-03	0.9941	5.859E-03
6.4	9.736E-07	3.702E-03	0.9963	3.704E-03
6.6	3.881E-07	2.339E-03	0.9977	2.340E-03
6.8	1.547E-07	1.477E-03	0.9985	1.477E-03
7	6.160E-08	9.324E-04	0.9991	9.325E-04
7.2	2.453E-08	5.885E-04	0.9994	5.885E-04
7.4	9.769E-09	3.714E-04	0.9996	3.714E-04
7.6	3.890E-09	2.344E-04	0.9998	2.344E-04
7.8	1.549E-09	1.479E-04	0.9999	1.479E-04
8	6.165E-10	9.332E-05	0.9999	9.332E-05

TABLE D-8. Tartaric Acid. *I* 0.1 M, 25°C. pK_a's: 2.82, 3.97 (*Continued*)

8.2	2.455E-10	5.888E-05	0.9999	5.888E-05
8.4	9.772E-11	3.715E-05	1.0000	3.715E-05
8.6	3.890E-11	2.344E-05	1.0000	2.344E-05
8.8	1.549E-11	1.479E-05	1.0000	1.479E-05
9	6.166E-12	9.332E-06	1.0000	9.332E-06
9.2	2.455E-12	5.888E-06	1.0000	5.888E-06
9.4	9.772E-13	3.715E-06	1.0000	3.715E-06
9.6	3.890E-13	2.344E-06	1.0000	2.344E-06
9.8	1.549E-13	1.479E-06	1.0000	1.479E-06
10	6.166E-14	9.333E-07	1.0000	9.333E-07
10.2	2.455E-14	5.888E-07	1.0000	5.888E-07
10.4	9.772E-15	3.715E-07	1.0000	3.715E-07
10.6	3.890E-15	2.344E-07	1.0000	2.344E-07
10.8	1.549E-15	1.479E-07	1.0000	1.479E-07
11	6.166E-16	9.333E-08	1.0000	9.333E-08
11.2	2.455E-16	5.888E-08	1.0000	5.888E-08
11.4	9.772E-17	3.715E-08	1.0000	3.715E-08
11.6	3.890E-17	2.344E-08	1.0000	2.344E-08
11.8	1.549E-17	1.479E-08	1.0000	1.479E-08
12	6.166E-18	9.333E-09	1.0000	9.333E-09

EXERCISES IN USING TABLES

Write the $\bar{\mathbf{n}}'$ expressions for dilute, 0.01 − 0.001 *M*, solutions of pure triprotic acids and of their pure conjugate bases. Look up the **pH** at which the table equilibrium $\bar{\mathbf{n}}$ is the maximum $\bar{\mathbf{n}}'$. Insert this **pH** into the **D** term of the $\bar{\mathbf{n}}'$ to get a corrected value to look up in the tables. Continue until agreement is found. For practice, guestimation between tabe entries will serve to familiarize one with the methods and strategies involved. Try ampholytes and buffers.

Examples

1. For 0.001 M citric acid, $\bar{\mathbf{n}}' = 3 - 1000\mathbf{D}$. It is not likely that this solution forms 0.001 M **H**. So, from the table try **pH** 3, $\bar{\mathbf{n}} = 2.43$. Now, $1000\mathbf{D} = 1.0$, so $\bar{\mathbf{n}}' = 2.0$. Try other table **pH** values, say, **pH** 3.2, $\bar{\mathbf{n}} = 2.29$, $\bar{\mathbf{n}}'$ is then $3 - 0.63 = 2.37$. We get closer. Continue.

2. What is the **pH** of a solution having Na_3PO_4 at 0.5 M and NaOH at 0.001 M? Use pK_w 13.80 and start with **pH** between 11 and 12, say, 11.8. This is $\mathbf{OH} = 0.01$. $\bar{\mathbf{n}}' = (C_{AB} - \mathbf{D})/C_A = (-0.001 + \mathbf{OH})/0.05 = -0.02 + 20(0.01) = 0.18$. (**H** is negligible.) At **pH** 11.8 the table for the phosphate system lists $\bar{\mathbf{n}} = 0.426$. Next, try $\mathbf{pH} = 12$ ($\mathbf{OH} = 0.016$) getting $\bar{\mathbf{n}}' = 0.30$ and $\bar{\mathbf{n}} = 0.32$. Continue if desired, or use the extrapolation formula with **OH** and $\bar{\mathbf{n}}$ values. Note that PO_4^{3-} ion is a good base and contributes significant **OH** (0.015 *M* adding to the 0.001 of strong base). The strong base alone would make **pH** 10.8.

3. A spreadsheet is shown for the secant intersection method (Appendix B) of finding the **pH** of mixtures in the tartaric acid system. The specific example in it is for the 0.001 *M* ampholyte, $HTart^-$, $\bar{\mathbf{n}}' = 1 - 1000\mathbf{D}$. The specific mixture affects the formula of only the $\bar{\mathbf{n}}'$. Therefore, to solve for other mixtures, enter the new formula for $\bar{\mathbf{n}}'$ in its cell in the first line and fill down. The automatic calculation of the table proceeds. A table of answers for several mixtures is given to check practice. Note use of formation constant direction in this case. The pK_a's are 2.82 and 3.97 at $\boldsymbol{I} = 0.1$ M, 25°C. Table 8.

Spreadsheet

TARTARIC ACID

			in $b, 1/K_{a2} = 9332$ in $c, 1/K_{a1} = 661$	I = 0.1 M, 25° 0.001 M HA^-	
H	$\bar{\mathbf{n}}$	$\bar{\mathbf{n}}'$	delta	**H** (int. − extr.)	**pH**
1.0000E − 03	1.3132	0.0000	1.3132		3
1.0000E − 04	0.5296	0.9000	−0.3704	2.9798E − 04	4
2.9798E − 04	0.8955	0.7020	0.1935	2.3004E − 04	3.52581
2.3004E − 04	0.8061	0.7700	0.0361	2.1445E − 04	3.63819
2.1445E − 04	0.7819	0.7855	−0.0036	2.1587E − 04	3.66866
2.1587E − 04	0.7842	0.7841	6.785E − 05	2.1584E − 04	3.66581
2.1584E − 04	0.7842	0.7842	1.276E − 07	2.1584E − 04	3.66586
2.1584E − 04	0.7842	0.7842	−4.526E − 12	2.1584E − 04	3.66586
2.1584E − 04	0.7842	0.7842	0	#DIV/0!	3.66586

Answers and functions to use: Exact equation **pH** values calculated with this spreadsheet are

C_A		0.1 M	0.01 M	0.001 M	0.0001 M
pH of:	H_2T	1.933	2.480	3.111	3.863
pH of:	NaHT	3.400	3.440	3.666	4.213

C_A		1 M	0.1 M	0.01 M
pH of:	Na_2T	8.885	8.385	7.887
$\bar{\mathbf{n}}$		1.22×10^{-5}	3.84×10^{-5}	1.21×10^{-4}

The $\bar{\mathbf{n}}'$ functions for the first 8 cases are, $[\bar{\mathbf{n}}' = (C_H - \mathbf{D})/C_A]$,

$\dfrac{2 - 10\mathbf{D}}{1 - 10\mathbf{D}}$	$\dfrac{2 - 100\mathbf{D}}{1 - 100\mathbf{D}}$	$\dfrac{2 - 1000\mathbf{D}}{1 - 1000\mathbf{D}}$	$\dfrac{2 - 10000\mathbf{D}}{1 - 10000\mathbf{D}}$

For the last three, Na_2T; $0 - \mathbf{D}$ $\quad 0 - 10\mathbf{D}$ $\quad 0 - 100\mathbf{D}$

In the first 8, $\mathbf{D} = \mathbf{H}$ to very close approximation. For the basic three, Na_2T, $-\mathbf{D} = 10^{(\mathrm{pH}-13.8)} - \mathbf{H}$, or practically **OH**. Change of the starting **pH** choice is needed for very different cases. The first 8 happen to be near enough to **pH** 3 – 4 for convergence to occur. The last three used a **pH** start above 7 and then converged. This demonstrates the robustness of the secant method for these functions. Inspection of the numerical effects of the **D** values on $\bar{\mathbf{n}}$'s is instructive. Very small **D** can be negligible in the first 8 because of the subtraction from 1 or 2. But, in the last, only **D** determines $\bar{\mathbf{n}}$.

Practice without computer can be achieved by calculating just $\bar{\mathbf{n}}'$ at **pH** points in the tartaric acid table 8 to find first the interval **pH** 3 to 4; then 3.6 to 3.8 using the deltas. Then using the interpolation formula (secants) between these gives a close approximation. This can be checked by using it with the nearest previous **H** in another interpolation and/or caculating the $\bar{\mathbf{n}}'$, $\bar{\mathbf{n}}$ and delta.

One might compare the results of various approximate calculations for these values. Buffers of any ratio can be solved simply by inserting the formula for the $\bar{\mathbf{n}}'$ into the spreadsheet column. For example a buffer made 0.0155 M in H_2T and 0.0377 M in NaHT has $\bar{\mathbf{n}}' = 1.291 - 18.80\mathbf{D}$. This comes from C_H, 0.0687 and C_A, 0.0532. Thus, one would know to start where $\bar{\mathbf{n}}$ is about 1.3, at **pH** about 3.0, using the full $\bar{\mathbf{n}}$ table for tartaric acid above. This **pH** gives $\bar{\mathbf{n}}' = 1.272$ indicating that the pH is too low (**H** too high). One might continue by trial and error, for practice, and sketch the $\bar{\mathbf{n}}$ curves to see how the convergence progresses. The spreadsheet gives **pH** = 3.0501 and $\bar{\mathbf{n}}$ = 1.27425.

KEY POINTS

Calculations discussed and displayed are listed here by subject area and in order of appearance in the chapters and are designated by section numbers. Those numbered items with asterisks show or describe spreadsheets that might be used. Spreadsheet and other calculational techniques are treated in an elementary way in Chapter I and in further detail in Appendix B.

MONOPROTIC ACID–BASE SYSTEMS

1. Elementary bar diagram methods for deduction of proton balances, strong and weak acids and bases (Section II-2). Some 10^{-7} M systems.

*2. Equimolar buffer of 0.001 M each HA and A^-. An α graph solution and spreadsheet solution (Section II-5). The p**H** of a buffer at 10^{-7} M is solved.

3. The p**H** of a mixture of formic, acetic, and propionic acids is shown. The p**H** effect of given additions of NaOH and the amount of NaOH to change this to a desired p**H** are shown. An acid, HA, plus a base, NaB, of another system is solved, acetic acid plus sodium formate (Section II-6).

*4. Weak acid–weak base compounds are shown: NH_4CN, and anilinium trichloroacetate are used for contrasting examples. The α diagrams and spreadsheets are described (Section II-6).

5. Weak base plus strong base (NaCN + NaOH) is a case having **OH** needed in the Charlot equation (Section II-7).
*6. Weak and strong acid: The **pH** of 0.100 *M* H_2SO_4. Spreadsheet interaction of ionic strength and **pH** is described to achieve valid results (Section II-7).
7. Dilution of weak acid solutions: common mistakes and the correct relations are shown (Section II-8).
8. Conjugate base solutions. Advantages of α and of log α plots and calculations are described. Use of the complete cubic equation (Section II-9).

POLYPROTIC ACID–BASE SYSTEMS

1. The α, $\bar{\mathbf{n}}$ and $\bar{\mathbf{n}}'$ relations and diagrams are shown using triprotic examples (Section III-1).
2. Bar diagrams for deduction of proton balances for ampholytes, acids, and buffers (Section III-2).
3. Quantitative aspects of α diagrams, a quick sketching method, and rough **pH** estimating methods (Sections III-4 and III-5).
4. Equivalent solutions (Section III-6): the deduction of useful analytical *C* terms from various starting materials (Table III-1) illustrates the use of C_{AB}.
*5. Numerical solutions of **H** equations. The secant method: graphical illustrations and examples in triprotic mixtures (Section III-7). Also Appendix B.
6. Illustrations of complete $\bar{\mathbf{n}}'$ curves for various cases (Section III-8).
*7. Ampholytes: the $\bar{\mathbf{n}}$ equation contrasted with the usual charge–mass–balance derivation. Calculations of **pH** of various cases to show difficulties (Section III-9).
*8. Mixed-system cases: (a) $(NH_4)_2HPO_4$ and (b) a citrate–oxalate mixture: the **pH** of 0.03 *M* sodium oxalate with 0.02 *M* citric acid (Section III-10).
9. Dilution effects in oxalate and in phosphate systems. Why high *C* limiting of **pH** occurs in some solutions and not in others (Section III-11).
*10. Various problems with bases, buffers, and log-log techniques. Large error of the approximate buffer equation for a citrate case (Section III-12).

TITRATIONS AND LABORATORY CALCULATIONS

1. Deduction of complete titration equations from the $\bar{\mathbf{n}}$ function: Plot of citric and phosphoric acids (Section IV-1, Fig. IV-1*a*).
2. Examples of the effects on titration curves of pK_a spacing and of C (dilution) in oxalic acid (Section IV-2).
*3. Weak acid–weak base systems: showing interesting curves and species shifts in the system citric acid–ethylenediamine (Section IV-3).
4. Full equation treatment of the carbonate error in NaOH solution used with a weak acid (Section IV-3).
5. Statistical effects (microscopic k's) on titration curves of H_3A and equivalent HA having the same macroconstants (Section IV-4).
*6. Linear titration functions obtained from $\bar{\mathbf{n}}$ equations (Gran method) applied to weak and strong acid mixtures and buffers (Section IV-5).
7. Examples of calculations for ionic strength control in titrations of various mixtures by NaOH (Section IV-6).
*8. Detailed calculation of the pK_a's of citric acid from an actual laboratory set of data from a glass electrode–calomel experiment (Section IV-7).
*9. Universal buffer: how to adjust $\boldsymbol{I}$ and added NaOH to obtain desired p**H**'s starting with a mixture containing phosphoric, succinic, and boric acids (Section IV-8).
10. Calculations and plots of p**H** effects of dissolving CO_2 in carbonate and phosphate buffers (Section IV-8).
*11. ML_n species calculations for given mixtures: Since high ligand concentrations are required for weak complexing systems (not just excess L), the amount of NH_3 left free will not be negligible. We get $\mathbf{L} = 0.0646$, $\bar{\mathbf{n}} = 3.541$ (bound L 0.0354) even though a 10-fold excess of ligand is available in the Ni(II)–NH_3 system which can make up to the 6-complex (Section V-2).
12. Reverse: solution design for desired ML species: How can one make a solution having 0.0500 M of the tetrammine ion and that would be insensitive to loss of that species? (Section V-2).
13. Close-spacing effects: The effect of close spacings of K values is especially common in ML systems. The Zn(II)–imidazole system is shown (Figs. V-2 and V-3).

*14. p**H** Effects for Basic Ligands: The Ni(II)–glycinate system. Take 100 ml of a solution having 0.0300 *M* Ni(II) and 0.1000 *M* HGly, a slight excess for formation of the ML_3 complex. Add NaOH (Section V-3 and Fig. V-4).

*15. Example: Find the pH of a solution made 0.0500 *M* in Ni^{2+}, 0.1000 *M* in glycine, (HGly), and 0.0300 *M* in NaOH. Here $C_M = 0.0500$ M, $C_L = 0.1000$ M, and $C_{AB} = 0.0700\ M$ (Section V-3).

16. Figure V-5. Behavior of $\bar{\mathbf{n}}$ curve slopes for log *K* sets: (i) 7, 4, 2, 1; (ii) 6, 5, 4, 3; (iii) 4, 4, 4, 4; and (iv) 3, 4, 5, 6.

17. EDTA–metal ion solutions: Polydentate ligand K_{f1} values and the interacting equilibria of Cu(II) with EDTA and one further ligand dependent upon p**H** are described briefly with references (Section V-3).

*18. Experimental determination of $\bar{\mathbf{n}}$ and the formation constants (Sections V-4 and V-5). Bjerrum's iteration formula is used and calculation details are shown in Section V-6 for ammonia complexing of Cu(II) and of Zn(II). Spreadsheet reduction of Berrum's data is shown in Section V-5 and by determinants in Appendix B.

*19. $\mathbf{F}_0$ or α_0 methods: Silver–allyl alcohol data are used in a Fronaeus–Leden method (Section V-7).

20. Polynuclear and mixed complexing: Ni(II)–lysine complexes. See discussion in Section V-8.

*21. Hydroxide complexes and polynuclear cases: The Cu(II)–OH^- system. Data proving polynuclear effects. Iterative spreadsheet method for α plots after constants have been obtained (Section V-9).

*22. What species and p**H** are found in pure water solutions of "neutral" Cu(II) compounds, say, $Cu(NO_3)_2$? The p**H** of Cu(II) solutions from 0.1 to 10^{-4} *M* are shown in a spreadsheet (Section V-9).

*23. The system Cu(II)–NH_3–OH^-. Published constants are used to plot α diagrams to show *C* effects. Take the given initial 0.0500 *M* $[NH_4^+]$ and examine three cases of total Cu(II): 0.001, 0.005, and 0.01 *M*. We add NH_3 and make the equilibrium $[NH_3]$ a chosen variable for calculation (Section V-10).

24. See the exercises at the end of Chapter V. Include a p**H** titration of a metal ion–EDTA mixture.

25. The solubility *S* of $RbClO_4$ in 0.0400 *M* $HClO_4$, assuming simple ions only, a near ideal, simple (complex-free) case. Also *K* and C_s cases (Section VI-1).

26. Experimental determination of K_{s0}: Treatment of data to reveal complexing (Section VI-2).
27. $CaSO_4$ solubility data: Treatment using a single complex fits the data well (Section VI-3).
28. Solubility with multiple complexing: AgSCN in KSCN solutions. Use of log species diagrams for saturated solutions (Section VI-4).
29. General solubility diagrams: Slopes of log S vs. log[ligand] reveal extent of complexing (Section VI-5).
30. Solubilities of MX with added ligand L which is not the common ion X: AgCl in NH_3 (Section VI-6).

*31. Solubility with proton reactions: Silver acetate, $CaCO_3$, tyrosine, and $MgNH_4PO_4$ examples. How P_{CO_2} affects S and p**H** in limestone systems (Section VI-7).

32. Polynuclear hydroxide solubilities: $Cu(OH)_2$ (Section VI-8). See Appendix B; where spreadsheet methods for several acid–base and ML cases are shown in detail.
33. A simple algebraic example is used to demonstrate graphically the secant method for the intersection of a cubic and a quadratic function (Fig. B-1).
34. Determinant method for solution of linear sets of equations (Gaussian substitution) is shown for the case of finding formation constants in the Cu(II)–NH_3 system using Bjerrum's data. Compare the spreadsheet iteration method for the same data in Chapter V.

ANNOTATED BIBLIOGRAPHY

*Bjerrum, J., *Metal Ammine Formation in Aqueous Solution*, P. Haase, Copenhagen, 1941, 1957. Details of experimental and calculation methods for many metal ion–ammonia and other amine complex systems. A clearly written account by a major originator in this field.

Ricci, J. E., *Hydrogen Ion Concentration*, Princeton University Press, Princeton, NJ, 1952. An extensive mathematical treatment of protonic acid–base equilibria. The style and notation are difficult, but almost every conceivable relation and situation is to be found somewhere in this book.

*Rossotti, F. J. C. and Rossotti, H., *The Determination of Stability Constants and Other Equilibrium Constants in Solution*, McGraw-Hill, New York, 1961. This is a detailed work on mathematical treatment of experimental data to get values of constants. Extensive graphical methods and precomputer computational approaches are given.

*Butler, J. N., *Ionic Equilibrium*, Addison-Wesley, Reading, MA, 1964. A pioneering monograph introducing clear, complete equilibrium mathematics to American chemistry. Extensive diagrams and calculation methods are shown.

*Jones, M. M., *Elementary Coordination Chemistry*, Prentice-Hall, Englewood Cliffs, NJ, 1964. Chapter 8, "The Determination of Stability Constants," in 65 pages presents a comprehensive but very clearly written account of experimental and calculational methods accessible to students. This book places solution equilibrium in context with the study of the chemistry of the elements.

* Works with asterisks use largely the same α, $\bar{n}$ notations of this book.

Bates, R. G., *Determination of pH*, 2nd ed., Wiley, New York, 1973. This is the authoritative work on measurements and standards by a long-time researcher at NBS.

*Guenther, W. B., *Chemical Equilibrium*, Plenum, New York, 1975. An earlier attempt to clarify and rationalize the field. It contains extensive computer driven graphical illustrations. Chapter 9 shows α, $\bar{\mathbf{n}}$ diagrams for 18 systems of *d*-metal ions with halides, CN-, SCN-, and ammonia. I no longer advocate the needlessly complex methods of deriving equations.

*Baes, C. F. and Messmer, R. E., *The Hydrolysis of Cations*, Wiley, New York, 1976. A comprehensive survey, often with reinterpretation, of results of investigation of protonic equilibria of aqua-metal ions in water. It is a good entry into a very difficult field of measurement and controversial deductions.

*Hartley, F., Burgess, C., and Alcock, R., *Solution Equilibria*, Wiley, New York, 1980. This book focuses on major calculation methods and measurements. Four detailed case studies lead the reader through calculations often comparing several approaches to getting constants. Representative simple systems are chosen.

Norris, A. C., *Computational Chemistry*, Wiley, New York, 1981. Extensive examples of using computers for numerical solutions to chemical problems. Good details on finding roots of higher power equations and of solving sets of simultaneous equations are given.

Stumm, W. and Morgan, J. J., *Aquatic Chemistry*, 2nd ed., Wiley-Interscience, New York, 1981. Detailed mathematical treatment of many complex natural systems. Extensive reference lists of geochemical, oceanographic, and pollution problems.

Pytkowicz, R. M., *Equilibria, Nonequilibria, and Natural Waters*, Wiley, New York, 1983 (2 vols.). This is an extensive and critical presentation of many difficult facets of aquatic chemistry. Problems met in experimentation and calculation are freshly treated. Ionic activities are well presented.

Connors, K. A., *Binding Constants*, Wiley, New York, 1987. This book treats measurement and calculations of association constants for a wide variety of interactions not often found in other monographs. Extensive details and explanations of equations are given. Many systems of organic and biochemical interest are treated.

Martell, A. E. and Motekaitis, R. J., *Determination and use of Stability Constants*, VCH Publishers, New York, 1988. Computer programs for getting constants and their reliability statistics are the focus of this book. The stress here is on cases having various and often numerous complications. This is for the advanced worker in the field. A disc of programs is included. See pp. 116–117 for the ultimate in α diagrams!

Collections

Volumes of the two editions of the *Treatise* contain many chapters on equilibrium mathematics displaying, especially, methods developed by Scandinavian pioneers in this field:

Kolthoff, I. M. and Elving, P. J., Eds., *Treatise on Analytical Chemistry*, Vol. 1, Wiley-Interscience, New York, 1959, Part I, especially Chs. 7–19; Vol. 2, 1979, Part I, Chs. 15–19.

Compilations of Equilibrium Constants

Sillen, L. G. and Martell, A. E., Eds., *Stability Constants of Metal–Ion Complexes*, Special Publications, No. 17, 25, The Chemical Society, London, 1964, 1971. This work lists available values with primary literature references. Temperature, medium (ionic strength), and experimental method are indicated. Choice among conflicting values is left to the reader.

Martell, A. E. and Smith, R. M., Eds., *Critical Stability Constants*, Plenum, New York, 1974–1989 (6 vols.). This work makes a choice of most likely values of constants and their estimated reliabilities for several media from the primary publications available. One must consult the references to see the methods and what decisions and choices have been made.

Kragten, J., *Atlas of Metal–Ligand Equilibria in Aqueous Solution*, Halsted Press. Wiley, New York, 1978. A collection of α_0 and log[M] plots vs. p**H** for metal ions, aqua alone (OH^-), and with other ligands. Note use of a reciprocal α function explained in the introduction.

Harned, H. S. and Owen, B. B., *The Physical Chemistry of Electrolyte Solutions*, 3rd ed., Reinhold, New York, 1958. Equilibrium and activity data for many electrolyte mixtures are given in Ch. 12–14 and in the Appendices. Solution thermodynamics is treated at an advanced level.

Ringbom, A., *Complexation in Analytical Chemistry*, Wiley-Interscience, New York, 1949.

INDEX

A list of major symbols is listed on pages 15 and 16 and on the front cover.
Key Points, beginning on page 301, groups major ideas by topic and order in the book.
Tables of α and $\bar{n}$ values for eight systems are found on pages 283–298.